The Balance of Nature?

The

*Ecological Issues in
the Conservation of Species
and Communities*

Balance of Nature?

Stuart L. Pimm

The University of Chicago Press

Chicago and London

The University of Chicago Press, Chicago 60637
The University of Chicago Press, Ltd., London
© 1991 by The University of Chicago
All rights reserved. Published 1991
Printed in the United States of America
03 02 01 00 99 98 97 96 95 94   3 4 5 6 7 8

Library of Congress Cataloging-in-Publication Data

Pimm, Stuart L. (Stuart Leonard)
    The balance of nature? : ecological issues in the conservation of
species and communities / Stuart L. Pimm.
        p.   cm.
    Includes bibliographical references and index.
    ISBN 0-226-66830-4 (paperback)
    1. Biotic communities.   2. Ecology.   3. Biological diversity
conservation.   4. Species diversity.   I. Title.
    QH541.P558   1991
574.5'247—dc20   91-3089
                                                                CIP

♾ The paper used in this publication meets
the minimum requirements of the American National
Standard for Information Sciences—Permanence of
Paper for Printed Library Materials, ANSI Z39.48-1984.

Contents

For Stephanie and Shama

Preface

What ecologists do, what they know is important, and what they speculate about are often different. The expected catastrophic extinction of species (already well under way in many places) will alter the planet's biological diversity so profoundly that, at known rates of speciation, it will take millions of years to recover it. Yet, few ecologists study extinctions. Indeed, very little ecology deals with any processes that last more than a few years, involve more than a handful of species, and cover an area of more than a few hectares. The temporal, spatial, and organizational scales of most ecological studies are such that one can read entire issues of major journals and see no hint of the impending catastrophe. The problems ecologists must face are so large: how do we contemplate processes that last longer than our research careers and that involve more species than we can count, over areas far too large for conventional experiments? The problems are also complex; understanding ecological processes at these large scales is far more of an intellectual challenge than is the stupefyingly tedious sequencing of the human genome. The problems are also more important. With complete certainty, I predict that there will be human genomes around in fifty years to sequence; with somewhat less certainty, I predict that there will be at least ten billion of them, dying from many causes each of which is orders of magnitude more important than the genetic causes the human genome sequencing will uncover. If we do not understand ecological processes better than at present, these ten billion humans will be destroying our planet more rapidly than we are now. When we contemplate this, it is no wonder we ecologists take ourselves off to beautiful, untouched environments and study fascinating species.

I have not been any different. Like nearly every other ecologist, I have undertaken small-scale, short-term studies of a few species in wonderful places. One of the places was the Hawaiian Islands. Here, perhaps 50% of the bird species were lost following their initial discovery by the Polynesians, 50% of the remainder were lost after ornithologists first visited the islands, and 50% of *that* remainder are critically endangered. Almost half a dozen species have completely or almost completely disappeared during the fifteen years I have been visiting there. My visits to Hawai'i have had a great impact on my research: nothing concentrates the mind quite like extinctions do.

Saving species is a noble aim, but are not the problems political and social ones? Of course they are, but there are major scientific questions too. Some are questions of a strictly practical nature: how do you get rid of pigs from areas where they are not native and are destroying the native ground cover? Other questions, however, require a deep understanding of the way nature works. How fast do populations recover from unusually low numbers, and why do some species vary more from year to year—and does this make these species more likely to encounter those low numbers from which recovery will be unlikely or impossible? Which species are extinction prone? Will alien species be able to invade native communities—and what will be their effects if they do? If these invasions cause extinctions, which other species will become extinct as a consequence? Which species play such important roles in the community that, if they become extinct, then many other species will be lost as a consequence?

These questions strike me as obvious, yet they are hardly major themes of our current literature: they are, equally obviously, on temporal, spatial, and organizational scales larger than that literature. In trying to address these and related questions in this book, I am simultaneously trying to broaden the scope of ecology and to provide some of the conceptual underpinnings to conservation biology. It follows that I do not consider conservation biology to be a sexy name for what ecologists have been doing all along; rather, it is a region of the possible ecology research space that is desperately important and sparsely populated with investigators.

Ecologists have thought about these issues before. If our studies are on rather small scales, our speculations are not. Off the record, ecologists are happy to talk about "stability," "integrity," "fragility," "balance," and a host of other terms that usually mean rather vague things about how natural systems behave over the long term and in the face of various stresses. It is out of these speculations that this book emerges. I have listened to hundreds of colleagues over the years and have attempted to define their terms and to summarize their hypotheses, exploring them theoretically and testing them with whatever data are available.

This book owes a great deal to those many colleagues. During the past decade or so, I have been invited to seminars and meetings in dozens of countries and so have had unusual opportunities to discuss the ideas and data in this book with a broad range of colleagues. The discussions have helped enormously in sharpening my ideas, correcting factual errors, and suggesting important literature. This is, however, a book of large scope: there must be a great deal of relevant ideas and data that I have omitted through ignorance or carelessness. I hope I will be told about what I missed and be forgiven for omitting what I ought to have known was there.

Though I cannot thank everyone, there are many who deserve special thanks. Much of the work in this book was done collaboratively with current or former graduate students (Michael Moulton, Andrew Redfearn, Anthony King, Gregory Witteman, and Hang-Kwang Luh) or by or with colleagues at the University of Tennessee (James Drake, Gary McCracken, and John Gittleman) and at other universities (Jared Diamond, Michael Gilpin, Roger Kitching, John Lawton, and Jake Rice). Parts of this book were written while I was a guest at various institutions: The School of Ecosystem Management, The University of New England, Australia (host: Roger Kitching); The Institute for Nonlinear Sciences, University of California at San Diego (hosts: Michael Gilpin, Ted Case, and George Sugihara); and The Centre for Population Biology, Imperial College, Silwood Park, England (host: John Lawton). Most of the book was written at The University of Tennessee, Knoxville, which has continually encouraged my research and travels.

I am indebted to a large number of colleagues who, despite very busy schedules, reviewed chapters, carefully, and sympathetically. In particular, I thank James Brown, Donald DeAngelis, Jared Diamond, James Drake, Torbjörn Ebenhard, Michael Gilpin, Peter Grant, Robert Holt, Stuart Kauffman, Pavel Kindleman, Mark Kot, John Lawton, Sam McNaughton, Donald McQueen, Michael Moulton, Lauri Oksanen, Andrew Sih, Michael Soulé, Catherine Toft, David Wilcove, Mark Williamson, and Peter Yodzis. Wynne Broon designed my figures.

During my many visits to Hawai'i, Sheila Conant and her husband David McCauley have met me at the airport at all hours of the day and night and allowed me to recover from jet lag while listening to the light rain and the 'Amakihi song of their Manoa valley home. They, Lenny Freed, Rebecca Cann, Jim Jacobi, and Charles and Danielle Stone have ensured that my trips have been both relaxing and rewarding and have made it no accident that Hawaiian experiences permeate this book. For over two years, I found it very difficult to write this book—or any other material. Many friends, in Tennessee and throughout the world, made those years tolerable through their support, their kindness, and their belief that this book would eventually be completed. In particular, Paul and Anne Ehrlich have continually provided friendship, delightful company, and unlimited enthusiasm. Susan Abrams provided not only friendship but infinite patience. My wife, Julia, did many of the tasks associated with this book, for which I thank her. I thank her even more for renewing my enthusiasm for this project.

1

Why "The Balance of Nature"?

What Ecology Is About and What Ecologists Study

Every ecologist knows what ecology should be. After all, every introductory textbook tells us. What we study, however, is often a different matter. So what *do* we study, for how long, and over what area? And, if we are community ecologists (who are supposed to study large numbers of species), how many species are we actually studying? The answers to these questions are not surprising. Look at any current ecology journal and you will see that our studies are often very brief: ten years is a long-term study. In the United States, such a study would require a minimum of four consecutive National Science Foundation grants, and such continued support is enjoyed by only a very few. Studies, especially experimental studies, are generally on a scale of square meters and rarely cover more than a few hectares. The typical community ecologist, as evidenced by his or her contributions to edited volumes on the subject, tends to study no more than a dozen species (Pimm 1986b).

Of course, there are exceptions. Biogeographers and students of diversity study many species over very large areas. These are nearly always studies of static patterns, for few data record the changes in species' ranges. And there are very long-term studies by paleoecologists, but these have a coarse resolution. Even pollen samples have a resolution of perhaps a century at best (Delcourt and Delcourt 1987). So where is the ecology of tens to hundreds of species over decades to centuries, across hectares to thousands of square kilometers? Should we care that this region of ecological research space is rather sparsely populated (figure 1.1)?

The most pressing ecological problems involved many species and their fate across decades to centuries, over large geographical areas. The catastrophic loss of species expected in the next several decades is predicted to rival that at the Cretaceous-Tertiary boundary (Ehrlich and Ehrlich 1981). Managers of natural areas must protect their fragmented communities against a spectrum of possible threats. They may have many species in their charge, manage ten to

1

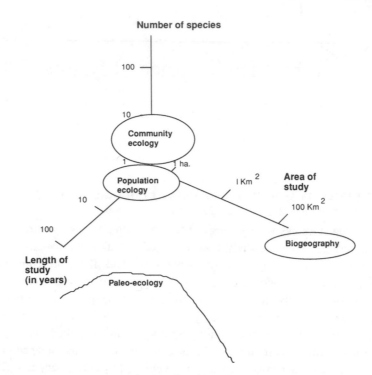

Figure 1.1 Caricature of the temporal, spatial, and organizational scales of ecological re-search. Most ecological research is done on short time scales and over small spatial scales, and even community ecology typically involves few species. The minimum time resolution of paleo-ecology is very much greater than the longest-term studies in the rest of ecology, and, although biogeography deals with large areas, it rarely deals with how ranges change over time.

thousands of square kilometers, and want to pass on their communities rela-tively intact to their successors a human generation in the future. In order to manage we must understand how populations and communities will change over the very organizational, temporal, and spatial scales about which we know least. I have deliberately written this book to address ecology on these scales. This is consciously a book about scientific issues in conservation biology. I will attempt to cover topics involving dozens of species, large areas, and es-pecially the time scales of decades that most ecology neglects. It is simul-taneously a book about what I think are the most challenging questions in ecology.

THE PROBLEMS OF SCALING UP

The problem of scaling up our understanding from the one-to-three year studies that are the staple of our science to the decades-to-centuries scales of our major problems may best be introduced by looking at what happens when we alter the density of a species. In the very short term, we can think of the processes that tend to restore the species' density to its former value. Every ecologists is familiar with the terms *density dependence* and *density independence:* these describe the change in a species' density relative to the density itself. Everything else in the universe is assumed to be more or less the same, while our species is recovering its former density. In the short term, this is a sensible view of things, because, indeed, "everything else" probably has not changed much. Over slightly longer time periods, we recognize that the initial change will affect the species' immediate predators and prey, then its competitors. Over longer periods, the predators and prey of these predators and prey will be affected; to put the matter simply, the effects will ripple throughout the community. So while in the very short term we may be content to view a species changing against a fixed background and concern ourselves with whether those changes are density dependent or not, over longer terms *we must understand population changes in the context of an entire food web.*

This idea that long-term population ecology must be community ecology is the first major theme of this book. The rate at which populations recover their former densities I call *population resilience.* Resilience depends, in part, on the species' birth and death rates, but it also depends on connections to other species—in particular, on the length of the food chain in which the species occurs (chapter 2). The variability of population density from year to year also depends on the species' birth and death rates (chapter 3), but it also depends on the diversity of predators exploiting—and on the diversity of prey exploited by—that species (chapter 4). Extinction depends on resilience and variability (chapter 7) and therefore on features of a species' ecology and on the ecology of the community of which the species is a part. The degree to which a change in one species' density causes changes in other species' densities I call *resistance.* Resistance depends so critically on interactions with other species that simple-minded predictions based on immediate interactions with predators and competitors produce "unexpected" answers (chapters 12 and 13)—and progressively more unexpected answers the longer the study lasts. If you look for long enough periods, the role of community structure in shaping dynamics is increasingly obvious. *A long-term theory of population changes has to be a community-based theory.*

My second major theme is that, over long periods, community structure is not static. Species are added to and lost from communities, at rates that depend on features of the species themselves and also on the features of the community's food web structure and on how that structure developed (chapters 8–11). Community structure changes, and as it changes it alters the framework against which population changes take place. Over the important temporal scales, *community theory is dynamic community theory.*

The problems of scaling up, from the small spatial scales at which we usually perform our experiments to the scales at which we encounter important problems, are directly comparable to those of extending our understanding across time. Indeed, the problems may be identical. A local change in density propagates not only over time but over space, affecting increasing numbers of species as it does so. When species densities are altered over large areas, these changes inevitably disrupt many species in the community. Understanding changes on large spatial scales also means developing a community-based theory. In short, to understand dynamics on longer time scales and over larger areas, we must develop community-based theories.

WHAT IS "THE BALANCE OF NATURE"?

The idea that most ecological studies are conducted at very special scales (and perhaps even the less important scales, from an applied view) is hardly a new one. For instance, the importance of interrelated scales is the subject of an influential monograph (O'Neill et al. 1986). But the recognition of a problem does not necessarily provide the tools for its solution. The tools we need to understand the longer-term, larger-scale ecology have been buried in our science for a very long time indeed. The origins of this book lie in a phrase familiar to all: *the balance of nature.*

What is this balance of nature that ecologists talk about? If ecologists now speak more of *ecological stability,* perhaps we are just substituting one phrase for another. *Balance* or *stability* implies some restoration following disturbance. The phrases indicate that the stability arises from "nature": ecological processes within populations, among the interactions between species in a community, and between the community and the physical environment. There is something unmistakably fuzzy about the terms *stability* and *balance of nature* as most ecologists use them (though, just because the terms are fuzzy, this does not mean that the underlying ideas are unimportant). For example, we use the singular *balance* and not *balances*—or *stability* and not *stabilities*—because ecologists once imagined that there might be a unitary process. It is now clear that ecologists mean many different things when they talk of stability.

This book defines the different meanings of ecological stability, in an effort to clarify these ideas and their meaning.

Why do we need one or more meanings of stability? Ecology also has a long tradition of discussing the temporal changes of single species. Studying the temporal changes of many species poses problems. When we are focusing on just one species over a few years, each increase or decrease in population can be described and discussed in exquisite detail. When we study more species, we need the means to describe those changes succinctly; the same is true if we have many years of data for just one species. How do we categorize or summarize the changes in densities to allow comparisons of many species in complex communities over long periods of time? Similarly, how do we describe and compare the dynamics of communities? The various meanings of stability are summary statistics, of course, that exist to enable us to describe these long-term complex patterns of change. If I have heard my colleagues correctly as they speculate about stability, the definitions may be fuzzy but the scale is clear: the speculations are an attempt to move toward an ecology of larger spatial, temporal, and organization scales.

I have particular scales in mind, roughly ten to one hundred species, time scales from ten to one hundred years, and spatial scales certainly orders of magnitude larger than most ecological field experiments—the characteristic scales of community ecology, if you will. Because most of our science is done at scales smaller than these, throughout this book I will often look at what population ecology tells us, contrast it with the results of larger-scale ecology, and argue that an understanding of both scales is essential. I will examine certain kinds of stability and the population, community, and ecosystem processes that affect them, but it will be the community processes that receive most attention. If I had chosen to write only about one kind of stability, say resilience, I would have considered a broader range of scales than those discussed in chapter 2 and, moreover, would have treated equally the various factors that affect resilience. This book emphasizes community effects and the scales they imply.

While I wish to explore ecology on neglected scales, I do not consider some scales at all. There is no discussion of very small scales where physiological or behavioral mechanisms may be responsible for "stability." There is no discussion of Lovelock's Gaia hypothesis (Lovelock 1982), in part because I do not understand it. It clearly deals with "the balance of nature"—but at a spatial scale (the entire planet) and at temporal scales much larger than my concerns. While there may exist global processes that could offset the catastrophic loss of species we are now experiencing, they are surely extremely slow. For example, from 50% to 100% of the vertebrate species have been lost from some

Pacific islands in the past ten to one hundred years. I know of no evolutionary biologist who thinks speciation could restore the original diversity within less than ten millenia. Similarly, while global climate changes may be somewhat limited by Gaia, they are not limited enough to prevent them from being a major concern for conservation biology.

BEGINNINGS: ELTON AND MACARTHUR

Despite the relative neglect of the scales I find most interesting, their study does have a history, and I want to review it briefly as the context of what follows. I am not a historian, and my review of the literature is perhaps an idiosyncratic caricature of a large and complex body of work; I do not expect anyone else would summarize this literature in the same way. But my review explains why this book is structured as it is and why I have treated some topics in detail and others only superficially. At the center of this book are the relationships between community structure (and particularly food-web structure) and different kinds of stability. The origins of this book lie in the ideas of Elton and MacArthur, who argued that community structure affects temporal changes in the community and its constituent species.

Two works of the mid-1950s are quoted repeatedly in modern ecology texts: Elton's (1958) book *The Ecology of Invasions by Animals and Plants* and MacArthur's (1955) paper "Fluctuations of Animal Populations, and a Measure of Community Stability." Both works contain arguments about stability, what determines it, and why its study is important.

Five Pages of Elton

Elton starts by saying (1958, 145):

> I will now try and set out some of the evidence that the balance of relatively simple communities of plants and animals is more easily upset than that of richer ones; that [the balance] is more subject to destructive oscillations in populations, especially of animals, and more vulnerable to invasions . . .
>
> For if this can be shown to be anywhere near the truth, it will have to be admitted that there is something very dangerous about handling cultivated land as we handle it now, and [it will be] even more dangerous if we continue down the present road of "simplification for efficiency."

Elton then presents evidence for his assertion that simple communities are less stable than complex ones. Interestingly, his first evidence comes from "the conclusions of mathematical speculation about population dynamics." Simple models of a predator and its prey or of a single host and its parasite have population densities that "fluctuate in numbers considerably, even without shocks from the outside like the vagaries of the climate" (1958, 146). Such fluctuations would mean that in simple communities populations would "never have constant population levels, but would be subject to periodic 'outbreaks' of each species."

Thus models of complex communities have stable populations while simple models do not. Notice that Elton did not present analyses of complex models; he simply *assumed* they would be stable. Elton's second line of evidence was from laboratory experiments, such as those done by G. F. Gause in the 1930s (Gause 1934), that made the same point. The evidence also had the same limitation: there were no laboratory experiments with really complex systems. "The third piece of evidence is that the natural habitats on small islands seem to be much more vulnerable to invading species than those of the continents. This is especially so on oceanic islands, which have few indigenous species" (1958, 147).

Elton's book was about invading species, and his evidence here is compelling. His fourth argument was that "invasions and pest outbreaks most often occur on cultivated or planted land—that is, habitats and communities very much simplified by man" (1958, 147). His fifth argument was similar. Insect outbreaks, Elton claimed, were a feature of simple temperate forests but not of complex tropical forests. His sixth argument seems to recapitulate the earlier one: insect pests are more likely to invade and become a nuisance in orchards, especially after spraying designed to eliminate the pests. The evidence for invasions into agricultural communities is persuasive. The evidence that insects sporadically achieve higher densities in temperate than in tropical forests was anecdotal at best, and insects may be more pestilential in agricultural habitats because they cause us more harm there than in habitats we do not exploit.

It is easy to criticize ideas that are thirty years old. The mathematical models and the laboratory experiments available to Elton looked only at simple systems. The data on invasions were compelling, but comparative studies of population fluctuations were nonexistent. Moreover, Elton discussed a range of very different phenomena: population variability, the potential abundance of a species, the rate at which a pest species might increase in numbers after removal of its predators, how readily communities may be invaded, and the extent of damage the invading species causes.

Yet, what Elton achieves is a remarkable statement of what ecologists should study. Ecologists should be thinking about a diversity of stabilities: population variability, population recovery, the ease of invasion, and the consequences of invasion. All are ways of characterizing both population and community changes, to facilitate comparisons. Moreover, Elton considers a factor called "complexity" that should determine "stability" and the importance of their interdependence: "Once the notion is grasped that [stability] is a property of the community . . . there is hardly any limit to the ways in which it could be introduced" (1958, 152). It is impossible to read Elton and not recognize that he identified important ecological problems involving large numbers of species, large temporal and spatial scales, and the context—i.e., community ecology—in which we must study them.

MacArthur on Fluctuations

MacArthur's (1955) paper was much more limited, for he dealt strictly with the theory of population fluctuations. Such fluctuations were considered by Elton, but his theoretical arguments were inadequate because analyses of complex communities were lacking. MacArthur's paper provided these arguments and began by carefully defining stability. "Suppose, for some reason, that one species has an abnormal abundance. Then we shall say that the community is unstable if the other species change markedly in abundance of the first. The less effect this abnormal abundance has on the other species, the more stable the community." MacArthur sought to relate this definition of stability to a measure of community complexity, "the amount of choice" that energy could take in its routes through the food web.

For all MacArthur's mathematical formalism, his argument was a verbal one: if a predator feeds on many species of prey, it will be less affected by the abnormal decline of one of them than is a predator that feeds only on that abnormally rare species. Indeed, the argument relating stability to complexity was nothing more than an admonition to not put all your eggs in one basket. Still, MacArthur's paper was an unmistakable statement that "stability" should depend on "complexity," and the paper was exemplary in providing careful definitions of stability and complexity.

THE LAST OF THE OLD MEETINGS

The twenty-second Brookhaven Symposium in Biology (Woodwell and Smith 1969) was published as *Diversity and Stability in Ecological Systems*. Some of the papers dealt explicitly with how environments that were physically variable or unpredictable would house communities with few species. But some of the papers looked at how aspects of community structure should affect features of dynamics. Both Lewontin (1969) and Margalef (1969) appreciated that "stability" meant different things; indeed, Lewontin's chapter was entitled "The Meaning of Stability."

Like every other ecology graduate student in the early 1970s, I had a well-thumbed copy of the proceedings of this symposium. In the nineteen contributions, MacArthur's paper is cited once—and only in passing. Elton's book is not cited at all. There is no illustration that has a measure of stability on the vertical axis versus *anything* on the horizontal axis. Watt's paper comes closest: he shows three populations and compares and contrasts their temporal changes (Watt 1969). In the volume as a whole, there is little effort to develop explicit theories of how community structure (or any other factor) might affect stability, and certainly no one was testing the ideas laid down a decade earlier by Elton and MacArthur. Lewontin's paper was prophetic, however. He carefully reviewed and explained the various mathematical definitions of stability. Theoreticians, at least, would now have something to work with.

A Theoretical Basis

In November 1970, Gardner and Ashby published a very short paper in which they noted that many different kinds of systems have many temporally varying components that are interconnected to greater or lesser degrees. They recognized that the dynamics of these systems might be highly nonlinear, but they decided to look at very simple, linear systems "as a first step." They built computer models that differed in (*a*) the number of interacting components and (*b*) in the latters' degree of *connectance,* or the extent to which these components were interconnected. For each combination of connectance and number of components, they created large numbers of such models with the parameters describing the interactions between components randomly drawn over an interval between minus one and plus one. Then they recorded the proportion of models that were stable *in the mathematical sense,* meaning that the values of the components returned to their equilibrium values after perturbation—that is, after the values were altered from the original.

The more components in the model and the greater the model's connectance, the *less* likely the models were to be stable. Indeed, there were "critical values for stability." As connectance increased in models with more than a few components, the transition from models being usually stable to usually unstable was a sharp one.

Thus, simple systems were *more likely* to be stable than were complex ones. This relationship, of course, was the direct opposite of the "increased complexity gives increased stability" argument that had dominated the literature thus far. Indeed, that argument had almost reached the point of becoming dogma. Watt (1973) took as one of his "core principles of ecology" the notion that "the accumulation of biological diversity . . . promotes population stability."

The Gardner and Ashby paper did not deal explicitly with ecological communities, but a few months later Robert May tackled directly the issue of stability in multispecies communities (May 1971). (Interestingly, Watt was the one who communicated May's paper to the journal that published it.) It is May whom we must thank for the vigor with which the topic of ecological stability has been pursued since 1971. May's paper began: "One of the central themes of population ecology is that increased trophic web complexity leads to increased community stability." He reviewed Elton's and MacArthur's ideas and explicitly investigated Elton's argument that complex models should be more stable than simple ones.

May's conclusions were uncompromising: "we consider a simple mathematical model for a many-predator-many-prey system, and show it to be in general less stable, and never more stable, than the analogous one-predator-one-prey system. This result would seem to caution against any simple belief that increasing population stability is a *mathematical* consequence of increasing multi-species complexity" (1971, 59).

The Consequences of Having a Method

May's (1973) *Stability and Complexity in Model Ecosystems* expanded on the theme of the 1971 paper and considered the relation between complexity and stability in randomly fluctuating environments as opposed to deterministic ones; how nonlinearities could produce cyclical oscillations; the role played by time delays; the importance of the strength of the interactions between species; and the idea that, even for a fixed number of interactions, some patterns of interaction would be more stable than others. The book's influence is best illustrated by the large number of modifications to the models that were proposed in the next decade. Theoreticians were quick to take up May's idea that the *exact* patterns of species interactions had a substantial impact on whether a model community would be stable.

Just what all these results mean is best explained in the addendum to the revised (1974) edition of May's book. Here, May contrasted *sets of* models that are unlikely to be stable with *sets of* models that are likely to be stable when the parameters in both sets of models are selected randomly over the same intervals. In the real world, parameters are constantly changing over space and time. Moreover, they are probably changing more in variable than in relatively constant environments. Therefore we are unlikely to find, in the real world, complex communities that correspond to "usually unstable" systems: such communities are too "fragile." The broad ranges of parameters consistent with "usually stable" simple models mean that corresponding communities in the real world will be "robust." We *are* likely to encounter such communities in the real world, for they can persist despite their constantly changing parameters. Notice especially that the comparison is not between *a* simple model and *a* complex model: any model is still either stable or unstable, and there are stable complex models and unstable simple ones. Rather, the contrasts are broad, sweeping ones that deal with large numbers of model communities and that consequently make predictions about large numbers of real communities.

Although May discussed simple versus complex models, his ideas are directly applicable to all community structures—including food-web shapes—that affect stability. May's argument becomes a prediction that the food-web structures we observe in nature ought to correspond to the structures that enhance the likelihood of stability in models. Using this argument, John Lawton and I and various of our colleagues derived a long list of predictions about food-chain length, the patterns of species feeding on more than one trophic level, compartments, and predator-prey ratios. We then showed that real food webs conform rather closely to these predictions. I reviewed this work in my book *Food Webs* (1982), and my principal conclusion was not only that food webs are simpler than one might predict but that the specific kinds of complexities that are most likely to make models unstable are rare in nature.

The next area pertinent to my story involves how food-web stability changes as species are progressively added to or subtracted from model food webs, a topic that arose naturally from the earlier analyses. The stability analyses discussed so far are of static structures: they discuss whether one food web is more likely to be stable than another. However, instability is equated with the loss of species, so what are the consequences of these losses to the species remaining in the community? This idea was the theme of a group of papers by Tregonning and Roberts (Tregonning and Roberts 1978, 1979; Roberts and Tregonning 1981). They looked at unstable communities, eliminated the doomed species, looked at the residual communities, decided whether they were stable, eliminated more species if they were not, and progressed until only stable communities remained. Communities gain species too; Post and Pimm (1983), Drake (1988), and others built models that assembled food webs from small numbers of species. These models were motivated by looking at how the chance of finding stable models varied across different structures and as the webs changed structures. The models quickly raised other issues—for example, how resistant to invasion a community might be.

HERESIES

In the middle of this explosive growth of the literature on theories of food-web stability came a contentious paper by McNaughton (1977), entitled "Diversity and Stability of Ecological Communities: A Comment on the Role of Empiricism in Ecology." McNaughton began by discussing the *Proceedings of the First International Congress of Ecology* (Anonymous 1974), in which a major theme was the relationship between species diversity and stability. McNaughton perceived that there was no consensus on what community properties diversity was supposed to stabilize: "the general tenor of the Congress papers is in marked contrast to those in (the Brookhaven) symposium on diversity and stability published just five years previously . . . Principal catalysts of the change in attitudes, based on Congress citations, seem to be mathematical models demonstrating that destabilization accompanies increases in number and connectance of system elements" (1977, 520).

McNaughton presented the results of an experiment in which he excluded African buffalo[1] from two grassland areas in the Serengeti National Park of Tanzania. One area had diverse vegetation, and the other had much less diverse vegetation. The effect of buffalo grazing was to reduce biomass by 69% in the less diverse area but by only 11% in the more diverse area, relative to areas from which the buffalo had been excluded. The more diverse community, therefore, was more stable, a result supported by other broadly similar experiments. He concluded with

> the hypothesis that plant community diversity stabilizes functional properties of the community to environmental perturbations . . . The weight of evidence resulting from explicit tests of the diversity-stability hypotheses (Margalef 1965, McNaughton 1968, Hurd et al. 1971, Mellinger and McNaughton 1975) suggests, not that the hypothesis is invalid, but that it is correct . . .
>
> Finally, how can we account for the marked instability of attitudes regarding the diversity-stability hypothesis (found in the *Brookhaven Symposium* and the *Proceedings of the First International Congress of Ecology*)? The explanation seems to be the appearance in the interim of a plethora of more rigorous models suggesting alternative conclusions. It is fascinating that there has been so much modeling and so little empirical testing during this period. The data reported here indicate that . . . verbal models (of stability and complexity) are more reliable generaliza-

---

[1]The scientific names of all species mentioned in the text are given in the index to species names, at the end of the book.

tions . . . than most of the "more elegant" mathematical analyses published to date. (1977, 522)

ARE THEORETICAL AND EMPIRICAL STUDIES INCOMPATIBLE?

Thus far, we have encountered three sects in pursuit of community stability. After a very slow start, *field ecologists* were marshalling increasing evidence to support their belief that more complex communities were more stable. They had few mathematical theories on which to hang their ideas. From 1970 onward, *mathematical ecologists* were building conceptually interesting models at a fast pace, but these models were not being tested. Third, there were the *pluralists,* who recognized the great diversity of ideas about complexity and stability but were not actively engaged in exploring these relationships either theoretically or experimentally.

Each sect made important points. From the field ecologists we learned that experimental tests of the relationships between stability and complexity were possible. From the theoreticians we learned that once stability and complexity were carefully defined, theoretical development could proceed rapidly. And from the pluralists we got the suggestion that, if theoreticians and empiricists were in disagreement, then perhaps they were looking at different parts of a large and complicated field.

An Attempt at Synthesis

In a 1984 review, I argued that further progress would come not just from recognizing the many meanings of stability and complexity (Pimm 1984a). Progress would require careful definitions of the quantities that would permit both theoretical explorations and empirical tests of the relationships between them, and from cataloging just what relationships had been proposed and tested. First, I suggested that theoretical and empirical ecologists had used the word *stability* to mean at least five different things.

A system is considered to be *stable* in the mathematical sense if, and only if, the variables all return to equilibrium conditions after displacement from them. By definition, either a system is stable or it is not.

Resilience is defined as how fast a variable that has been displaced from equilibrium returns to it. Resilience could be estimated by a return time, the amount of time taken for the displacement to decay to some specified fraction of its initial value. Long return times mean low resilience, and vice versa. Resilience is measured as a rate of change.

Persistence is how long a variable lasts before it is changed to a new value. Systems that often change could be described as having high turnover, so turnover would be the reciprocal of persistence. Persistence is measured as time.

Resistance measures the consequences when a variable is permanently changed: how much do other variables change as a consequence? If the consequent changes are small, the system is relatively resistant. Resistance is measured as a ratio of a variable before and after the change, and so it is dimensionless.

Variability is the degree to which a variable varies over time. It will be measured by such statistics as the standard deviation or coefficient of variation, of consecutive measurements of those things that interest us.

Theoreticians had considered stability in its strict mathematical sense. Systems are either stable or not, so, given May's argument that what we most often observe in nature will be stable systems, the theoretical results make predictions about *the kind of communities we observe and those that we do not.* The other definitions of stability do permit comparisons of different existing systems, however, and these were the concerns of field ecologists. Elton had explicitly considered variability. His discussion of the ease with which species could invade species-poor communities is a discussion of persistence. Species-rich communities, he asserted, would persist longer, because they would be less likely to be invaded than species-poor communities. MacArthur considered variability too, although his arguments equating the degree of change to food-web structure could also be an argument about resistance. McNaughton's discussion was wide-ranging, but his results are clearly about resistance—the extent to which biomass changes when a grazing herbivore is added to the community. So, in general, the results I have discussed are about very different definitions of stability; no wonder there was little agreement.

To further complicate matters, the different studies I have discussed look at stabilities at different levels of ecological organization, which is why I have had to use the words *a variable* in my definitions of stability. Studies of whether interacting sets of species will be stable consider whether the densities of all those species return to equilibrium. To ask how long a community persists before it is invaded is to ask how long community composition lasts. Such a discussion ignores fluctuations in the abundance of species and looks only at the species list itself. In contrast, McNaughton placed his emphasis on how much total biomass changed and not on which species were involved in the changes. Therefore we can discuss the stability of populations, of community composition, or of community biomass.

The research I have discussed so far also tried to relate the different definitions of stability to different features of community structure—"complexity" in particular. Complexity has been taken variously to mean the number of species in a system, the degree of connectance of the food web, and the relative abundances of a species in the community. And why stop here? Many other different features of community structure could bear on the five definitions of stability.

In short, my review argued that ecologists were looking at the combinations of five definitions of stability, three definitions of complexity, and three levels of organization—a total of forty-five possible questions about the relationships between community complexity and stability. Rarely did two ecologists look at the same question, although when empirical and theoretical ecologists *did* look at the same relationships between complexity and stability there was remarkably good agreement. I shall spend the rest of this book making the same point.

How This Book Is Organized

The history of the complexity-and-stability debate helps explain this book's organization. The book has four sections—one each dealing with resilience, variability, persistence, and resistance. Why only four? There is no section on stability in the mathematical sense, in part because I have tackled that subject elsewhere (Pimm 1982) but principally because the results of studies of mathematical stability appear throughout this book, as I address the other meanings of stability.

Chapter 2 addresses resilience, and chapters 3–5 cover population variability. Chapter 6 discusses the very complicated population dynamics that can arise and contribute to highly variable densities. Chapter 7 looks at extinctions, which are one consequence of high population variability. Communities are persistent only if they do not lose species through extinction and do not gain species through invasion. Chapters 8–11 deal with whether communities will persist because they are not invaded. Finally, chapters 12–15 deal with resistance.

There are many questions concerning stability and its causes that are not discussed in this book. Of course, some topics will be covered implicitly. A feature that emerged repeatedly is the interdependence of many of the results. For example, resilience affects variability, and variability affects the chance of extinction and so affects how long a community's species composition persists. There is no other way to discuss the material on stability than to break it down into its constituent pieces, of course. But once we have done this, it is important to notice that the various meanings of stability are *functionally* related. For instance, I have not explicitly asked how long a population persists. If by "persists" we mean how long it lasts before becoming extinct, then I will answer the question partially in chapter 7. On the other hand, if by "persists" we mean how long a population's density remains within some arbitrary range of values, then, obviously, the more variable the population, the shorter this persistence. So answers to that question would be found in the chapters on population variability.

There are some topics I have not addressed at all. For example, studies of resilience and variability have tended to be about *population* resilience and variability. There is no reason why ecologists should not study the resilience or variability of *species composition*. I have not addressed the resilience of the total density of a set of species. Similarly, my discussion of variability has been equally restricted. How variable is either species composition or the total density of a set of species over time?

This book is far from being a complete review of stability. It is a selective and personal account, meant to be stimulating rather than encyclopedic. If you feel there are important issues I have neglected, then this book will have had one of its desired effects.

The Approach to Be Used

There are perfectly good reasons why most ecology has been done on very few species, for a few years, and on small spatial scales. For such scales we can build models that include plausible descriptions of the ecological mechanisms involved, and we can test those models either with carefully controlled experiments or with data collected specifically for that purpose. Stated bluntly, the criticism of the approach used here is that the models are terrible and that the data are even worse. Many may find the whole approach to be profoundly disturbing. The obvious place to discuss such a criticism is at the end of this book, when I can refer to specific examples—and, indeed, this issue is the subject of chapters 16 and 17. The general features in the criticism are worth mentioning here, however, if only to warn you what to expect.

The frequent criticism that theoreticians simplify things is both correct and unnecessary. Of course we have to simplify things—in order to be able to understand them. To be comprehensible, every theory must make assumptions, and these assumptions must surely be false, for the very reason that the models are simpler than is nature. But for the scales to be discussed in this book, the models presented will seem unusually inadequate. This is because they seem much less *mechanistic* than are models in other areas of ecology. The population ecologist will look at the equations in this book (they are of the Lotka-Volterra variety) and find them totally lacking in details of population biology.

These critics will be exactly right, but this book is not about single species. For instance, the community models contain terms that are gross caricatures of how species behave: predator-prey encounters are modeled in the same way as are random collisions of molecules in an "ideal" gas—that is, one for which even the physicists must make unrealistic assumptions. The optimal foraging models of behavioral ecology deal with the details of how predators encounter prey, and the resulting models for the dynamics of one (sometimes two) species are appealing. I suggest they are appealing because the terms of the models can be readily identified with behaviors we can observe. Of course, these models will not be mechanistic to the physiologist, who is concerned with how the brain obtains the optimal solution and who observes at a much smaller scale. One scientist's mechanism is another's phenomenon.

I do not mean to suggest that we should accept inadequate models or that we should not relentlessly pursue better models. But an effective model for ten species over ten years will not have the same intrinsic appeal as does a model for what individuals will do over the next week. Surely, the effectiveness of a model has to be judged against its predictions about nature and not against the elegance of its formulation. (However, I do appreciate that a theory's internal consistency or elegance may be persuasive in other areas of science.) We only worry about the models' assumptions when the models make the wrong predictions. Indeed, models are often most informative when they *do* make the wrong predictions. When the models get it right, we feel that the simplifications only left out features that were not sufficiently important to alter the results.

That it is the data that provide the crucial tests for the models may not be that controversial a conclusion. The kinds of data available are more of a problem, for the data that I shall present are rarely experimental in the familiar sense of that word. No one can do neat, controlled experiments on the organizational, temporal, and spatial scales involved. Worse, the data I discuss in this book were never collected for the purposes to which I shall put them. Rather, we have to be opportunistic, interpreting what data we have and doing our best to find those data that fortuitously provide the means to test the models.

In short, unfamiliar models and data must be used, many ecologists will not like them, and all ecologists (I include myself) are not entirely happy about using them. The familiar methods of ecology will not work on the scales to be discussed here. Yet we do not have a choice in moving toward longer-term, wider-scale, multispecies ecology, for that is where the important applied problems lie.

2

Resilience

WHAT IS IT AND HOW DO WE MEASURE IT?

I have defined population resilience to be the rate at which population density returns to equilibrium after a disturbance away from equilibrium. Figure 2.1 shows two model populations that have the same equilibrium densities but clearly differ in their resilience. Many ecologists will know that there is another definition of resilience; Holling (1973) defines resilience to be how large a range of conditions will lead to a system returning to equilibrium. In his definition, highly resilient systems will almost always return to equilibrium, whatever happens to them; systems that are not resilient will often be fundamentally changed after a perturbation, perhaps by losing species or perhaps by moving

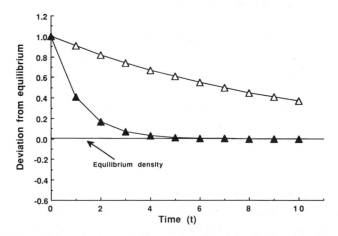

Figure 2.1. Two model populations returning to equilibrium density (set at a reference level of zero) from the same initial displacement at + 0.5. One population (open triangles) is much less resilient than the other (closed triangles). Theory suggests that returns will be approximately exponential decays for simple models. These are exponential decays, and the return times are 10 (top) and 1.1 (bottom). For details, see main text and appendix to chapter 2.

to a new equilibrium involving very different species' densities. Holling's definition is an interesting one, but I will not consider it here.

In this chapter I shall show how population resilience depends on a variety of factors, each likely to be important on different temporal and spatial scales. Resilience, as I have defined it, is a central feature of population dynamics, and comparable measures of resilience can be defined for measures describing communities and ecosystems. It is not, however, an easy quantity to estimate. How might we quantify resilience? The answer comes from the ubiquity of exponential decays. Exponential decay arises when the disturbance from equilibrium decays at a rate linearly proportional to the magnitude of the disturbance. Consider that cup of tea sitting on your desk; it is losing heat at a rate proportional to how much hotter is the tea than the surrounding environment. (Sitting here in Tennessee, my glass of iced tea is gaining heat at a rate proportional to how much colder is the tea than the surrounding environment.)

The single constant that relates the rate of change to how far the system is away from equilibrium is one measure of the system's resilience. An intuitively more appealing measure of resilience involves the time it takes for the disturbance to decay to some specified fraction of its initial value—the longer this time, the less resilient the population. We could measure the time it took for the population density to return to half its initial displacement from equilibrium. Rather than use such a half-life, convention defines return time as the $1/e$th life: e is the base of natural logarithms, and $1/e$ is approximately 37%. The reason for this choice is given in the appendix to this chapter.

The utility of these physical models to ecology is that how fast population densities return to their more usual values may depend in a simple way on how much the densities have been perturbed. Unusually small densities may lead to high birth rates and so to a fast rate of return, while unusually high densities may be quickly followed by very high death rates. In some population models, the rate of return to equilibrium can be shown to be linearly proportional to the deviation from equilibrium, and the return to equilibrium will match the exponential ideal.

As a theoretical example, consider a population that in year t has a density X_t and where the numbers the following year can be modeled by the equation:

$$X_{t+1} = X_t + X_t \cdot r \cdot (1 - X_t/X^*). \tag{2.1}$$

This is one of the simplest models of population growth: the population grows up to its equilibrium density, X^*, and declines as it increases above it; the rate of approach to equilibrium depends on r, and the return time is $1/r$, provided certain assumptions are met. I shall discuss these assumptions presently. (Details of these mathematically well-known results are found in Pimm 1982).

If we have values of population density, X, in successive years, we can recognize that equation (2.1) is a quadratic function of the form $y = bx + cx^2$, with x being X_t (the density in a given year) and y being X_{t+1} (the density in the next year); b and c are constants. Standard regression techniques allow the estimation of these constants—and, from them, the return time—provided we make a few assumptions.

A practical example of this method is given in figure 2.2. After a severe decline in abundance caused by the very hard winter of 1962—63, the song thrush populations in Britain recovered. The return and the approximate location of the equilibrium are obvious. The statistical recipe of the previous paragraph estimates the equilibrium density as 113 and the resilience as 0.62. The estimated return time of about 1.6 years matches what we might have estimated "by eye" for this population.

There are a number of complications in estimating return times. The first is that, even for a simple model, the exponential decay process is only an approximation—and one that becomes progressively better as the population density approaches its equilibrium density. Only as the population density moves close to equilibrium is its rate of change linearly proportional to the population's deviation from equilibrium. A further complication comes from multispecies models. For a model with n species, the decay to equilibrium is a sum of n exponential decays, provided, yet again, that the densities are all near

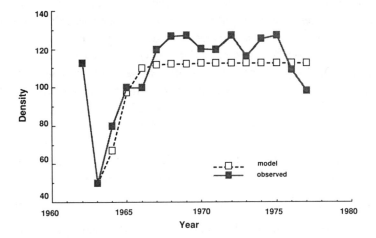

Figure 2.2. Population density of the song thrush in English farmlands (solid squares). Units of density are relative and are scaled to 100 in 1966. Statistical modeling fits the curve $X_{t+1} = (1.62)X_t - (0.62/113)(X_t^2)$ to these data (dashed line, open squares). The equilibrium density is estimated to be 113, the resilience is estimated to be 0.62, and the return time is estimated to be $1/0.62 = 1.61$ years.

equilibrium. (Details of this well-known mathematical result are in chapter 2 of Pimm 1982 and in the appendix to this chapter). There will not be a single resilience (or return time); an estimate of return time will depend on just how long a period we chose to measure it. This is not very helpful when we are trying to compare populations. We can use the slowest of these n exponential decays as a measure of return time. The fast exponential decays, however, may be numerically more important; that is, the density may show very large rapid changes in density, followed by a much more gradual, long-term trend after these transients decay. DeAngelis (1991) discusses more general measures of resilience that allow comparisons of systems where transients may sometimes be important numerically.

Another difficulty in calculating a single return time is that the processes (such as births and immigration) that lead to populations increasing toward their more usual densities are not those (such as deaths and emigration) that lead to populations decreasing toward their more usual densities. A hypothetical species may be upwardly resilient yet downwardly sluggish: I shall return to this problem later and present data to show that this is not generally true. In short, measuring resilience may be difficult, and simple, single estimates may provide only approximate and flawed means of characterizing recovery processes.

RESILIENCE AND THE LIFE-HISTORY CHARACTERISTICS OF INDIVIDUAL SPECIES

What affects resilience? How fast a population can grow when it is below equilibrium must depend, in part, on how fecund the species is; how fast a population declines when it is above equilibrium will depend on how quickly the individuals die. Population resilience will also depend on how the species interacts with other species, for predation and competition must slow down the potential growth. Moreover, growth may depend on how readily energy or essential nutrients become available within the ecosystem. If nutrients, for example, are tightly bound by the ecosystem, then how quickly they are released may be important in determining how quickly populations can recover. So there are three kinds of factors that affect resilience, and they correspond roughly to the three familiar levels of ecological organization: population, community, and ecosystem.

I start with the population effects and ask the obvious question, Do species differences in reproductive rate translate into differences in population resilience, or are they masked either by the interactions with other species or by ecosystem-wide features of nutrient flow? Here are two studies that answer this question.

Reproductive Rate and Resilience in Birds

The Common Birds Census (CBC) scheme for farmland has been organized by the British Trust for Ornithology since 1961 and provides an annual index of percentage change of bird populations on a sample of about one hundred farms throughout Britain (Marchant et al. 1990). The CBC technique measures percentage changes from year to year more accurately than it measures absolute densities, and for this reason an arbitrary index level of 100 in a reference year (often 1966) is used as the scaling factor for the index.

For most farmland species, CBC data are available from 1962. The abundances of resident birds were depressed by a rather cold winter in 1961–62 and by an excessively severe winter in 1962–63, so that the populations of these species were severely reduced from their levels before these winters. Migratory species that left Britain for the winter were not affected in this way. Figure 2.2 shows the effects both of these winters and of the subsequent recovery on the song thrush population. After about 1967, this species showed relatively restricted fluctuations about a fairly obvious equilibrium level. Although cold winters are a recurring feature of the environment of resident species in Britain, they occur at relatively long intervals (the previous winter as severe as 1962–63 was 1947–48).

O'Connor (1981) used as a measure of resilience the rates of recovery from the effects of the 1962–63 winter. His resilience index was derived, for each species, as the percentage change in the species' population size between the summers of 1963 and 1964. Since most resident bird populations in Britain were greatly reduced by the experience of the severe winter, most farmland populations were increasing from very low densities, and the observed population increase should be fairly close to the maximum possible. Values of this resilience index should (and indeed do) correlate closely with resilience estimates I have calculated by using the technique illustrated by figure 2.2, which uses more years of data. As a measure of reproductive effort, O'Connor calculated the seasonal egg productivity of each species by multiplying clutch size by the number of successful clutches laid in each season.

The resilience index was examined in relation to egg productivity and also in relation to survivorship. There was no significant relationship between survivorship and resilience, suggesting that the ability of these species to recover from the effects of hard winter was not determined by their survivorship. Capacity for the population recovery, however, was correlated with seasonal egg production (figure 2.3). Those resident species that had the greatest annual

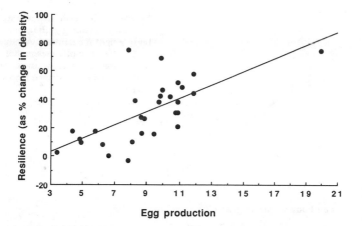

Figure 2.3. Resilience increasing with greater annual egg production, in resident British birds. Resilience was estimated as percentage change in the densities of farmland birds between 1963 and 1964, and egg production was calculated as the number of eggs laid per year. The years 1963 and 1964 were chosen because in these years the resident species were recovering from severe declines due to the 1962–63 harsh winter (see figure 2.1). The increase in resilience with egg production is significant at $P < 0.01$. (Redrawn from O'Connor 1981.)

production of eggs were the species that recovered their original population densities most rapidly, while species with lower egg productivity recovered more slowly.

Body Size and Maximum Population Change

While fecundity depends on many factors, across all animals the effects of body size are very pronounced. Figure 2.4 shows that for three orders of magnitude increase in body weight there is approximately a tenfold decrease in reproductive rate. Small-bodied species not only have larger numbers of young per generation than do large-bodied species; they tend to have more generations per year. (Peters 1983 provides a detailed review of these results.) So do these differences in yearly reproductive rate translate into differences in resilience? Another approximate method of calculating resilience is to look at the maximum relative change in a population—that is, the maximum of X_{t+1}/X_t—or at its logarithmic equivalent—that is, the maximum value of $\log X_{t+1} - \log X_t$. Populations may fluctuate considerably from year to year yet sometimes may not change very much between any two years. Nevertheless, given enough years, we may catch the population well below its normal levels, and the subsequent increase will give us an idea of how fast the population can change when it gets the chance.

Now, resilience also involves the factor of how fast a population *decreases*

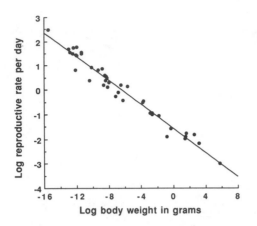

Figure 2.4. Intrinsic rate of natural increase r declining with increasing body weight for animals. A value of r = 2.0, for example, means that the population can increase one hundred fold in one day. (Redrawn from Fenchel 1974.)

when it is unusually abundant, so we could also look at the minimum of X_{t+1}/X_t. As I have pointed out, this measure depends on how fast individuals die, whereas increases depend on how fast individuals can reproduce. So there need not be any simple relationship between "down" resilience and "up" resilience; and certainly they need not be identical, as they are in most models. Seeing whether up resilience and down resilience are similar across widely different species is an interesting question in its own right.

Across widely different species, these minima and maxima statistics have an obvious problem. The more years of data we examine, the more likely we will encounter an extreme value. But, if we use a fixed number of years for all species, these ratios give us a way of comparing how fast populations can change. Andrew Redfearn and I have assembled data on maximum ratios of year-to-year changes in density (X_{t+1}/X_t) for more than two hundred species of widely different body sizes. (These data will be discussed in more detail in chapter 3). Because these ratios are likely to increase with the number of points used in their calculation, the data in figure 2.5(a) are based on fourteen points—that is, fifteen consecutive years of population estimates. Fifteen years was chosen as a compromise. There are many population studies that were conducted for fewer than fifteen years, but the estimates based on them are likely to be less reliable. And, as the length of studies increases much beyond fifteen years, the number of studies available drops precipitously.

There is considerable scatter in figure 2.5 (suggesting that resilience may depend on many things), but small-bodied species (those with higher reproductive rates) have greater resiliences than do larger-bodied species. I have also plotted the greatest decrease in abundances, in figure 2.5(b). Smaller-bodied species also can decrease relatively faster than can larger-bodied species.

There are two conclusions. Given (1) that small-bodied species increase faster from year-to-year and (2) the close relationship between body size and yearly reproductive rate, there is likely to be a strong relationship between yearly reproductive rate and resilience. (Although I do not have data for what we really need—the reproductive rate per generation, the number of generations per year, and the observed year-to-year increases—the correlations between each of the three variables and body weight are strong ones.) Species with high reproductive rates are likely to have highly resilient populations despite the many other factors that affect resilience.

Second, up resilience and down resilience are probably closely correlated across all these species; large-bodied species both go up and go down more slowly than small-bodied species. This is despite the differences between the causes. How fast a population goes down depends on a variety of factors that I will discuss at length in the next chapter.

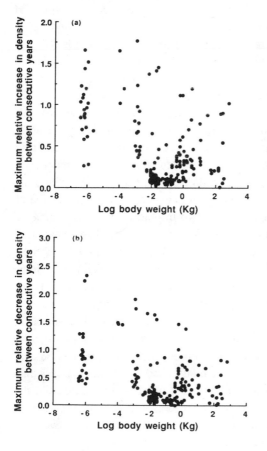

Figure 2.5. Both maximum (a) and minimum (b) relative changes in density between years, declining with increasing body weight for 212 animal populations. The relative changes are the maximum or minimum values of $\log (N_{t+1}) - \log (N_t)$. Thus, in the upper graph, a value of 2.0 means that the population had a maximum *increase* of 100-fold between years; in the lower graph, a value of 2.0 means that the population experienced a maximum *decrease* of 100-fold between years. Only fifteen years of data are used in the estimates, since extreme values such as maxima and minima are likely to increase with sample sizes. Regressions are significant at $P < 0.0001$ in both cases.

Food-Web Effects on Resilience

When we disturb the densities of a set of interacting species, no species density can return to equilibrium until all the others have done so. The resilience of each of the species in the community will depend on the least resilient species. Consider this example, which suggests that food-chain length may have a strong affect on resilience. Suppose we reduce the density of phytoplankton in a lake. This will reduce the density of their zooplankton predators—and, eventually, that of the fish that feed on the zooplankton. The phytoplankton might have the capacity to return quickly to their former densities. But while the fish densities are recovering (slowly), zooplankton will be unusually abundant, and this will keep the density of the phytoplankton down.

The Effects of Food-Chain Length on Resilience

To examine the effects of food-chain length on resilience, John Lawton and I (Pimm and Lawton 1977) modeled a variety of food chains and food webs. We used Lotka-Volterra models to construct food chains of varying length and investigated how long the constituent populations took to return to equilibrium after a disturbance. For each of the three food webs shown in figure 2.6, we created two thousand models. (The details and limitations of these and other models in this book will receive an extended discussion in chapter 16). We chose randomly, over intervals that we thought seemed ecologically sensible, the parameters describing the interactions between species. Figure 2.6 also shows the distribution of the return times we calculated for these models. As the number of trophic levels increases, so do the return times. The time units are arbitrary but are the same as those chosen to express the rate of change of the population densities in the model. For two trophic levels, nearly 50% of the models had return times less than or equal to 5 time units. The comparable percentage falls to 5% for three trophic levels and to less than 1% for four. Also indicated in the figure is the percentage of models in which return times exceed 150 time units. With two trophic levels this is very small (0.1%), but it is equal to 9% and 34%, respectively, with three and four trophic levels.

An alternative approach is to have only one species per trophic level (as in the case of the four-species, four-trophic-level model). For a two-species model with two trophic levels, the distribution of the return times can be analyzed analytically. Here, about 70% of the models had return times of less than 5 time units. Hence, models with both different numbers of species and the same number of species per level, as well as models with the same number of species, lead to the same conclusions: as the number of trophic levels increases, so do the average return times.

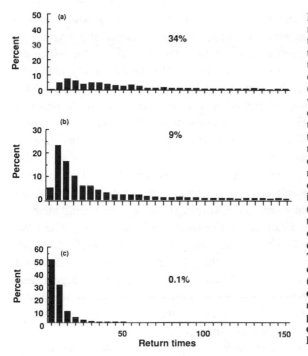

Figure 2.6. Distribution of return times from nearly two thousand models of three food-web configurations: (a), four trophic levels, one species per level; (b), three trophic levels, with two species at the lowest trophic level; and (c) two trophic levels, with three species at the lower trophic level. The number on each part of the figure represents the percentage of models with return times in excess of 150 time units. (The time units are arbitrary and are those used to quantify the rates of change of the species densities.) The return times were calculated as the reciprocals of the largest eigenvalue of each model's interaction matrix; for details, see appendix equation (A2). (Redrawn from Pimm and Lawton 1977.)

Testing the Theory: Resilience in Arthropod Pests after Pesticide Use

How might we test the idea that long food chains cause the population densities of the individual species to be less resilient? Unfortunately, my approach to answering this question is indirect, but it is the only example I know of. Consider an insect pest population increasing toward a high density that involves far more pests and far too few of its hosts for our liking. A species with a resilient population will reach unacceptable high levels faster than does a species that is not resilient. Population resilience can thus be a measure of *instability* and also a measure of how often we may have to control the pest by chemical means. There is an interesting paradox here. For species that we deem desirable, a high resilience means a quick return after some disturbance. For these species, high resilience would be equated with *stability!* In a general way, it is their high resilience that makes insects such pests. Small size is coincident with high population growth rates (figure 2.4). With regard to insects, it is also clear why some groups cause a greater problem than do others. Aphids, for example, have particularly high population growth rates. It is not, however, on such intrinsic factors that I wish to concentrate. Rather, I ask, How does food-web structure modify a species' rate of return to equilibrium?

In the specific terms of insect outbreaks, this means asking how food-web structure affects the rate at which pest densities reach unacceptable levels.

Food-web theory suggests that, the longer the food chain, the lower the resilience of the system's constituent species will be. Consequently, removing trophic levels should mean that pest outbreaks will occur faster (Pimm 1984a) than they would in the presence of predators. (The final density of the pest will probably be higher, as well.) A way to evaluate these ideas may also give some insight into why insecticide applications often induce pest outbreaks. Insecticides clearly kill many pests, but many predators are often killed as well. The predators, moreover, cannot recover until the pests come back. This suggests that after insecticide application there will be a "time window" during which pests have very few predators and so, according to the theory, should increase more rapidly. Andrew Redfearn and I (Redfearn and Pimm 1987) asked the question: At comparable densities, after being treated with insecticide applications, do pests increase more rapidly than do untreated controls? (I should add that it has often been documented that insect pests can reach higher densities after spraying with pesticides, but this does not necessarily mean that they are increasing faster at any particular density. Although this idea seems reasonable, it has not been tested experimentally.)

Redfearn and I restricted our attention to phytophagous aphids and mites on the above-ground parts of cereal, vegetable, and orchard crops. Why choose these groups of arthropods? We must assume that the population changes are a reflection more of reproduction than of emergence and immigration. Although we cannot exclude emergence and immigration of aphids and mites, the high reproductive capacity of these species, plus other features of their natural history, means that the changes in numbers we observe will more likely reflect the contribution of population growth.

We extracted data from articles appearing in the *Journal of Economic Entomology* during the years 1957–76. In total, we made 112 comparisons between treated and untreated populations from thirty-four data sets and from twenty-three articles. From each data set we extracted pest densities on different dates, for the control and for each treatment type. The recorded times represented the time period, in days, between the sampling date and the last treatment date. We grouped chemicals according to whether they were (*a*) applied as granules or liquids into the soil and then covered ("soil-applied" pesticides) or (*b*) applied as sprays, dusts, or foliage granules aboveground ("aboveground" pesticides). It was assumed that pesticides in both categories would reduce the densities of herbivorous species, whereas those in the second category would also severely affect predatory species.

To estimate resilience we had to make a number of assumptions contained in the following scenario: Let us imagine herbivorous insect populations grow-ing exponentially in an agricultural ecosystem such as a field or an orchard. The logarithms of density plotted against time will give a straight line for an exponentially increasing population. We can estimate the slope of these data by using regression analysis. The larger the slope, the faster the population is increasing, and the faster the (unwanted) end point will be reached. Therefore, we may equate the slope with resilience. An example of this method, compar-ing a treated population and an untreated population, is provided by figure 2.7. Second, if population growth is not simply exponential—that is, if log density does not increase linearly with time—then we must use a more complicated model. (Populations may behave in this fashion if, for example, resource lim-itation were restricting population growth at high densities.) Redfearn and I estimated how fast the densities were growing at one particular density, the same for both treated and untreated populations.

We found that aboveground pesticides accelerate population growth of aphids but not of mites. The results are in table 2.1. This accelerated growth was not due entirely to the fact that these insecticides might release the aphid populations from some density-dependent factor, such as resource limitation, which slows the growth of the untreated populations at their generally high densities. We reached this conclusion because our results were not altered by assuming density-dependent growth and by then comparing estimates of the instantaneous growth rates of increase of treated and untreated populations at equal densities. Furthermore, it is unlikely that the accelerated growth caused by aboveground pesticides was entirely a consequence of the removal of her-bivorous species competing with the aphids. These would have been attacked by soil-applied pesticides, which had no significant effect on the population growth rates of the aphids. I conclude that these data support the theoretical

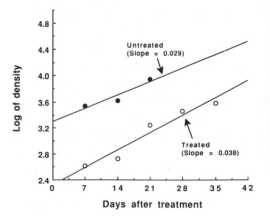

Figure 2.7. Effect of the insecticide carbofuran on green peach aphids on peppers. Both the population treated with the insecticide and that left untreated increased through time, but notice that the treated population increases faster than the untreated population. In the treated system, it is likely that predators were absent while the aphids were increasing. (From data in Hale and Shorey 1971.)

TABLE 2.1   The Effect of Soil-Applied and Aerial Pesticides on the Population Growth Rates of
              Aphids and Mites (after Redfearn and Pimm 1987)

| | Proportion of Populations Showing Higher Rates of Increase After Spraying[b] | | | | | |
| Model Used to Calculate | Soil-applied Pesticides | | | Aerial Pesticides | | |
Population Growth Rate[a]	Overall	Aphids	Mites	Overall	Aphids	Mites
Logarithmic increase	35/52 (67%)	31/37 (84%)	4/15 (27%)	53/60 (80%)	44/46 (96%)	9/14 (64%)
Nonlogarithmic increase	21/51 (41%)	16/36 (44%)	5/15 (33%)	45/55 (82%)	41/45 (91%)	4/10 (40%)

[a]Two models were used to calculate the rate of population change over time. The first assumed exponential growth—
that is, a linear relationship between the logarithm of density and time. For such a model, the slopes of the regression lines
were compared. The second model assumed a quadratic relationship between the logarithm of density and time, thus
allowing for the population to grow more slowly at higher densities. For this model, slopes were compared at a specific
density midway between the average density of the treated and untreated populations.
[b]Thus, for example, 35 of 52 populations (31 aphid plus 4 mite populations, of a total of 37 aphid plus 15 mite
populations) showed a higher rate of increase after the spraying than did the (unsprayed) controlled plots. Notice that the
effect of spraying is most marked in the aphids treated with aerial sprays, and is much less obvious when soil-applied
pesticides are used.

prediction that population resilience can increase as trophic levels are removed
from a community.

ECOSYSTEM EFFECTS ON RESILIENCE

Finally, resilience can be related to such ecosystem-level factors as energy flow
and nutrient cycling. In an important book, DeAngelis (1991) shows that system resilience can be related to two fundamental structural concepts: the energy
flow through the system and the mean number cycles a unit of matter makes in
the system before leaving it. A simple index which combines both of these is
the transit time of a unit of matter through the system. DeAngelis shows that
transit time should be an important determinant of resilience. Strictly considered, the DeAngelis work deals with the resilience of matter (carbon, phospho-
rous, or other elements), but the availability of these elements must, at some
level, affect the densities of individual species.

DeAngelis reviews three sets of studies. First, in an attempt to understand
underlying similarities and differences among ecosystem types, O'Neill (1976)
used data from six diverse ecosystems: tundra, tropical forest, deciduous forest, salt marsh, spring, and pond. O'Neill set parameters for a nonlinear
energy-flow model for each of the six ecosystems. The six models had the same
compartmental structure and were subjected to a standard perturbation. O'Neill
found that recovery times decreased as energy input per unit of standing crop
increased. The tundra system had the longest recovery time. The pond, with a
low standing crop and high biomass turnover, had the shortest recovery time.
O'Neill related his results to a suggestion by Odum and Pinkerton (1955).

They defined *power capacity* as the quantity of energy processed per unit of living tissue. They hypothesized that greater power would result in a greater capacity to counteract change. In other words, more power, greater resilience.

Second, DeAngelis discusses, as I have done earlier in this chapter, the effects that food-chain length has on resilience. The result—that, as the number of trophic levels increased, so the average recovery time also increased—DeAngelis argued, is entirely compatible with O'Neill's ideas. The flow of energy or biomass through the system has an important effect on resilience. The higher the flux (that is, the flow per unit of energy or biomass), the more quickly the effects of the perturbation are swept from the system—and so the higher the resilience. Long food chains may be expected to decrease this flux. Of course, there are likely to be other factors that will affect the flux, and these also will account for system-to-system differences in resilience.

Finally, DeAngelis discussed the recycling of nutrients. Some nutrients may be held very tightly by a system and may be recycled many times before being lost. Others may be washed out of the system much more rapidly. Jordan et al. (1972) argued that models of nonessential elements tend to be more resilient than models of essential elements. They suggested, as an explanation, that essential nutrients, such as calcium or phosphorus, tend to be tightly cycled. A disturbance to these elements tends to diminish only slowly. Minerals that are not essential (cesium, for example) are lost from the system at a high rate, and the disturbance disappears quickly. A similar argument was made by Pomeroy (1970), who pointed out that coral reefs and rain forests are examples of systems with tight nutrient cycles. When systems of this type are disturbed, recovery may be slow, because there is little through-flow of nutrients. A high degree of nutrient recycling allows an ecosystem to maintain a high level of biomass, even when the subsidy from the environment is small. When a perturbation removes some fraction of that biomass (along with the nutrient), it takes a long time for the system to recover, because the nutrient input is small compared with the stock of nutrient removed.

The basic factor determining the resilience of nutrient-cycling models seems to be the degree of recycling, while the factor determining the recovery in food-web or energy-flow models (with no feedback) is the energy or biomass flux. DeAngelis pointed out that these factors are really similar. They both relate to the rate at which a given unit of material or energy is carried through the system from the point where it enters to the point where it leaves. DeAngelis continued by formulating these intuitive ideas mathematically. He shows that it is possible to define a single index, transit time, that incorporates both power capacity and recycling. Transit time is the expected time that a unit of energy or matter remains in the system—that is, the time from input to output. Transit time is strongly related to recovery time: the longer the transit time, the longer the recovery time.

SUMMARY AND CONCLUSIONS

Population resilience is the rate at which a population's density returns to its equilibrium level after being moved away from it. We can estimate resilience either as this rate or as its reciprocal—that is, *return time,* the time it takes for the population's displacement from equilibrium to decay to some specified fraction of its initial displacement. Convention sets this fraction at $1/e$, or about 37%.

We might expect resilience to depend on many factors and at different levels of ecological organization. How fast a population recovers from a severe decline in density could depend on its reproductive rate: the more young produced, the faster the population can recover its former level. There is evidence for this effect. In birds, species that produce more eggs per year had densities that recovered more quickly from the effects of a severe winter. Across all animals, small-bodied species have more young than do larger-bodied species. Population densities of small-bodied species can be shown to increase more rapidly from year to year than do large-bodied species. Small bodied species also decrease more from year-to-year than do large-bodied species. Although I have not directly related year-to-year changes in density to reproductive output, it seems likely that high reproductive rates will translate into high population resilience.

How fast a population can increase will also likely depend on the species' interactions with other species in the community. A species' predator can slow down the recovery of the species, because the predator takes some of the species' reproductive effort. The predator's predator also has an effect, because, until it has recovered its normal density, the predator will be unusually abundant, and the predator's prey will be unusually rare. Given this, long food chains may be expected to reduce the resilience of their constituent species. I present evidence to show that this occurs for aphid populations in agricultural ecosystems.

Finally, we might expect that resilience will depend on how quickly the nutrients necessary for a species' growth become available. How quickly the nutrients become available will depend partially on food-web structure but also on abiotic processes.

So which of the three sets of factors (population, community, and ecosystem) is the most important one? O'Neill et al. (1986) argue that this question is inappropriate, for the factors operate at different—though interrelated—spatial and temporal scales. The population data that show that resilient species have high reproductive rates, i.e., that resilience is determined at the population level—are for relatively short time periods (typically, year-to-year changes). Even here, there is much scatter about the relationships, suggesting the importance of other factors. Toward the other extreme of spatial and temporal scale are the recoveries of plant and animal communities after glacial retreats, when the rates of climatic amelioration and of nutrient release from rocks must play a predominant role in setting the rate of plant colonization. At some intermediate scale we may identify all the factors I have discussed. How fast the plant and animal communities recover after a forest has been felled will depend on how the nutrient cycles in the soil have been changed, on interactions between species (predators may not become established until their prey are present), and on population characters (large-bodied—hence long-lived and slowly reproducing—species will probably appear after small-bodied—hence short-lived and quickly reproducing—species.) There is no single answer to what determines resilience, therefore. Rather, there is a recognition that many factors operate and that which factors predominate will likely depend on the spatial and temporal scales we are considering.

APPENDIX

Consider the equation for the exponential return to equilibrium:

$$X_t - X^* = (X_0 - X^*)e^{-k \cdot t}. \tag{A1}$$

X_t is the population density at time t, X_0 is the initial population density, X^* is the equilibrium density of the population, and k is a constant. $X_t - X^*$ is the current value of the displacement of the density from equilibrium, and $(X_0 - X^*)$ is the initial value of this displacement. The ratio of these two quantities is the proportional decay, and it can be seen to be $e^{-k \cdot t}$. So when t has the value $1/k$, the disturbance has fallen to e^{-1} (approximately 37%) of its initial value. The reciprocal of the rate constant k is a simple measure of resilience (there are more complex ones) and has units of time. It is an inverse measure of resilience (the longer the time, the lower the resilience), and it is called the population's *characteristic return time* (or *return time*, for short).

For multispecies models, for a typical species X_i the return to equilibrium can be expressed as

$$X_t - X^* = \Sigma(c_{ij} \, e^{-k_j \cdot t}). \tag{A2}$$

The c_{ij} are constants, and the k_j are constants called *eigenvalues*. The more negative the eigenvalues, the faster that part of the sum decays. So eventually the return to equilibrium will be dominated by the largest (that is, least negative) eigenvalue. For multispecies systems, we can define return time as the reciprocal of this largest eigenvalue (Pimm and Lawton 1977). For return time to be an appropriate measure of resilience, we must now consider not only population densities close to equilibrium but also sufficiently long periods of time, so that the decay to equilibrium is dominated by the largest eigenvalue. Over short periods, the return to equilibrium may be dominated by large, transient changes, which correspond to the other eigenvalues. Details of these mathematically well-known results can be found in chapter 2 of Pimm (1982).

A Note on Temporal Variability

The temporal variability of population density differs greatly between species as a comparison of the two populations of figure VAR. 1 shows. Identifying such gross differences in variability is easy, but discussing the factors that affect variability is not. This is because there are various processes that interact to determine population variability. The processes are themselves complicated, and they have complicated interactions with each other. By way of introduction, I return to the example of the song thrush in English woodlands (figure VAR.1a and chapter 2). The density of the song thrush declined dramatically between 1962 and 1963. The cause was an unusually hard winter that gripped England from the last few days of 1962 until well into 1963. The mean temperature for January and February of 1963 was 5°C lower than that for the average of those months during 1962–79 (Cawthorne and Marchant 1980). After its decline, the song thrush population then recovered, returning to pre-1962 levels after about five years. This example is quite typical of populations that experience irregular reductions in density because of the vagaries of the environment. The cause of the density changes may not be winter temperature—it could be variations in rainfall in a desert or changes in circulation patterns in the ocean—but the principles are much the same. The song thrush example is a simple one, but it poses a long list of questions:

1. Do different populations encounter different levels of environmentally induced mortality? In other words, do some populations vary more or less than the song thrush because the environment is more or less variable?

2. Over time, how does the environment change? Even if the variability of, say, winter temperatures remains constant, the longer we measure the environment, the more extreme conditions we are likely to observe. But does the environment remain the same—or does it change, leading to even more extreme conditions than we might otherwise expect?

3. Other species of birds were affected by the cold winter of 1962–63;

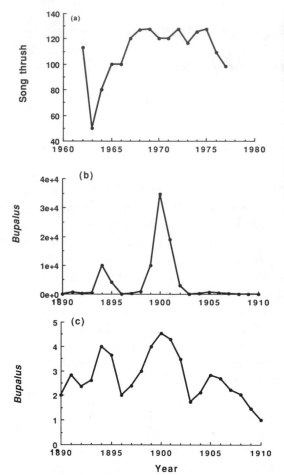

Figure var.1. (a), Densities of the song thrush in English farmlands. Units of density are relative and are scaled to 100 in 1966. (b), Densities of moth *Bupalus* in German forests, on an arithmetic scale; numbers are per hectare of forest floor. (c), *Bupalus* densities on logarithmic scale. Whereas the song thrush experiences a twofold variation in density over about twenty years, *Bupalus* densities vary over nearly four orders of magnitude during the same period.

populations of some species declined much more than those of others. How do species differ in their susceptibility to environmental changes? Some species might have been better able to withstand the low temperatures. Other species might have feeding habits that made them more or less susceptible to the low temperatures. Another common English bird is the moorhen, which feeds around the edges of small ponds or streams. During the 1962–63 winter these habitats were frozen solid for long periods; it is not surprising that the moorhen suffered a particularly large decline in numbers (Cawthorne and Marchant 1980).

4. Environmental changes also have indirect effects on species; how do these affect population variability? The hard winter of 1962–63 not only

killed the song thrush directly but probably killed the species on which the bird fed, as well. Some species would be reduced more than others, and the total drop in food supply would depend on the sum of various losses in prey species, as well as on the birds' ability to switch to relatively more abundant prey species. The indirect effects are further complicated because the hard winter killed off some of the song thrush's predators and competitors too. Reduced competition and predation might be expected to have had an effect on the song thrush's density, but these effects would surely have taken some time to become apparent, as they were indirect. In short, in addition to direct effects on the population, there were a complex of effects that came indirectly through the food web.

Suppose the density of a prey species is strongly regulated by the density of the birds; then a bad time for the birds would be a good period for the prey. Then the prey species' numbers would look like the mirror image of the birds'—fairly constant numbers punctuated by irregular high densities that occur when the birds are rare.

5. The song thrush population eventually recovered from the hard winter 1962–63. High resilience might reduce variability: the population that recovers more quickly will remain at low—and hence unusual—densities for a shorter time than will a population that recovers slowly. A consideration of the hypothetical prey species of the previous paragraph shows that the effect of resilience on variability can be complicated, for that prey species of the previous paragraph will achieve higher numbers (and so have higher variability) when its predator, the bird, is absent—if the prey species has a *higher* resilience.

6. After five years, the song thrush population had recovered from its losses suffered during the hard winter, and it remained at a fairly constant level. Under what circumstances should we expect this kind of pattern? The population could have overshot the predecline level and then declined. A highly resilient population, through very quick growth and a mortality rate that lags behind population recruitment, might show other, more complicated changes in density that are driven in part by its interactions with other species. All these features would increase the population's variability.

Items 5 and 6 show that resilience will play an important part in determining variability but that the interaction between variability and resilience is likely to be very complicated. Moreover, food-web structure affects both sensitivity of a species to environmental disturbances (chapter 4) and its resilience (chapter 2).

To deal with basic problems posed by each of the six sets of questions, in the next chapter I will examine how the features of individual species (including the resilience of their populations) affect variability. In chapter 4, I will look at the effects of food-web structure. Finally, in chapter 5, I will consider the nature of environmental variation in both space and time—and how organisms differ in their responses to this variation.

3

Temporal Variability and Individual Species

INTRODUCTION

This chapter will consider how features of a species' natural history affect the variability of the species' density from year to year. I am going to begin with an empirical result—rather than present the data last, as a test of already developed theories. The empirical result involves the relationship between body size and temporal variability. Andrew Redfearn and I assembled 202 previously published population data sets of terrestrial animals that were counted at least annually and for at least fifteen years. The results appear here for the first time. The data comprise 116 bird studies, 43 mammal studies, and 43 insect studies. The majority of the bird data were collected by the British Trust for Ornithology's Common Bird Census (CBC). Many of the insects are British moths and aphids collected by the Rothamsted Insect Survey. Most of the remaining data were used in two other comparative studies of population dynamics (Tanner 1966; Petersen et al. 1984). Finally, nineteen miscellaneous studies were assembled by my graduate ecology class in 1986. Redfearn and I calculated two indices of variability: (*a*) the coefficient of variation (CV), which is the standard deviation of the densities divided by the mean density, and (*b*) the standard deviation of the logarithms of the densities (STDL). There are a number of difficulties in calculating year-to-year variability and, particularly, in comparing such a wide range of species. I discuss these issues at length in the appendix to this chapter.

   Both indices of variability showed the same result. Small-bodied species were significantly more variable than were large-bodied species (figure 3.1). Ecologists have long suspected that insects, for example, have more variable densities than do birds or mammals (Williamson 1972; Southwood 1977). Attempts to demonstrate taxonomic differences in variabilities have not always been successful (Connell and Sousa 1983), but Schoener (1985) is correct in arguing that previous attempts to demonstrate these differences relied on too few species or a nonrepresentative selection of species. I concur with Gaston

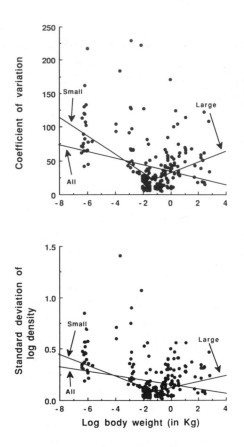

Figure 3.1. Two measures of population variability versus body weight. All estimates of variability are calculated from fifteen consecutive years of annual density estimates, N_t. The coefficient of variation is of the N_t, and the standard deviation is of the logarithms of the N_t. (Both measures are discussed in the appendix to chapter 3.) Regression lines are for all the species, for small species weighing less than 0.1 kg (such species might be expected to have life spans of about a year or less), and for large species weighing more than 0.1 kg. Note that, for species weighing more than 0.1 kg, variability increases significantly with body size. All the regressions are significant at $P < 0.01$.

and Lawton (1988a, 1988b) in arguing that this relationship between size and variability is not inevitable, both because its explanation is not obvious and because it is certainly complicated. Why should smaller-bodied organisms be more variable? Body size may be directly responsible—or correlations between body size and reproductive rate, longevity, or other factors may influence variability. In this chapter I shall tease apart the various explanations.

BODY SIZE AND RESISTANCE

During unusually harsh conditions, the rate at which a population declines will determine, in part, how variable the species' density will be. I define resistance as the degree to which a population (or community) is unchanged when the environment changes (chapter 1). Consequently, small-bodied species may have less resistant population densities in the face of inclement weather. Small-bodied species may be more temporally variable, because they suffer greater declines in abundance than do large-bodied species (see figure 2.5b).

Support for these ideas comes principally from studies of vertebrates. Lindstedt and Boyce (1985) review the mechanisms by which large-bodied species of mammals may be more resistant to harsh seasons than are small-bodied species. They noted two broad-scale patterns. Within individual mammal species, body size increases with the increasing seasonality of the environment, and, across mammalian faunas, large-bodied species tend to predominate the farther north is the fauna. As an explanation for these patterns, Lindstedt and Boyce note that mammals store energy primarily in the form of adipose tissue. When food is limited, metabolism shifts to a reliance on stored fats. Pitts and Bullard (1968) measured fat stores as a function of body size. As body size increases, the amount of stored fat (M_f) becomes a greater fraction of body mass (M_t). Calder (1984), analyzing their data, found a nonlinear relationship such that M_f was proportional to $M_t^{1.19}$. However, the energy an animal uses per unit time also increases with M_t. Does this mean that a large bodied mammal has more or less M_f relative to its energy needs than does a small-bodied mammal? Under normal conditions, energy used per unit time increases in a nonlinear way with body weight and is proportional to $M_t^{0.75}$. The survival time of mammals without food ought to be proportional to the ratio of M_f to energy needed—that is, $M_t^{0.44}$. Put simply, with increasing M_t, M_f increases more rapidly than energetic needs. Larger mammals should be able to withstand starvation longer than smaller mammals. Energy needs under conditions of severe cold need not necessarily scale with M_t in the same way as do needs under "normal conditions." Calder argues that under extreme conditions small-bodied species, because their greater surface-to-volume ratio exacerbates heat loss, are at even more of a disadvantage than this previous discussion would suggest.

Root (1988) finds that empirical patterns for wintering birds match those for mammals: for about 60% of birds that winter in North America, the northern limit of the winter range corresponds to a particular line of average winter temperature. Very roughly, for this group of species, the smaller the species' body size, the warmer the northern limit of their range. In general, small-bodied species do not winter as far north as large-bodied species.

Direct evidence for these physiological speculations that large-bodied vertebrates should be more resistant to inclement weather and so suffer small declines in bad years also comes from the data on British birds. As noted in chapter 2, the 1962–63 winter was an unusually severe one, with average January and February temperatures running about 5°C below the 1962–79 average. The next most severe winter was that of 1978–79, when the average temperature was about 3°C below this average. Cawthorne and Marchant (1980) provide a fascinating discussion of the effects that this winter had on the breeding densities of species in 1979. From 1978 to 1979 the small-bodied species decreased in densities whereas large-bodied species were relatively unaffected (figure 3.2). Cawthorne and Marchant document various other factors that affected the decline in density. Declines were greater among species in the more open farmland habitats than among species in relatively sheltered woodlands, and declines also were greater at higher elevations.

It is not inevitable, however, that large-bodied species should do better than small-bodied species. The result depends on the relative rates at which the costs (energy losses) and benefits (energy storage) change with body size. It is far from clear whether invertebrates and ectothermic vertebrates will follow the patterns for birds and mammals, which are endotherms and cannot easily reduce their energy needs. Among certain groups of insects, there appears to be no clear relationship between body size and how long individuals can withstand either a loss of food or unusually cold temperatures (J. H. Lawton, personal communication.) Stress may force insects into diapause. Lawton has also pointed out to me that, for many insects, starvation or cold may be the wrong stress to consider, in any case. For example, the highly variable aphids may decline more than larger-bodied insects, because the aphids are more likely to be killed by the physical action of heavy rain than are the larger-bodied species. The environment in general may be more punishing to small-bodied species.

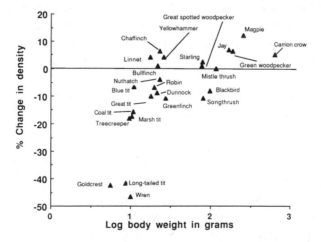

Figure 3.2. Small-bodied birds versus large-bodied birds: decreases during harsh winters. The data are the percentage changes in woodland bird populations in England between 1978 and 1979, plotted against body weight. The winter of 1978–79 was an unusually harsh one. (Redrawn from Cawthorne and Marchant 1980.)

In sum, how body size affects population resistance in all but mammals and birds is far from clear.

LONGEVITY

The second reason why population densities of large-bodied species should vary less involves longevity (Connell and Sousa 1983; Gaston and Lawton 1988a, 1988b). The yearly counts of individuals making up the estimates of temporal variability will not be independent of each other if the individuals are long lived. At one extreme, the same individuals may be counted in each year of the study. For example, a hypothetical population of adult redwood trees might show no variation over fifteen years of a study, because we would be counting the same individuals year after year.

Long-lived species will appear less variable than short-lived species when we use yearly estimates of density, because large-bodied animals live longer than small-bodied animals. Peters (1983), in his appendix VIIIb, summarizes a number of previous studies, mainly of vertebrates (but including that of Blueweiss et al. [1978] covering all animals), that demonstrate the relationship between size and longevity. We might ask whether the observed decrease in temporal variability with body size is explained entirely by the fact that large-bodied species live longer. One way to answer this question is to calculate temporal variability from generation to generation, rather than from year to year, and to see whether these new calculations show the same decrease in variability with increasing body size (Connell and Sousa 1983). This is not easy to do, especially if one wants to equalize the number of lifetimes that go into the calculations. There is a simpler alternative. Animals weighing more than 100 grams (= 0.1 kg; all birds and mammals in figure 3.1) will surely have *generation times* on the order of a year or more. The long generation times mean that the counts each year will contain individuals from overlapping generations, and the effect of longevity on reduced variability should be particularly pronounced. It is surprising that, for the species larger than 100 grams, temporal variability actually *increases* significantly with body size (figure 3.1). For species smaller than 100 grams, temporal variability is significantly less for those with relatively larger bodies. Indeed, among these smaller species, regression analyses show that temporal variability decreases more markedly than it does for all the other species combined (figure 3.1). This is a very odd result—although, as we shall soon see, it is not incompatible with existing theories. The result probably suggests that longevity does not have a powerful effect on variability. If it did, variability would continually decrease with body size, even for species with very large body sizes. Of course, the result may be the consequence of limited data. Notice, however, that by far the most numerous studies in figure 3.1 are those for species with body weights of 10 grams– 1 kg, and it is their unusually constant densities that cause the odd result. The data are not easy to dismiss, but they should certainly be confirmed by other compilations of population studies.

RESILIENCE AND TEMPORAL VARIABILITY: MODELS

Resilience—the rate of population recovery—also depends strongly on body size (see figures 2.4 and 2.5); the population densities of large-bodied species show low rates of increase. Small-bodied species not only produce more young per generation; they typically have more generations per year than do large-bodied species. When we consider the effects of resilience on population variability, however, we encounter a dilemma—two seemingly reasonable approaches giving diametrically opposite results.

Why High Resilience Leads to Low Variability.   May (1973) has suggested that resilient species may vary less than those that return to normal levels more slowly. This runs counter to the observation that larger-bodied species generally have lower variabilities, although it might explain why, for species larger than 100 grams, variability does increase with body size. To help understand May's argument, the examples given in Figure 3.3a are two simulations of a single-species model. The growth rate of each population is disturbed at random once in each generation. The variance of these random disturbances is the same in the two simulations, but their effects are very different. The simulations differ in the rates at which the populations return to equilibrium. In one simulation there is a long return time; in the other there is a short one. In the former case, particularly when the population level is small, it cannot easily recover, while in the latter case the population recovers more quickly, fluctuates less, and does not reach the low levels that, in the real world, might mean extinction of the species.

This seems like a plausible scenario, but it has some subtle features. Notice that the populations are subject to often violent mortality over a very short period of time. Indeed, the "random disturbances" act instantaneously on the population. This is a quite reasonable assumption for a population such as the song thrush of figure VAR.1, where we can easily identify both the episodic nature of major mortality and the obvious subsequent recovery. The mortality of the 1962–63 winter was rapid, and many of the birds probably died during

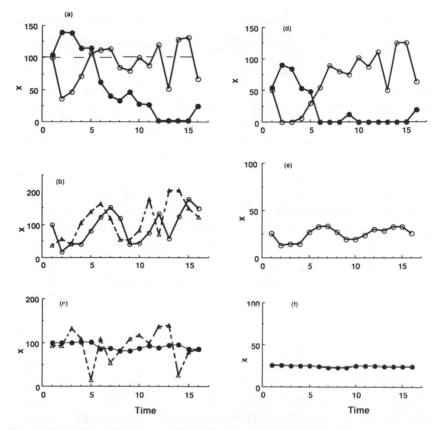

Figure 3.3. Addition of random variation, showing decidedly different effects on population variability that depend on how the variation is added to the model and on how resilient are the populations. The results are sample simulations of a discrete version of the logistic growth equation (see text). In (a) the equilibrium density capacity (dashed line) is constant; in one simulation the reproductive rate is high (open circles), and in the other it is low (closed circles). Randomly distributed terms are added to the species densities each generation. The population with the *higher* reproductive rate varies *less*. In (b) and (c) the equilibrium density changes randomly (dashed lines with triangles); in (b) the species' reproductive rate is higher than that in (c). The species in (b) varies *more* than the species in (c). In (d), (e), and (f) simulations have parameters identical to those in (a), (b), and (c), respectively, except that an additional trophic level has been added to the systems. The new equilibrium density for the species is 50 in all three simulations. The changes in equilibrium density for (e) and (f) are exactly as those in (b) and (c) and are, therefore, not shown. In (d) the more resilient of the two populations varies *less*, and in (e) the species varies *more* than its counterpart in (f). Comparisons of (d), (e), and (f), with (a), (b), and (c) show that it is resilience—and not just reproductive rate—that determines variability. Resilience depends on reproductive rate but also on the number of trophic levels in the model community. (Redrawn from Pimm 1982.)

the first few days of the prolonged cold. The period of recovery was much longer, spanning a period of several years.

Why High Resilience Leads to High Variability.   High resilience can lead to high variability for one of two reasons, either of which could provide an explanation for the results shown in figure 3.1. First, if mortality lags behind the rate of population growth, populations with sufficiently high resilience may first overshoot and then undershoot equilibrium. As it declines, the population continues to decrease below equilibrium because the birth rate now lags behind the rate of mortality. Such effects can lead either to simple two-point cycles or to the extremely complicated dynamics I discuss in chapter 6.

Second, let us suppose that the equilibrium density is itself variable. A species with a high rate of return to equilibrium will track the fluctuations in this density. Such a species will have a more variable density than does a species that responds more slowly, sitting it out until more favorable conditions prevail (Luckinbill and Fenton 1978; May et al. 1978; Turelli 1978; Whittaker and Goodman 1978). The return times in figures 3.3b and c are exactly those of the simulations in 3.3a. In figures 3.3b and c, the density of the species with the higher rate of return to equilibrium (figure 3.3b) fluctuates more than the density of the species with the lower resilience (figure 3.3c).

So far, our examples consider only a single trophic level. The return times of the model populations differ because the species differ in their reproductive rates. It is easy to show that the return time of the system—not the reproductive rate per se—determines the population's variability. Figures 3.3d–f repeat exactly the same simulations shown in figures 3.3a–c, except that another trophic level, a predator, has been added to the systems. The effect of the predator is to increase considerably the return times of both systems (chapter 2). The effects are what we would expect: where random variation is added to growth rates (figure 3.3d), fluctuations are much higher than previously (figure 3.3a). When the equilibrium density of the prey species varies randomly (figures 3.3e and f) we see only a slight variation in density, compared with results shown in figures 3.3b and c.

However the changes are brought about, long return times mean less variation in densities if random variation is added to equilibrium densities and more variation in densities if random variation is added to growth rates. A mathematical analysis of this topic is provided by Turelli (1978) and Pimm (1984b).

What does it mean *ecologically* if random variation "is added to equilibrium densities?" Imagine reducing the equilibrium density of a species by half. How will the increased mortality of the species act? Clearly, it will be spread out over several generations as the population approaches the new and lower equilibrium, even though the greatest mortality will occur initially. In contrast to the example of the song thrush and of random variation added directly to a species' density, when variation "is added to equilibrium densities" the mortality occurs over a time period longer than the species' generation time.

In short, the two arguments about the way resilience affects variability are not strict alternatives. Whether increasing population resilience causes increased or decreased variability will depend largely on the nature of environmental disasters—in particular, on their duration and frequency. We would expect all populations to closely adjust their densities to very-long-term environmental changes. Over shorter periods, highly resilient populations will increase and decrease more quickly in response to environmental variations than will less resilient populations, and so the latter populations will vary less. Finally, over intervals dominated by episodic catastrophes causing severe population crashes, highly resilient species will vary less. Overall, the data on body size and variability that are presented in figure 3.1 are consistent with the idea that more resilient populations are more variable—and for the species weighing more than 100 grams the data are consistent with the alternative! With a bit of speculation, we might even understand why the least variable species have intermediate body sizes.

In sum, for the aphids-to-elephants range that the data of figure 3.1 encompass, there are many explanations for the decrease in variability. Explanation of the patterns of variability in terms of decreasing resilience would be more compelling if there were more detailed studies of how resilience affects variability; it is to these studies that I shall now turn.

RESILIENCE AND TEMPORAL VARIABILITY: EMPIRICAL STUDIES

Laboratory Studies of Protozoa

There are experimental data that support the idea that more resilient populations are more variable. Luckinbill and Fenton (1978) experimentally manipulated the equilibrium density of species of protozoa by altering the supply of bacteria on which they fed. Luckinbill and Fenton compared two species of ciliates: *Paramecium primaurelia* and *Colpidium campylum*, the smaller of the two. Both species were cultured in Cerophyl medium inoculated with the bacteria *Enterobacter aerogenes*. Population numbers were determined by the quantity of bacteria supplied.

Estimating Rates of Return. To estimate how quickly the populations returned to equilibrium, six replicate populations of each species were grown to

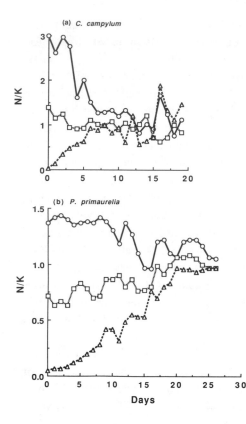

Figure 3.4. Populations of two ciliate species as they return to an
equilibrium set by their food supply. After they had been growing
with a constant level of their bacterial food supply, populations
were either set to very low (triangles) or to very high densities
(circles) or were left unchanged
(squares). *C. campylum* (a) returns
to equilibrium much more quickly
than does *P. primaurelia* (b). Numbers, N, are scaled to their long-
term average, K, so all lines eventually approach $N/K = 1$. (Redrawn from Luckinbill and Fenton 1978.)

their saturation limit in Cerophyl medium for fourteen days. At that point, two
of the replicate cultures were retained untreated as controls. Experimental populations were then displaced from their equilibrium. Two of these had cell
densities reduced to near zero, and two were increased to approximately twice
the density of populations before displacement. Converging experimental populations were sampled daily over the following twenty-eight to thirty days.

Density changes in the displaced populations are shown in figure 3.4. Experimental populations converged to the same approximate density as did controls. Equilibrium density was estimated as the average density that populations
maintain between convergence and the termination of the experiment. The species demonstrate different rates of return to equilibrium, and, indeed, the
smaller, faster-growing species, *C. campylum*, was the more resilient.

The Effects of Resilience on Variability.   Luckinbill and Fenton then investigated the effect of varying the food supply. They asked: "Do slowly regulating
species lag behind and thus vary more widely, while fast regulators closely

track the changing equilibrium? Or does instability enhance the tendency of strong regulators to over-compensate, causing wider oscillations?" (Luckinbill and Fenton 1978).

Because the level of bacteria determines protozoan population numbers in these systems, the period during which equilibria remain constant can be varied by simply changing the rate at which food is supplied to the systems. These changes can be made either more or less frequently. The relative abilities of the two species to maintain constant population sizes can then be compared in several environments, ranging from relatively constant to highly variable.

Luckinbill and Fenton prepared different concentrations of bacterial food by modifying the standard preparation of Cerophyl. Bacterial concentrations were varied, from a high level (full strength medium) to a lower level (dilute medium), for each species. When there was a change from one level of food supply to another, the protozoans were concentrated by low-speed centrifugation and were resuspended in the desired dilution of medium. That same concentration of medium was resupplied to daily samples of the populations until the equilibrium was shifted again.

Food levels alternated between high and low densities, at three frequencies. At the lowest frequency, alternating high and low food levels were held constant for twenty-one consecutive days, a period that assured that the populations reached the numbers determined by the level of the food supply. At the intermediate frequency, bacterial levels were alternated at two-week intervals. In the most rapidly varying environment, equilibria were alternated at one-week intervals.

The average densities for four replicate cultures of each species under this treatment are shown in figures 3.5. The species with the shorter return time, *C. campylum,* experienced oscillations of increasing amplitude as the food levels were alternated more frequently, and it became extinct after more than 175 days. On the other hand, *P. primaurelia,* with a comparatively sluggish return time, enjoyed more even population levels as the environment became more variable.

Luckinbill and Fenton drew three conclusions from their study. First, different species respond differently to changing environments: the resilient species became unstable when tracking a fluctuating equilibrium, whereas the less resilient species achieved relatively even population densities. Second, the comparatively large size and greater individual consumption of bacteria by *P. primaurelia* sustained this species through periods of depleted food supply in the experiment in which the food supply was rapidly fluctuating. Third, the differential effects of environment are determined more by the frequency of environmental fluctuations than by their magnitude.

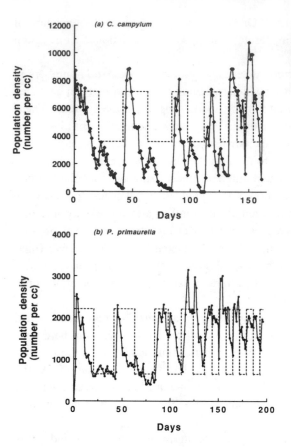

Figure 3.5. Population fluctuations in two species of ciliate protozoa (solid lines) as their bacterial food supplies (dotted lines) are changed. The more resilient species (a) fluctuates much more than the less resilient species (b) and often approaches extinction. (Redrawn from Luckinbill and Fenton 1978.)

With these conclusions, Luckinbill and Fenton showed that more resilient populations can be more variable. Moreover, since their experimental design cleverly controlled the protozoans' food supply and thus its fluctuations, their data clearly show the mechanism involved. It is still important to ask whether this possibility will hold in the world outside the laboratory. Therefore, I turn next to field studies.

Noctuid Moths

Spitzer et al. (1984) showed that species of moth with the greatest potential reproductive rate were the more variable. The moth data were from a light-trap situated near České Budějovice, in south Bohemia, Czechoslovakia (Rejmanek and Spitzer 1982; see chapter 4). The light-trap was operated nightly from April to November 1967–79, with the exception of 1969.

To estimate resilience, females of the most abundant species were collected and dissected, and the number of eggs they contained were counted. Because of possible earlier oviposition, these numbers might be underestimates of egg numbers, so Spitzer et al. calculated the means of the three highest values for each species. The species with the highest numbers of eggs per female were also the more variable (figure 3.6).

Bird Populations

The second field study uses the data from the British Trust for Ornithology's CBC (see chapter 2). I analyzed the available data up to and including 1977 and used the coefficient of variation as the measure of variability (Pimm 1984b). Estimating the rate of return to equilibrium presents more difficulty. A species' rate of return to equilibrium depends on its reproductive parameters, but it also depends both on the number of trophic levels in the system in which the species finds itself and on the characteristics of the other species in the system (chapter 2). Consequently, I chose another approach to estimate the return times. (That approach is detailed in the first section of chapter 2 and in figure 2.2.) Subsequent to my analysis, it became clear that for these bird species there is a relationship between return time and reproductive rate. O'Connor (1981) found that a measure of reproductive effort (eggs per season) was closely correlated with another measure of resilience (the rate at which the population increased from 1963 to 1964 following the severe 1962–63 winter; see figure 2.3). O'Connor's estimates of resilience are in turn closely related to the estimates of resilience I obtained.

The CBC data were collected in two habitats: farmland and woodland. The species include two trophic levels: insectivores (almost entirely migratory) and gramnivores, with varying tendencies toward omnivory (largely resident). Densities of the migratory species tended to vary more (chapter 5). These dif-

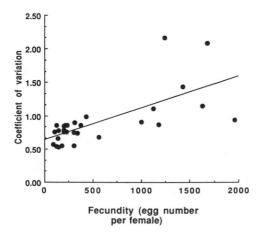

Figure 3.6. Species of moths in southern Bohemia, which vary more when they are more fecund (and so when their populations are likely to be more resilient). $P < 0.05$. (Redrawn from Spitzer et al. 1984.)

ferences, however, were insufficient to alter the qualitative results I shall now present.

For both farmland and woodland populations, the data show that, as resilience increases, the variability of the population density *decreases* significantly (figure 3.7). All but two of the estimates of resilience (willow warbler and blackbird in woodland) were small enough to suggest that the populations should return to equilibrium without overshooting it.

This result is consistent with the increase in variability with increasing body size (and so decreasing resilience) for large-bodied species (see figure 3.1). Moreover, the increase in variability with increasing reproductive rate for the moths matches the overall pattern for small-bodied species. Thus, both detailed studies support the strange result that species of intermediate body size will be the least variable.

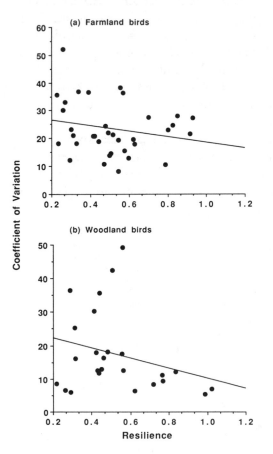

Figure 3.7. British birds' population variabilities, declining with increasing resilience in farmland (a) ($P < 0.05$) and woodland (b) ($P < 0.05$) populations. The coefficients of variation are plotted against resiliences calculated by the method illustrated by figure 2.2.

The Contribution of Cycling to High Variability

One way in which high resilience can cause high variability is by the generation of cycles. This motivates two questions: Are species whose densities do cycle more variable than those whose densities do not cycle? Are highly resilient species generally those that cycle? In other words, does high resilience cause cycling, and does cycling cause high variability, and, if so, is this two-stage process sufficient to explain the overall increase of variability with increasing resilience?

To see whether highly resilient populations are those that generally cycle, we can examine the populations in figure 3.1 that obviously cycle. There are some obvious limitations in doing this. There were few cycling species. Populations of many species may show simple cycles, but these may not be apparent in only fifteen years of data. Most of the species for which I recognized obvious cycles (lynx, for example) were studied for much longer than fifteen years, thereby making the cycles more obvious than they would have been otherwise; nor can I exclude the possibility that populations show deterministic changes of density far more complex than cycles. (I shall return to this complicated subject in chapter 6).

Populations with obviously cycling densities were not restricted to those small-bodied (and thus more resilient) species. Indeed, they tended to be medium- to large-bodied species of mammals—voles and lynx, for example—though there were some insects that showed obvious cycles. Cycling species did tend to be more temporally variable than were noncycling species of the same body size. In chapter 4, I shall discuss results showing that vole populations whose densities cycle have levels of density variation higher than those of vole populations whose densities do not cycle. My tentative conclusion is that, if high resilience causes cycling which contributes to high temporal variability, the effects are independent of and *additional to* the main correlation between body size and temporal variability. Small-bodied species are highly variable even when no cycles are apparent.

OTHER FACTORS

The species contributing to the relationship depicted in figure 3.1 are diverse both trophically and taxonomically, and there must be many other correlates of body size affecting temporal variability as well. Within taxa, large-bodied species tend to be more polyphagous (Wasserman and Mitter 1978; Gaston and Lawton 1988a, 1988b). Small-bodied species may be more variable because their diets are more specialized than those of large-bodied species. Does it seem reasonable to suggest that mammals are less variable than insects because they are sufficiently more polyphagous than insects? I cannot see a way of sensibly comparing the diets of elephants and aphids; the differences in their natural history are just too great. In figure 3.1, however, the smallest species (aphids) are herbivorous, as are the largest species (mammals). Therefore, the decline in variability with increasing body size is not confounded by gross trophic differences.

SUMMARY AND CONCLUSIONS

Across all animal species, densities of small-bodied species tend to be more variable than those of large-bodied species. This result may have been expected, but its explanation is not obvious; there are several explanations that may account for this tendency. Strangely, above 100 grams, large-bodied species become more variable than small-bodied species.

During inclement conditions, the densities of small-bodied vertebrates decline more than the densities of large-bodied species. In mammals at least, differential declines may occur because, as body size increases, so do the fat stores available to the species relative to its energetic needs.

Large-bodied species live longer; we may be counting the same individuals from year to year. Long life acts to reduce population variability, but this effect is probably not a powerful one. For species clearly living more than one year (those over 100 grams), densities of larger-bodied species are *more* variable.

Resilience is greater in small-bodied species (chapter 2), but the effects of resilience on variability are complicated. High resilience can contribute to high variability because highly resilient species can track environmental fluctuations more closely than can less resilient species. However, highly resilient species can recover more quickly from short, sharp declines in density and this will make highly resilient species less variable. Detailed examinations of the relationship between resilience and variability provide evidence for each of these diametrically opposite effects.

APPENDIX: THE MEASUREMENT OF POPULATION VARIABILITY

In choosing a measure of population variability, we require both that it be a good representation of the data on which it is based and that it enable us to compare variabilities of different populations. Such criteria are obvious, but meeting them is far from easy. An excellent review of the problems in calculating variability is provided by McArdle, Gaston, and Lawton (1990). While many of their concerns were already known to statisticians, ecologists have sometimes overlooked them. As a consequence, some previous and important results about variability (Gaston 1988; Gaston and Lawton 1988a, 1988b) had to be retracted by McArdle et al. (1990).

For a measurement to be a "good representation," it needs to utilize the information provided by all the observations and not be unduly sensitive to a small number of them. No single measurement of population variability will always meet this requirement. Some populations, such as *Bupalus* (figure VAR.1c), usually have low levels, with occasional excursions to very high levels. These rare, high levels contribute strongly to either the standard deviation of population numbers or the CV (the standard deviation divided by the mean). The standard deviation of numbers may vary greatly, depending on whether these rare, high levels were encountered. As a consequence, the standard deviation may increase the longer the population is sampled. If we transform the population numbers, by taking their logarithms, the distributions of these logarithms of densities may be distributed more symmetrically. (Compare the two graphs for *Bupalus,* for example.) If this is the case, the standard deviation of the logarithms (SDL) will provide a better measurement of variability. This logarithmic transformation places more weight on variations at low densities and places less weight on variations at high densities. For these reasons Williamson (1972, 1984) has argued strongly for the need for logarithmic transformation of population sizes. In the case of the song thrush (figure VAR. 1a), however, the logarithms of density data are strongly influenced by the rare, very low densities caused by hard winters. For such species, CV will be a better measure than is SDL.

For some species we should use population sizes, for other species the logarithms of the population sizes. In perhaps the majority of cases, however, both measures will be closely correlated, and we can use either. Of course, where the standard deviation of numbers are relatively different than SDL of numbers, we can see the relative effects of unusually high or low population sizes. Wherever possible I shall use either CVs or SDLs, whichever seems more appropriate, when comparing similar sets of species; and I shall use both measures when comparing very different kinds of species.

Wolda (1978, 1983) has measured variability in another way that has attractive features for comparing populations. He calls it "annual variability" (AV). This index is based on differences of the logarithms of population sizes in consecutive years. If the logarithms of population density are near symmetrical in their distribution, then so are the differences in the logarithms. As Williamson (1984) shows, there is a simple relationship between AV and the SDLs of the population sizes. Consider a plot of the density of the population in one year versus the density of the population in the next year: such a plot would have one point fewer than the number of years of data available to us. If we now define r as the correlation coefficient between the points in this plot—that is, the population densities in these consecutive years—then

$$AV = SDL \cdot [2(1 - r)]^{(1/2)}. \tag{3.1}$$

For short-lived species this *autocorrelation* coefficient between populations measured at yearly intervals will often be small. The density one year will have little effect on the density the following year. Under such cases, AV and SDL will have similar values. For a population growing exponentially (the world's human population comes close!), r will be close to 1. This means that, although SDL will be large, AV will be close to zero, because equation (3.1) contains a term in $1 - r$. What this means is that AV looks at variability *about* trends in the densities—such as, for example, the exponential growth of the human population.

The logarithmic transformation has an obvious difficulty: it cannot be used if there are zeros in the data. This is often the case, especially when species are quite rare and particularly when species vary greatly from year to year. Ecologists typically add one to the numbers and then take the logarithms. This is a mistake, for the result is to greatly underestimate the true variability. Consider a population which has true relative densities of 10^{-3}, 1, 10^{-2}, 10^{-3}, 10^{-1}, 10^{-1}, 1, and 10^{-1}; such densities look a little like those of *Bupalus,* and this is deliberate. Of course, we cannot estimate the true densities when they fall to low levels, so we may actually report those densities as 0, 1, 0, 0, 0, 0, 1, and 0, respectively. Adding 1 to each, we get 1, 2, 1, 1, 1, 1, 2, and 1. Far from concluding that this is a very variable population, we consider it to be very constant from year to year. This bias is discussed in more detail by McArdle et al. (1990). Such a bias can lead the unwary to conclude that rare species are less variable than common ones; rare species are more likely to have their densities recorded as zero. Rarity is often correlated with other variables that are of interest to us. The potential for being misled is considerable.

In addition to the problems already discussed, there is the issue of population abundance. One of the reasons for taking the logarithms of density or using the CV is to make the estimates of variability independent of the overall abundance. CV is independent of the mean, for example, only if the standard deviation increases in direct proportion with the mean. The pioneering work of L. R. Taylor and his colleagues (see, for example, Taylor et al. 1978, chapter 4) shows that this is often true. But it is not *always* true, and so we must be alert for overall abundance affecting estimates of variability. The obvious solution is to test for this directly, correct the estimates of variability accordingly, and then look at how other factors affect the corrected estimates of variability.

Finally, there is the issue of spatial scale. As McArdle et al. (1990) argue, variability is likely to depend on the spatial scale over which we sample the population. The problem is that, in general, we have no idea what that scale is. The song thrush densities in the figure are representative of changes over an area much larger than the area of the approximately one hundred farms from which the data were actually gathered. My guess is that the data reflect a scale of thousands of square kilometers, but certainly the data are over a large scale. Given this, it is worthwhile to contrast results over known and very small scales—for example, populations in small islands.

The Data of Figure 3.1

The data in figure 3.1 pose a number of problems because they involve comparisons across such a broad range of species. There is no sensible way to correct these data for the effects of population abundance on variability, except to note McArdle et al.'s (1990) argument that, for the majority of common species, the measures of variability I used will not depend strongly on abundance. (Redfearn and I excluded species for which density estimates were often zero and which would necessitate adding 1 to the densities before taking logarithms. We also standardized the length of years included in the calculation of variability: fifteen years for each species.) Variability increases the longer we measure it—a point to which I shall return in chapter 5.

An additional problem with these data involves the very different spatial scales over which they were collected. The CBC bird data reflect changes over much of England, and the aphid and moth data samples reflect changes over large but unknown areas. Other data are much more local. The best comparisons of data on clearly different scales are for birds on small islands (Pimm et al. 1988) versus the countrywide CBC data (tabulated in Pimm 1984b). When contrasted with the wide range of variabilities found in figure 3.1, the variabilities (estimated as CVs) are very similar in both data sets. Similarly, Ehrlich's (1965) study of a butterfly population limited to an area of a few hectares yields an estimate of variability in general agreement with the data in figure 3.1. While I am sure that effects of spatial scale will be found and probably will be

important, the available evidence suggests that they are not sufficiently important to swamp the effects which I have discussed in this chapter. Uncertainty
will remain, however, because comparisons of variabilities on different scales,
such as this comparison between island and mainland birds, have not been
made for other taxonomic groups.

4

The Effects of Food-Web Structure

THE IDEAS OF MACARTHUR AND WATT

The temporal variability of a population depends, in part, on food-web structure—that is, how the species interactions are organized within a community. MacArthur (1955) argued that population densities of species should be more stable in communities with more complex food webs. By "stable" I think he meant "less variable" (see chapter 1). If so, his idea provides a good starting place for discussions of food-web structure and temporal variability, but it focuses on only a small part of a complex problem. Watt (1964) derived a prediction diametrically opposite to MacArthur's idea on the relationship between temporal variability and food-web complexity. To resolve these apparent contradictions, I must develop a general model of how food-web structure does affect temporal variability. With the theoretical predictions in hand, I shall present a small number of studies that attempt to test them.

MacArthur's Idea

When MacArthur argued that more complex communities were also more stable, he relied on careful definitions of both complexity and stability. Complexity was the number of pathways energy took to reach a given population; stability was the extent to which population density changed if one of those pathways failed. The argument is no more than the admonition "don't put all your eggs in one basket." If you drop your only basket, you will break all your eggs; if there is only one energy pathway, the energy supply is highly vulnerable. The crux of this argument is about maintaining a reliable food supply in an unpredictable world.

This simple idea can be illustrated by the comparison of two predatory species, one monophagous and the other polyphagous (figure 4.1a). These predators may be herbivores feeding on their plant "prey" without any change in the argument. In the kind of complex system envisioned by MacArthur, each predator would feed on many different species of prey, with much overlap between

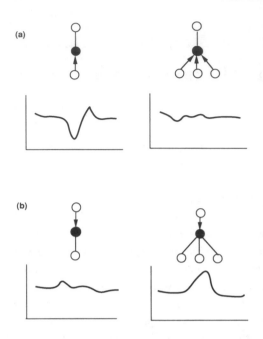

Figure 4.1. Two opposing arguments for the possible relationships between population variability and the degree of polyphagy: (a) MacArthur's suggestion that monophagous species (left) should be more variable than polyphagous species (right) because they will decrease most when one prey species declines in abundance and (b) Watt's contrasting argument that polyphagous species may be more variable because they may be able to reach higher numbers during periods when their predators fail to control their densities. Circles represent species, and lines represent trophic interactions between species, such that species higher on the page feed on species beneath them. The species for which we measure the population variability is shaded. Arrows indicate which trophic interactions are most important in determining the density of the species of interest.

predators. In a simple system, the food chains would be more or less distinct. The specialized (monophagous) species is likely to vary more than the more generalized (polyphagous) species. If one prey species of the polyphagous predator becomes rare, then individuals can switch to alternative prey. In contrast, the monophagous predator would become rare if it were deprived of its sole species of prey.

Actually, prey switching is not necessary for this argument that polyphagous species should be less variable than monophagous ones. Suppose that the polyphagous species is one of those insect herbivores composed of separate subpopulations, each of which specializes on a different host plant; Fox and Morrow (1981) provide some examples. When one plant species becomes unavailable, those individuals that feed on it may be doomed, but only a fraction of the herbivores' total population is likely to suffer.

The essential features of this argument are three: (1) the argument is about the reliability of the factors that determine population density; (2) the factors that control population density are the species' prey or resources; and (3) high variability is caused by severe reductions in abundance.

Watt's Idea

Watt (1964) tested MacArthur's idea by using data on the temporal variability of Canadian moths, but it is Watt's ideas—and not his data—that I want to discuss first. Watt argued that polyphagous species could be *more* variable than monophagous species. He suggested that species typically may be kept at levels below those set solely by the availability of their food supply, by the attentions of their predators, parasites, diseases, competitors, or generally inclement weather. Now, suppose that occasionally these various enemies are themselves rare or that the weather, for once, is favorable to the prey species we are considering. What will determine how abundant the species will become before the population declines to a more usual level? Two parameters seem preeminent: (1) how fast the population can increase—that is, its resilience (chapter 2)—and (2) how much of the habitat the species can exploit.

Both parameters involve the response of the population after the failure of the factors that normally control the population. Let us suppose that we are comparing species that have broadly similar rates of increase. What affects how much of the habitat a species can exploit? A monophagous species can only exploit a large proportion of its habitat if the habitat is a monoculture of its food supply. A polyphagous species is likely to be able to exploit a large proportion of its habitat because of its diverse feeding habits. Therefore, populations that reach the highest relative densities will be *either* monophages in monoculture *or* polyphages. With the economically important exceptions of species in monocultures, populations of polyphagous species will be more variable than monophagous species (figure 4.1b). This is the opposite of the relationship suggested by MacArthur. Yet, like MacArthur's argument, Watt's has three essential features: (1) the argument is about how the population responds when the factors controlling it fail; (2) the factors that control the population are the species' enemies; and (3) high variability is caused by large increases in abundance.

When we compare MacArthur's and Watt's ideas, we see they have nothing in common. MacArthur was discussing population crashes and the reliability of the factors that control abundance, while Watt was discussing population outbreaks and concentrated on the consequences of those factors failing to provide reliable control. MacArthur implied that the factors controlling a population were resources, and Watt felt that the factors controlling a population were those that kept the population rare. Although this sketch is an oversimplified presentation of MacArthur's and Watt's ideas, it is clear that a complete understanding must combine both views, because they deal with different aspects of the problem. To gain a more complete understanding of how food-web structure affects temporal variability, we must consider all the elements presented by MacArthur and Watt: the reliability of the factors that control population density and the response of the population when those controlling factors fail.

In sum, prey diversity can affect variability in opposite ways. Increased prey diversity can *reduce* variability by lessening the size of population declines, or increased prey diversity can *increase* variability by leading to greater population increases. But what of predator diversity? Neither MacArthur nor Watt discussed it, but their principles should still apply. Increased predator diversity might *reduce* the variability of their prey species, because diverse predators might provide a more reliable control of those prey densities. A species preyed on by many species of predator may be much less likely to escape the attentions of those predators than is a species exploited by only a very few species of predator. Increased predator diversity may reduce the likelihood of unusually high population densities.

Can increased predator diversity *increase* variability? In order for this to occur, increased predator diversity would have to be responsible for declines in prey abundance that are greater than those that occur when there are only a few species of predators to depress the prey abundance. As Mark Williamson (personal communication) has pointed out, the first law of ecological modeling must be that it is possible to provide a theoretical model of almost any circumstance; and I am sure this is no exception. At this stage, such an effort would be pure speculation, and I prefer to leave that to the reader.

How Do We Test These Ideas?

If we are going to examine the complex relationships between temporal variability and food-web structure, we shall need to measure a number of food-web features, including the diversity of prey exploited and the diversity of predators that exploit the species of interest to us. We also may need to know the influence of abiotic factors, and we must surely need to know whether a population is limited more severely by its resources or by its enemies.

Complex though these relationships may be, there are data that we can use to begin to tease them apart. What we can expect to measure in the field are the temporal variabilities of a set of species, the different numbers of predatory species that exploit it, and the different numbers of prey species it exploits. We will probably not find data on a species' other enemies—its diseases and parasites—although for insects we may find data on the number of parasitoids (parasitic insects that lay their eggs on or near the host and whose larvae consume it). No one has yet assembled data on the temporal variability of a set of species, the diversity of their prey, and the diversity of their predators. However, we can look both at a number of studies that relate the temporal variability in population density to the diversity of prey exploited and at other studies that relate temporal variability to the diversity of predators. To analyze these data, we obviously need two kinds of information: long-term population data and the numbers of species of prey or predators involved.

If we are looking at temporal variability and either prey diversity or predator diversity, we must also make an assumption. The number of prey species (N_1) that our species of interest exploits must be independent of the number of species of predator (N_3) that exploit our species of interest. The subscripts on the N's reflect the relative trophic positions; our species of interest is at trophic level 2. The independence means that the correlation between N_1 and N_3 is zero.

Suppose that the more species of prey the different species at trophic level 2 exploit, the more species of predator they suffer: N_1 and N_3 are positively correlated. Then what if we compared the temporal variability of both a monophagous and a polyphagous species and found that the polyphagous species was less variable? We might conclude that the polyphagous species was less variable because it had a more reliable food supply and that both species were limited primarily by their food supplies. However, the polyphagous species might be limited by its predators, and the greater diversity of predators might be responsible for its numbers being more reliably controlled.

Do polyphagous species suffer the attentions of more predatory species? I have asked this question before (Pimm 1982), but existing food-web data were insufficient to provide an answer. Two recent studies of insects (Hawkins and Lawton 1987; Stiling 1987) were unable to detect a significant correlation between the number of species of prey a species exploits and the number of species of predators it suffers.

Temporal Variability and the Diversity of Prey

It is difficult to count the numbers of prey species on which many predatory species feed. However, we have good data for herbivorous insects, which tend to be relatively specialized, so that entomologists have been able to record the plant species on which they feed. For species at higher trophic levels, say, insectivorous birds, we may have good descriptions of the habitats they occupy, but we do not know details of diet. Therefore, the examples in this section all involve herbivorous insects.

Canadian Moths

Watt (1964, 1965, 1973) analyzed data from the Canadian Forest Insect Survey (CFIS) (McGugan 1958; Prentice 1962, 1963). The data are yearly counts of immature stages of relatively large species of moths, collected by hand picking and by beating trees. The yearly totals represent the summed collections from areas in various parts of Canada. The survey was intended to provide only approximate indices of relative abundance, and the data were collected quite haphazardly. However, the data set is a large one, so there is the possibility of selecting the best-studied species for more detailed analyses.

Watt analyzed 552 species of moth and found that those feeding on a wide variety of host tree species were more variable in their densities than were more specialized species. This finding contradicts MacArthur's (1955) argument. Watt (1964) suggested that his results were best explained by the amount of available habitat—rather than by the number of host species per se—because he also found that certain specialized species were also highly variable. The spruce budworm, for example, is very variable yet specialized. It lives in habitats where its principal hosts, balsam fir and white spruce, occur in stands with few other tree species. When predators, parasites, or diseases are no longer limiting factors, the moth population can increase very rapidly indeed. Watt also found some interesting differences between species of moths with different lifestyles. Those that fed gregariously were generally more abundant and more temporally variable than were solitary feeders.

Confounding Factors. Two decades of hindsight suggest some methodological difficulties with Watt's analyses. Because of these difficulties and because we were fascinated by the differences between Watt's ideas and MacArthur's, Andrew Redfearn and I (Redfearn and Pimm 1988) decided to reanalyze the original data and to seek other data sets to test the rival ideas. We began by considering factors that might confound Watt's analyzes. An independent discussion of the confounding factors inherent in studies of variability appeared in the same year as our study (Gaston and Lawton 1988a, 1988b), with a further discussion taking place the next year (Owen and Gilbert 1989). All the discussants raised similar concerns. In particular, Watt's finding that variability in density increased with degree of polyphagy could be a spurious one, because both variables might depend on abundance.

Abundance-polyphagy. A rare species may appear to be more specialized than it really is, simply because its rarity will lead to our finding it on few host plants. This is a particular problem with the CFIS data, for the same data were used to estimate degree of polyphagy and abundance. A species collected only ten times could, at most, exploit ten species of tree. A species collected one hundred times could be found on as many as one hundred species of trees.

Abundance-variability. Occasional high numbers (reflecting either truly high abundance or unusually active collecting) will contribute to high variability and will also inflate the mean abundance. Taking logarithms of the densities and estimating variability by using the standard deviation of these logarithms (see chapter 3) will remove some of variability's dependence on mean abundance. Unfortunately, for very rare species, taking logarithms will not be possible unless we add 1.0 to the likely zero estimates of density. This leads to underestimates of variability for such species (see the appendix to chapter 3). The possibility exists that uneven collecting effort will lead to spurious increases in variability with increasing abundance. (I should explain why I have not included Gaston's work [Gaston 1988; Gaston and Lawton 1988a, 1988b] in the examples I shall soon present: the estimates of variability in these studies are biased by the inclusion of many zero estimates of densities, and McArdle, Gaston, and Lawton [1990] have withdrawn their conclusions.)

In the CFIS data we expect high estimates of abundance to be associated with high estimates of polyphagy and also with high estimates of variability. Watt's observation of increasing variability with increasing polyphagy may be no more than a consequence of the fact that each of these variables depends on apparent abundance.

Statistically, the presence of these confounding factors poses no difficulty. If either abundance and variability or abundance and degree of polyphagy are related purely because of sampling artifacts, we can factor out that part of the relationship between variability and polyphagy that is due to their mutual correlation with abundance, by using a partial regression analysis. The difficulty is not statistical but conceptual. If apparent dietary specialization or low variability is caused by poor sampling, we should correct it. The difficulty is in the theoretical argument which states that dietary specialization might decrease average density by making population crashes more severe. If the theory is correct, abundance is *not* a confounding variable, and estimates of polyphagy or variability should not be corrected for it. The question then becomes whether we should statistically remove the effects of abundance. There is no answer, since a correlation analysis does not give the direction of causation. Redfearn and I analyzed the data both ways, looking at the relationship between degree of polyphagy and variability directly and, where there were relationships between these variables and abundance, looking at these variables' correlation corrected for abundance.

A Reanalysis of the Canadian-Moth Data. In our analysis of the data on Canadian moths, we calculated variability, abundance, and polyphagy for 151 species. Our data set was much smaller than Watt's, for several reasons. Instead of using the CFIS data to obtain abundance, variability, and degree of polyphagy, we used Tietz's (1972) data as an independent source of dietary information. Tietz provides a list of genera on which the moth larvae have been found. We did not use the diversity of plant species used by the moths because, to our knowledge, these data were not available. In all probability, the diversity of species used will closely correlate with the diversity of genera used by the various moth species. We also restricted our analyses to moth species that were recorded sufficiently frequently for us to be confident of the abundance data; that is, we avoided the pitfalls of logarithmically transforming data including zero-density estimates.

We recognized differences between gregarious (sixteen species) and solitary species (the remainder), as as well differences between families. Most of the solitary species belong to one of four families: Geometridae, Lymantridae, Noctuidae, and Notodontidae.

Within these data, the degree of polyphagy clearly increased with increasing abundance, but the relationships between variability and abundance were much weaker. For analyses both corrected and uncorrected for the effects of abundance, temporal variability usually decreased with increasing degree of polyphagy, but there were no significant regressions. We could not confirm Watt's result that temporal variability increased with the degree of polyphagy. Indeed, there was only one such positive relationship between variability and degree of polyphagy, and that occurred in the family Lymantriidae. For the gregarious species and families Geometridae, Noctuidae, and Notodontidae, variability decreased with degree of polyphagy.

A Limited Analysis of the Canadian Moths.   Watt and Craig (1986) analyzed data on five medium-sized Canadian-moth species. All five species are solitary leaf defoliators, have similar ranges, and share Douglas fir as their principal host. They differ from each other in how many other species of tree they exploit as alternative hosts. Species with few other hosts tended to be more variable that those with many alternative hosts. Again, there was not a statistically significant relationship between variability and degree of polyphagy.

British Moths

Redfearn and I next considered data on twenty-five species of British moths collected by the Rothamsted Insect Survey, between 1966 and 1982 (see, for example, Taylor et al. 1978, 1979, 1980; Tatchell et al. 1983). The data are yearly counts of moths of economic interest that were collected in light traps. We used data from fifteen sites spread across Britain and summed densities across sites because records at each site were usually too sporadic for us to be confident of the density estimates.

We could find no source that gave detailed records of the number of host plants used by the larvae of these species. Instead, we made a qualitative assessment of the degree of polyphagy for each species (on the basis of its description in the field guide by Skinner [1984]) and assigned each species to one of three groups of increasing polyphagy. Unlike our analyses of the Canadian-moth data, our estimates of variability in British moths were based on adults and our estimates of degree of polyphagy were based on larvae.

We found no relationship of either variability or degree of polyphagy to abundance, so there was no need to correct the data for abundance. The most specialized species were significantly more variable than either the most polyphagous group or the intermediate group, but there was no difference between the intermediate group and the most polyphagous group. Again, we noticed that variability declines with increasing polyphagy. The data are far from satisfactory, however, for they included only three species belonging to the most specialized group, and only one of these species, *Operophtera fagata*, was particularly variable. Without this species, there would have been no significant differences between any of the groups. Worse, the various species of moths occurred in different habitats: some fed on trees, which are likely to provide relatively constant resources from year to year, while others fed on grasses, which may vary far more from year to year.

Czechoslovakian Moths

The last stop on this world tour of moth population data brought us to Rejmanek and Spitzer's (1982) study of noctuid moths. They examined seventy-four species of moth caught at a light trap located near Čcské Budějovice, in southern Bohemia, over a twelve-year period. They grouped the feeding habits of the larvae into four categories: those that fed on (1) one genus, (2) one family, (3) two or three families, or (4) more than three families of plant. The data showed both no relationship between polyphagy and abundance and no relationship between variability and abundance; therefore, once again, we could use uncorrected data. The most polyphagous species were the most variable, and, for a change, the result was highly significant statistically (figure 4.2a).

Rejmanek and Spitzer also found that temporal variability differed between species of different habits and habitats. Migratory species were more variable than nonmigratory species, and species in agricultural ecosystems were more variable than species using shrubs and trees. (Similar results both in studies of migratory and nonmigratory birds and in studies of forest and grassland insects will be presented in chapter 5). These confounding factors do not alter their overall result, however.

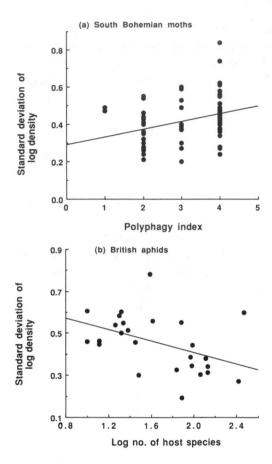

Figure 4.2. Population variability (a) increasing significantly ($P = 0.006$) with increasing polyphagy for seventy-two species of South Bohemian Noctuid moths analyzed and collected by Rejmanek and Spitzer (1982) and (b) decreasing significantly ($P = 0.02$) with increasing polyphagy for twenty-five species of British aphids analyzed by Redfearn and Pimm (1988) and collected by the Rothamsted Insect Survey at Silwood Park, England. Variability is measured as the standard deviation of the logarithms of densities, calculated over fourteen to sixteen years for the aphids and over twelve years for the moths. For the aphids, polyphagy is estimated by the logarithm of the number of recorded host plants. For the moths, polyphagy is scored as the following categories: 1 = on one plant genus; 2 = on one family; 3 = on two to three families; and 4 = on more than three families of plants.

British Aphids

The Rothamsted Insect Survey also includes data on aphids, data based on monthly counts of winged aphids of economic importance that were taken from aerial suction traps located in agricultural land. The data are reported from many sites, but Redfearn and I confined our attention to six sites located along a north-south transect across Britain. We estimated temporal variability and abundance at each site, and our analyses cover twenty-six species. Tatchell et al. (1983) provided us with data on the total number of British plant species on which each aphid species had been found. The distribution of these numbers of host plants tends to be very skewed: many species have relatively few host plants, but some aphid species feed on an very large number of host species. For this reason, we used the logarithm of the number of host plants as our measure of polyphagy.

Aphid life histories divide this group into "alternating" and "nonalternating" (technically termed *heteroecious* and *autoecious,* respectively). Alternating species spend the autumn, winter, and spring on some "primary" (usually woody) host on which they may lay eggs. They spend the summer on "secondary" hosts, usually from families different than that of the primary host and usually herbaceous. Nonalternating species do not switch hosts seasonally. We determined the life style of each species on the basis of descriptions by Tatchell et al. (1983) and Blackman and Eastop (1984). Clearly, defining a sensible estimate of polyphagy for alternating species is hard: should we use the diversity of primary or of secondary hosts? Which is more likely to limit the aphid numbers? Whichever we choose also suggests when we should count the aphid numbers: in late spring if we are looking at primary hosts, in late summer if we are looking at secondary hosts. Separate counts were not readily available to us. In what follows I shall only consider the results for nonalternating species. In nonalternating species, polyphagous species tended to be less variable than specialized species, at all six sites. Variabilities corrected for abundance yielded three (of six) statistically significant decreases in variability with increasing polyphagy, and uncorrected data yielded four significant decreases in variability (figure 4.2b).

Hoverflies

Owen and Gilbert (1989) described a fifteen-year study of the hoverflies (Syrphidae) sampled in a suburban garden. Their paper discusses many of the ecological factors affecting variability, as well as many of the potential pitfalls in analyzing the data. I plot some of their data in figure 4.3, which shows for three categories of trophic specializations, the temporal variability as the standard deviation of the logarithms of density versus the mean of the logarithms of density (SDL). Two features are apparent. The variability index SDL appears unusually small when mean density is low. The cause of this may be the reasons explained in the appendix to chapter 3: adding unity to zero densities before taking logarithms leads to a gross underestimate of variability. The second feature is that even when we exclude the rarest species, the two common trophically specialized species and the three common trophically generalized species appear more variable than the ten common species that are trophically intermediate. This is the same conclusion that Owen and Gilbert reach, although they apparently obtained it in a slightly different way. Theirs is an intriguing result, but the data are probably too few for us to draw a firm conclusion.

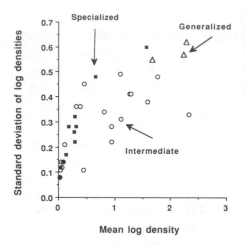

Figure 4.3. Standard deviation of the logarithms of density (SDL) versus abundance for species of hoverflies classified as being trophically specialized, intermediate, and generalized. The estimates of variability, SDL, clearly depend on abundance—and in a nonlinear way. For those species sufficiently abundant for this dependence to not matter (say, above a mean log density of 0.5), the trophically intermediate species seem to be the least variable, though there are few data in this region of the graph. (From data in Owen and Gilbert 1989.)

Herbivorous Insects: Summary and Discussion

The results are of three kinds: (1) significant decreases in variability with increasing degree of polyphagy, (2) significant increases, and (3) results that are not significant.

Significant decreases in variability are found in the data for the nonalternating species of aphids. At all six sites variability decreases with increasing polyphagy, and at three of these sites the decreases are statistically significant with corrected data; at four sites the decreases are significant with uncorrected data. *A significant increase in variability with polyphagy* is found in the Bohemian-moth data. The most variable hoverflies are those with intermediate trophic specialization, but the result is based on very few species.

Generally unconvincing results are found in the rest of the data. Watt's original analyses of Canadian moths fail to correct for possible confounding factors. For the various subsets of these data, Redfearn and I found three decreases and two increases in variability with increasing polyphagy (none were statistically significant), and, for a carefully selected subset of species, Watt and Craig (1986) found a decrease (nonsignificant). The decrease in variability with polyphagy seen in the British-moth data is based on only one point.

Perhaps the obvious feature common to all these results in the "generally unconvincing" group is the quality of the data. Canadian-moth density estimates were collected haphazardly. The British-moth data have better density estimates, but our groupings of the estimates of polyphagy were subjective by force of circumstance. Discarding data that do not fit one's preconceptions is a very dangerous maneuver, which is why I have discussed a number of studies that did not find significant results. Suppose you tentatively accept my argument that the "generally unconvincing" results are simply the consequence of inadequate data. We then face an interesting question: Why should moths show increasing temporal variability with increasing degrees of polyphagy while aphids show the opposite relationship?

As yet, I do not know the answer to this question. It certainly is not because of gross differences in habitats, as the aphids and most of the moths are agricultural species. Perhaps it could be because the aphids are food limited while the moths are predator limited; if this is the case, the aphids should be characterized by occasional very low densities, the moths by occasional very high densities. To test this idea, we can look at the relative skewness of the data. The population data on the moths are not available, but the summary statistics are, and they allow making comparative plots of coefficient of variation (CV) versus SDL. Such plots tell us whether population lows or highs are more important, because CV is more sensitive to highs while SDL is more sensitive to lows. As figure 4.4 shows, in this respect there are no differences between the moths and the aphids.

TEMPORAL VARIABILITY AND THE DIVERSITY OF PREDATORS

Canadian Moths

Using CFIS data, Andrew Redfearn and I (Redfearn and Pimm 1991) analyzed forest lepidoptera and the parasitoids that attack them in southern Ontario. We argued that there are at least two hypotheses for why, the greater the number of parasitoid species that attack a host species, the less variable should be its density. The more species of enemy a species suffers, the more reliable

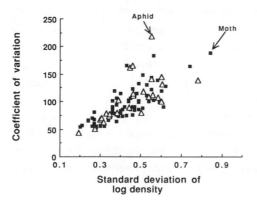

Figure 4.4. Coefficients of variation of density (CV) versus the standard deviations of logarithms (SDL) of density for moths and aphids. Relatively high values of CV indicate unusually high population densities, whereas relatively high values of SDL indicate unusually low population densities. MacArthur's hypothesis stressed unusually low densities as major contributors to population variability, and Watt's hypothesis stressed unusually high densities as major contributors to population variability. There are no obvious differences between the moths and the aphids.

those enemies may be in providing control; a high number of enemies may prevent sporadic outbreaks of the prey. In addition, the lower a population's resilience, the less variable its numbers may be (chapter 3); and more enemies may mean a lower resilience. Alternatively, species with very variable populations may only be able to support very few predators; many species may not be able to cope with the frequent rarity of their prey.

The data comprised thirty-eight species from thirteen families for which there were ten years of data. There were no zero estimates for these species, though there were for many other species that we excluded. The analyses are also complicated by (a) the many factors that confound studies of variability (factors that I have discussed earlier) and (b) some other factors that were unique to the present study. The myriad details are in our paper. The number of species of parasitoids reared from these host species were obtained from the literature. Most of the host species (twenty-two species) feed externally on the host plant, but ten species build simple structures or are in some way more concealed. (For the rest we were unable to find the relevant information.) We might expect these two groups (which we called *open* and *concealed* species, respectively) to differ in how parasitoids affected their dynamics. We found that, given the differences between these two groups, an increase in the numbers of species of parasitoid attacking the host species leads to the host species being less variable. This result was expected, but the correlation does not imply causation. As explained earlier in this section, it could be the high variability that causes the low diversity of parasitoids. The idea that high numbers of parasitoids cause low variability stems in part from the greater reliability of population control to be expected from more parasitoids. According to this argument, as the diversity of parasitoids increases, so the likelihood of an unusually high density in the host population should decrease. That is, when there are few parasitoids the populations densities may be quite strongly skewed toward higher densities, whereas when there are more parasitoids the skewness of population densities should become lower. We tested this idea by calculating the skewness values, noting that skewness statistics are generally thought to require large sample sizes in order to provide reasonable estimates (and on only ten observations are likely to be less than reliable). The result, however, was in the expected direction (skewness of population densities declined with increasing parasitoid diversity), but it was not significant ($P = 0.1$). The conclusion is that species with less variable population densities were those with more species of parasitoids, and the tentative conclusion is that the correlation is caused by the effects of the parasitoid diversity on variability, rather than vice versa.

Fennoscandinavian Voles

A convincing demonstration of how a diverse set of predators reduces the temporal variability of their prey involves the work of Hansson, Henttonen, and their colleagues. I find the example convincing because of its details, for it is the details that are so obviously missing in the previous study of Canadian insects. Of course, it is an example that I have very carefully selected on the simple criterion that it matches what I expected—although I do not know of any obvious counterexamples. What we would all like are studies that have both generality and detail; such studies will always be needed, and for obvious reasons they will always be rare.

Clethrionomys and *Microtus* voles are small mammals abundant throughout Scandinavia, Finland, and elsewhere. The data I shall discuss are not as simple as a plot of vole population variability against the diversity of their predators. Rather, the story is a detailed investigation into vole population variability and its likely causes. There are a number of results.

The temporal variabilities of *C. glareolus* and *M. aggrestis* densities increase with latitude (figure 4.5). When the effects of latitude are factored out statistically, temporal variability still increases with the duration of snow cover. As might be expected, the duration of snow cover also increases with latitude, but it also increases as one moves toward the center of the continent. Continental vole populations vary more than those at even very high latitudes in maritime areas, such as those on islands off the Norwegian coast.

High variability in these vole populations is associated with multiannual cycles. All the populations show intraannual cycles as the birth of young, generally in spring, increases the population, and as mortality, generally in winter, decreases it. Statistical analyses discussed more fully in chapter 6 can identify the dominant period of population cycling—ten years, for example, in the

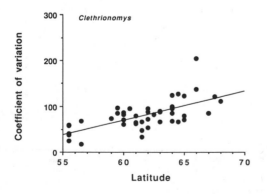

Figure 4.5. Coefficients of variation of density of populations of the vole *Clethrionomys*, increasing with the increasing latitude of the population. (From data in Hansson and Henttonen 1985.)

lynx. Such analyses of several species of *Clethrionomys* (Henttonen et al. 1985; table 4.1) show that the most variable populations are those with cycles longer than 2.5 years; these are the northerly and continental populations. Southerly and maritime populations do not cycle at these relatively long periods and vary less. This cyclic tendency of the northerly populations, as well as its association with high variability, is an important part of the results. Were the northerly populations simply more variable, then we might look for explanations in terms of more variable climatic conditions farther north (see chapter 5).

There are other species of small mammals in these communities. These cycle too, but in the north and not in the south. Moreover, in the north but not in the south, the cycles are roughly synchronous—species reach unusually low levels or unusually high levels within a year of each other (Henttonen 1986; table 4.2). In reviewing this work, Hanski (1987) contrasts two communities: a northerly one in Finnish Lapland and one near Moscow in the Soviet Union (figure 4.6). (Notice that in figure 4.6 *Microtus* always peaks a year after *Clethrionomus*; this may be because of differences both in the growth rates and in the competitive abilities of the two genera [I. Hanski, personal communication]. This complicates the story but does not alter its main features.) Not only

TABLE 4.1   A comparison between the Standard Deviation of the Logarithms of Density (SDL) and Spectral Analyses of the Population Data, for Thirteen Populations of the Vole Genus *Clethrionomys*

Population and Species	SDL	Wavelength Range (in years)			
		<1.5	1.5–2.5	2.5–5.5	>5.5
Populations with low variability:					
1. *C. rutilus*	0.17	30	**32**	18	20
2. *C. rutilus*	0.17	**47**	29	23	0
3. *C. rutilus*	0.18	**40**	28	26	6
4. *C. rutilus*	0.22	**31**	23	28	18
5. *C. gapperi*	0.35	25	**39**	18	18
6. *C. glareolus*	0.33	16	**31**	24	29
7. *C. glareolus*	0.32	**46**	25	21	8
8. *C. glareolus*	0.26	**51**	36	12	2
Populations with high variability:					
9. *C. glareolus*	0.85	30	19	**39**	11
10. *C. rutilus*	0.78	23	19	**45**	12
11. *C. rufocanus*	0.77	9	29	**47**	15
12. *C. rufocanus*	0.74	19	24	**39**	18
13. *C. rutilus*	0.64	30	15	**39**	15

Note.—The spectral analysis gives the percentage contribution of wavelengths of different periods, and the percentage in boldface gives the dominant period. Thus, for population 1, the dominant period occurs at 1.5–2.5 years. The data show that populations in which the cycling is at most, <2.5 years, vary much less than those populations with longer-term cycling. (From Henttonen et al. 1985).

TABLE 4.2   Correlation Coefficients between the Year-to-Year Densities of Pairs of
              Vole Species Counted in Long-Term Studies.

	Number of Studies in Which Correlation Coefficients Were		
Region	Negative	Positive but Not Significant	Positive and Significant
Northern Fennoscandinavia	0	5	10
Transition	0	2	3
Russia	1	5	0
West-central Europe	0	1	0

Note.—High correlations mean that the numbers are varying synchronously, typically because the species
are cycling in or near synchrony. The data show that northern, continental populations are more likely to cycle
than are more southerly populations. The correlation analyses used Spearman's rank correlation. (With thanks to
H. Henttonen for permission to reproduce these unpublished data.)

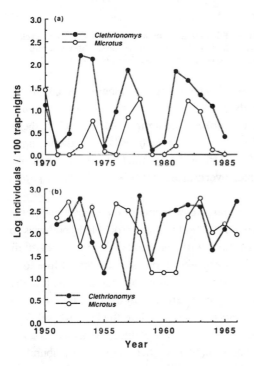

Figure 4.6. Two examples of abundance fluctuations in cooccurring
species of vole. (a), *Clethrionomys
glareolus* (solid line) and *Microtus
oeconomus* (dashed line) in Finnish Lapland. These two species
(plus four others) cycle almost in
synchrony. (b), *C. glareolus* (solid
line) and *M. arvalis* (dashed line)
in the Tula region (near Moscow).
These two species do not obviously
cycle synchronously. Data are
from Sadovskaja et al. (1971) and
Henttonen et al. (1987). (Redrawn
from Hanski 1987.)

small rodents but also insectivorous shrews cycle synchronously in the north
(Henttonen 1985; figure 4.7). Data on shrews in the south are less extensive,
but shrews there do not appear to cycle (Hansson 1984). Some birds, such as
grouse, that cycle in densities in the south do not cycle synchronously with
voles (Linden 1988).

   In the north, voles reach much higher densities. In peak years they deplete
their food and cause considerable damage to tree seedlings. (Hansson and

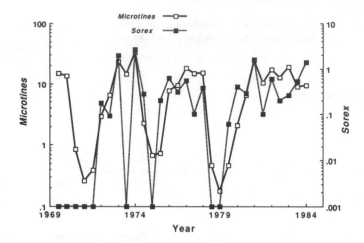

Figure 4.7. Densities of microtine rodents (solid line) and shrews (dashed line), which cycle synchronously at Pallasjarvi, Finnish Lapland. Note both the logarithmic scales and that different scales are used for the two groups of species. (Redrawn from Henttonen 1985.)

Zejda 1977; Hansson and Larsson 1980). There is little evidence of such effects in the south (Hansson 1987). Finally, the high variability of northern populations is caused, in part, by very low densities (see figure 4.6). Moreover, these low densities are often associated with "summer declines": in northern populations, densities continue to decline during the summer at a time when the vegetation has recovered from winter overgrazing (table 4.3). Southern populations do not have summer declines.

Hansson (1987) argues that these results reject some of the many hypotheses proposed to explain vole cycles. He considers the most probable hypotheses to be those that invoke interactions either with the voles' food or with their predators. Furthermore, he argues that the observations suggest that, while food limitation may (and eventually must) limit increasing vole populations, it is the interactions with predators that lead to the differences in degree of cycling. In the north, the voles are exploited by a specialized predator, the least weasel. The weasel can feed on voles and on other small mammals during long winters and deep snow cover. The weasel population increases when the voles are abundant in the north but not in the south (Hansson 1987). The cyclical changes are driven by the weasel, for once the vole growth rate has slowed, the weasels are still abundant. The abundant weasels then force the voles (along with other small mammals, including shrews and birds [Järvinen 1985]) to low levels, before the weasels too decline. The vole population will even decline during the summer. In short, the dominant predator in the system responds to changes in vole numbers with a change in density that results in the predator's density lagging behind the density of its prey.

In the south, there are many more predators, the predators are more generalized in their food habits, and there are more species of prey to exploit. There are birds and mammals that are specialized vole predators in the south (table 4.4), but, as Erlinge et al. (1983) show, these specialized species take only about 20% of the voles killed. Generalized predators take the rest. What is especially interesting is that, while the generalized predators are so important to the voles, the reverse is not true. Rabbits constitute about 50% of the diet of the generalized predators; voles constitute less than 15%. In contrast to conditions in the north, the shorter winters and reduced snow cover of the south make mammals available to predators for most of the year.

Certainly, for predators in the north there are other prey beside voles, but

TABLE 4.3   Population Variability as Standard Deviation of the Logarithms of Density (SDL) of Two Species of Vole at Different Latitudes

Species and Latitude		Mean SDL	Summer Decline	
			No	Yes
Clethrionomys glareolus:	<59°N	0.22	6	1
	59–61°N	0.47	4	3
	>61°N	0.52	4	13
Microtus aggresis:	<59°N	0.30	6	0
	59–61°N	0.53	2	6
	>61°N	0.62	1	9

Note.—Northern populations vary more than southern populations, and variable populations are those observed to decline during the summer, when food is generally abundant. (From Hansson and Henttonen 1985). Thus, for example, the first row of data mean that, of seven populations south of 59°N, 1 had summer declines, and 6 did not; and the mean SDL of these seven populations was 0.22.

TABLE 4.4   Percentage of Field Voles Consumed by Predators that either Do or Do Not Specialize on Small Rodents, Averaged over Two Years in the South of Sweden

Predator Species	Percentage of Voles Killed
Facultative vole predators:	
Fox	17
Domestic cat	16
Badger	4
Polecat	2
Common buzzard	32
Tawny owl	6
Vole specialists:	
Stoat	9
Kestrel	8
Long-eared owl	6

Note.—Facultative rodent predators take many more voles than do vole specialists. (From data in Erlinge et al. 1983.)

these other prey are neither sufficiently abundant nor available for enough of the year to compensate for the loss of voles when the voles have become scarce. In the south, the diversity of relatively abundant prey to the generalized predators means that, in response to changing vole numbers, the predators have changes in their diet rather than just changes in their density.

In sum, the data show that northern communities are simpler: they have fewer predators, and what predators there are must be more specialized to cope with long winters and deep snow. These specialized predators are less reliable controls of voles because they are dynamically dependent on them. In the south, the diverse communities of generalized predators provide a more reliable and effective control of the densities.

There is one final point to be made. The "cause" of the cycles appears to be more a property of the nature of the food webs than of a particular aspect of a species' ecology. The cycles and the resultant high variability seem to be a consequence of a simple food-web structure, low variability a consequence of a more complex one. Henttonen, Hansson, and Hanski seem to be among the first ecologists to look at cycles in community terms—looking for food-web correlates rather than explanations involving just the species' natural history.

ABUNDANCE: THE RELATIVE CONTRIBUTIONS OF PREDATOR AND PREY DIVERSITY

There is one study that does evaluate the relative effects of both the diversity of prey a species exploits and the diversity of predators the species suffers. It does so in an indirect yet plausible way, and it compares how species density changes over space and not across time. Nonetheless, the results, which concern insect behavior, are relevant to our previous discussion.

Risch and his colleagues summarized some 150 studies of nearly two hundred herbivorous insect species that are pests in agricultural communities (Risch et al. 1983). An early and typical experiment of this genre is Root's (1973) measurement, in two treatments, of the density of insects that feed on collards (*Brassica oleracea*). In monoculture the collards were grown alone, and in mixed plantings the collards were grown with many other species of plants. Root found that the densities of insects on the collards in the mixed plantings were generally much lower than those on collards in the monoculture.

Risch et al., in reviewing comparable experiments for a wide range of insects and crops, found that Root's results are typical. Across different experiments, 53% of the species reported were less abundant in the more diverse systems, 18% were more abundant in the monoculture, and 29% showed either no significant differences or inconsistent differences (figure 4.8).

The review argues that these results can be explained by one of two hypotheses. The "predation hypothesis" argues that mixed plantings will have a greater range of natural enemies and that these will keep the herbivores at lower levels. This greater diversity is generally an assumption, albeit a reasonable one. The "resource concentration hypothesis" proposes that the diverse plant community in the mixed plantings may directly reduce the ability of an herbivore to find and utilize its preferred host plant.

Risch et al. assembled studies that contain both monophagous and polyphagous insects and annual and perennial plants. These groups differ in the proportion of the species that show changes in abundance across treatments. The review uses these differences to resolve the two hypotheses. Both monophagous and polyphagous herbivores are likely to encounter a greater diversity of predators in the mixed-species plantings and so should be more reliably controlled there. If there are differences between monophagous and polyphagous species, these must reflect the fact that the monophagous species cannot exploit many of the plants in the mixed-species plantings.

Indeed, there are differences (figure 4.8). It is only the monophagous species that are generally much rarer in the mixed plantings: 61% are more common in monoculture, and 10% are more common in the mixed plantings. The differences are reversed and not so marked in polyphagous species: 27% of the species were more common in monoculture, but 44% were more common in mixed plantings.

The first conclusion is that food diversification—and not predator diversification—affects abundance. Risch et al. (1983) proceeded to test this conclusion by repeating Feeny's (1976) argument about the differences between perennial and annual crops. Annual crops may be more susceptible to herbivores because the former are short lived, which means that predators have only a short time to find the plants and the herbivores they host. Long-lived perennials are more likely to support predator-limited herbivores. For these and other reasons, perennial plants are more likely than annual plants to host predator-limited herbivores.

Yet, when we compare annuals and perennials, we do not find that the effects of predator diversification are more important for perennial plants (figure 4.8). For insect species on annual plants, there is a greater reduction in abundance in mixed plantings versus monoculture (59% of monophagous species were more abundant in monoculture, and only 3% were more abundant in mixed plantings.) On perennial plants the comparable figures are 67% and 24%.

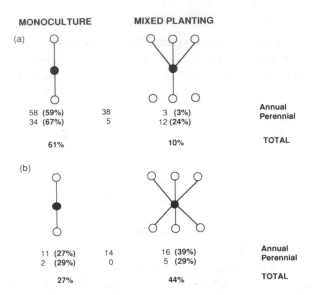

Figure 4.8. Summary of the data of Risch et al. (1983). Densities of monophagous species
(a) and polyphagous species (b) were compared both in single-species plantings of their
hosts plants (left) and in mixed-species plantings (right). Shown are the numbers of studies
(and, in parentheses, the percentages) in which the species of interest (solid circles) had
higher densities. Numbers in the middle, without associated diagrams, are those studies in
which there were no consistent differences. In mixed-species plantings, there are expected to
be more predators. Only polyphagous species can exploit more species of plant in mixed-
species plantings. These data suggest that it is the diversification of food, not the diversification of predators, that affects densities. See text for details of this argument.

What, if anything, do these results say about temporal variability? One has
to suppose that temporal changes in both predator diversity and plant diversity
would effect changes in abundance in the same way as do these spatial differences. Greater predator diversity in the mixed plantings is generally only an
assumption. If we concede these assumptions, the data suggest that polyphagous species should fluctuate less than monophagous species. (This matches
the result for aphids but not the result for moths.) It also argues that the effects
of plant diversification are more important than the effects of predator diversification.

SUMMARY AND CONCLUSIONS

MacArthur was one of the first to argue that food-web structure would modify
temporal variability. After the loss of a single prey species, species that fed on

few species would be more likely to become very rare than would species that were more polyphagous. Watt, in contrast, argued that, after the loss of the predators that controlled their densities, polyphagous species would be able to increase faster and so become more abundant than monophagous species. Both arguments are plausible, and they are not truly alternatives. MacArthur concentrated on food-limited species and on the reliability of their food supply. Watt concentrated on predator-limited species and on the consequences of the failure of that limitation. A comprehensive theory of food-web structure and population variability must take into account the arguments of both MacArthur and Watt.

Teasing apart the theoretical predictions on population data is difficult. We would need population data over many years for many species, as well as a knowledge of the diversity of both the species' predators and the species' prey. To my knowledge, such data are not available. But, for herbivorous insects, there are population data on species for which we have some knowledge of diet. Not all of these data are very compelling, and some were collected very haphazardly. For the data that seem reasonable, I have shown that increasing polyphagy is associated with a decrease in variability in some species of aphid yet is associated with an increase in variability in some species of moth. Perhaps MacArthur's argument applies to the aphids and perhaps Watt's argument applies to the moths, but it is not clear yet why this should be so; nor can I detect in these data any more detailed features that would support this idea. Predator diversity also has an effect. Studies show that the insect species with more predatory species vary less and are less likely to reach unusually high densities than are species with fewer predatory species exploiting them. For voles in northern Europe, there is evidence that increased predator diversity reduces variability. In the far north, where there is only one important vole predator, vole densities cycle and, as a consequence, are relatively more variable. Farther south, there is a diversity of relatively more polyphagous predators, and the vole densities do not cycle and are less variable.

My final thoughts are these. Temporal variability can be shown to depend on food-web structure; however, the relationships are complicated and not the same in all groups of species, and there are some important caveats. I have chosen these groups so that they are in similar places and counted over the same temporal scale. The data were probably collected over similar spatial scales, although we cannot be certain that the same sampling technique samples different species on exactly the same spatial scales. How large an area is sampled by a light trap, for example, will depend on the behavior of the different species of moths. The choice to equalize scales as much as possible is deliberate. It means that the effects of spatial and temporal variability of physical factors (discussed in the next chapter) should be relatively constant across the species I am comparing. It also means that various effects which could confound the statistics are minimized. The species discussed in this chapter are taxonomically related ones, so species differences in variability ought to be reduced. Nonetheless, nearly all the studies discussed in this chapter involve the removal of possibly confounding factors, and this removal is technically difficult and inevitably leaves open the possibility that other important confounding factors have been overlooked.

5

The Variability of the Environment

ENVIRONMENTAL VARIATION ACROSS SPACE

At any location, how much a population varies from year to year will depend on how variable are the environmental factors there. In some locations, these factors may be more variable than they are at others. Two examples are sufficient to illustrate (*a*) the differences in environmental variability across space and (*b*) the effects these differences are likely to have on population variability. My first example involves desert and grassland rainfall. These habitats are relatively dry, and rains often fall sporadically, the amount varying from year to year. In a given year, places just a few kilometers apart may receive very different amounts of rain. The thunderclouds that bring the rain deposit it over relatively small areas. The effects of desert rains are well known, and year-to-year differences in rainfall have obvious effects on species, at all trophic levels (Raitt and Pimm 1976). Less obvious is that different desert and grassland areas differ in how reliable are the rains from year to year. Figure 5.1 shows the variation in yearly rainfall for two desert and two grassland sites in the western United States. Months in which rainfall is scarce tend to have higher year-to-year variation in rainfall. Similarly, sites that receive less rain are also more variable from year to year.

For a second example, there is McGowan and Walker's (1985) contrast of two oceanic areas. The first area is the Central Pacific Gyre—a vast body of water that gently circulates, ensuring constant mixing. The temperature of the water drops from a surface temperature of about 25°C to about 3°C at a depth of 1,000 m. Along this gradient, there is a corresponding shift in salinity. For a given temperature, the range of salinities is slight. Nearer shore are areas in the California Current, where turbulence and mixing of warm and cold waters leads to more variable temperatures and a much greater range of salinities. These differences in the variability of both temperature and salinity have a major effect on the community structure, for they affect what species are to be expected in any plankton sample. Species composition in the Central Pacific

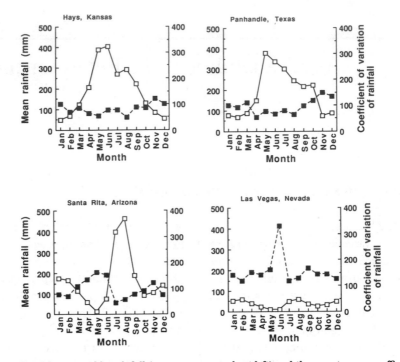

Figure 5.1. Mean monthly rainfall (open squares, scale at left) and the year-to-year coefficient of variation of monthly rainfall (solid squares, scale at right) for four sites in the United States. Hays (Kansas) is in a relatively wet prairie, Panhandle (Texas) is in a dry, short-grass prairie, Santa Rita (Arizona) is in the Sonoran Desert region, and Las Vegas (Nevada) is in the Mojave Desert region. Both within years and between sites, low rainfall is associated with high variability of rainfall. Data are from the U.S. Goverment's thirty-year summaries.

Gyre is remarkably predictable. Plankton samples taken two to four years apart (and hence from very different parts of the water) typically share 50% or more of their species in common. In contrast, half the samples taken, less than a day apart from the California Current have fewer than 50% of their species in common. This low similarity in species reflects the very different water conditions encountered in the California Current over short time periods.

We should not be surprised if the variabilities of species densities in different areas could be easily related to these differences in the variabilities of physical factors (Gaston and Lawton 1988a). Yet I do not have any examples where known differences in the variabilities of physical factors have been used *to predict* differences in population variability. While it is easy to provide after-the-fact explanations of differences in population variabilities in terms of the differences in physical factors, they can be misleading. For next are two well-conceived examples which attempted to relate population variabilities to differences in the variability of physical factors. These two examples involve comparisons of tropical with temperate insects and of migratory with nonmigratory birds; neither found the expected differences.

Insect Populations

Wolda's studies (1978, 1983) are among the few comparative studies of population variability. In 1978, Wolda used a variability index I discussed in the appendix to chapter 3, and, in 1983, he looked at the standard deviation of the logarithms of population size (SDL). (Population sizes were measured annually; no one doubts that the wet tropics are less seasonal than the arctic and that they therefore have populations that vary less within each year.) The conclusions of Wolda's two studies are similar. I will consider the latter in more detail because it is rich in data. Wolda analyzed seventy-seven data sets for which there were annual censuses in six or more years. Each data set consisted of several populations: sixty-one of the data sets contain information on twenty or more populations, and eleven contain information on more than fifty populations. Each data set is a taxonomic group—usually a family—in a specific location. Comparing the mean variabilities within these data sets is thus the best way to look at differences between habitats and taxa.

One of Wolda's objectives was to test Elton's 1958 idea that tropical populations fluctuate less than do temperate ones (chapter 1). Ecologists frequently assert that the tropics are more stable than temperate regions; Wolda is one of the few who have tested this assertion. Wolda's data show that species of tropical insects are only slightly less variable than are temperate species. The larger number of studies of temperate species shows both larger and smaller variabilities than do studies of tropical systems. Wolda concludes that "tropical insects are about as stable, no more and no less, as temperate insects, even insects in relatively undisturbed tropical forest."

Wolda's data certainly show that there is nothing very special about the variabilities of tropical versus temperate insect populations. The data also show that there are large, statistically significant differences between taxa. The tropical-temperate comparison is made difficult because so few insect families were counted in both places. Mosquitoes were counted in both places, and the tropical species were much less variable in their densities than were the temperate species. However, when short-term studies are included (Wolda 1978), this result does not hold. Wolda found that mosquito densities from Canada were not more variable than those in Jamaica. Some tropical mosquito populations "can be very variable indeed" (Wolda and Gulundo 1981).

At the level of insect orders, there are fourteen studies of temperate Homoptera that can be contrasted with five of Wolda's own studies of Homoptera in the tropical forest of Panama. The tropical populations are again less variable. Again, this result does not hold up when short-term studies are included. Wolda (1978) found that Homoptera from various sites in the humid temperate zone were not more variable than those from Panama.

In Wolda's data the differences between the temperate and tropical Homoptera could easily reflect differences between forest habitats (the tropical data) and agricultural systems in which the aphids and the temperate species of delphacids occur. Species listed as "pests" in Wolda's paper, plus other groups of clear economic importance (for example, aphids and delphacids), do have very high variabilities. Pests may be pests precisely because they sporadically reach very high numbers and so are highly variable (chapters 2 and 4). Wolda did find that insects in relatively dry areas (grasslands plus agricultural systems) generally had larger variabilities than did species in wet habitats. This result appears across and within families. For instance, in the various studies of carabids (reported originally by den Boer 1977), the woodland populations are all less variable than the populations in heather, grasslands, and pasture habitats. These differences between "dry" and "wet" habitats may be because physical factors are more variable in dry habitats, but between wet and dry habitats there must be many ecological differences that could also explain the results.

Wolda cites just one study that relates population variability to the variability of a physical factor. Malicky (1956) recorded the emergence of twenty-five species of caddisflies in two streams in Austria. In one, the temperature varied from 0°C to 19°C, and the standard deviations of log densities (SDL) of these species averaged 0.61. The temperature of the other stream varied only from 6.0°C to 6.3°C, and the SDL of the twenty-five species averaged 0.47. I am not certain whether recognition of the many factors that bias estimates of SDL would alter the conclusion of Malicky's study.

In summary, there are differences in the variabilities of populations in different habitats, and these can be explained a posteriori by differences in the variability of the physical environment. While different environments certainly differ, data showing the effects on population variability are scarce. Finally, there appears to be little evidence to support the frequent assertion that tropical habitats are less variable from year to year than are temperate habitats.

British Woodland Birds

The example of the song thrush (figure VAR.1a) suggests an obvious comparison. Bird species that either are resident in or migrate locally within western Europe suffer from occasional hard winters that greatly reduce their numbers. In contrast, migratory species that winter far to the south (often south of the Sahara) should not be affected by hard winters. I confidently predicted that migrant species' breeding densities would be less variable than those of species that remain in or near Britain year round, because hard winters are a major contributor to the variability of many of the latter species (Pimm 1984b). As the data in figure 5.2 show, I was wrong. For whatever reasons, migrant birds tend to have slightly higher, rather than slightly lower, population variabilities. Recall from chapter 3 that migrant moths in Czechoslovakia are also more variable than species that do not migrate. It appears that migration is generally a very risky behavior.

ENVIRONMENTAL CHANGE OVER TIME

We must accept that the physical factors of the environment affect populations and do so at a variety of scales. Ten thousand years ago ice sheets were very much more extensive than they are now, and, as a consequence, plant and animal communities were very different from their present forms. At much shorter time scales, we can observe the effects that a hard winter has on bird populations, for at least several years afterward. Because of these observations, it seems sensible to ask both how the physical environment changes over time

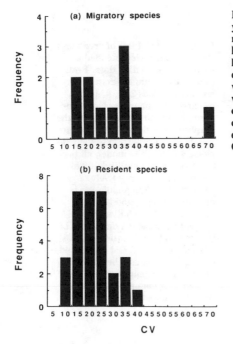

Figure 5.2. Coefficients of variation (CV) for year-to-year densities of migratory (a) and resident (b) bird species in English farmlands. Despite the resident species suffering large reductions in 1963 (see figure 2.1), resident species are not more variable. The very variable migrant species is the whitethroat, a warbler, which suffered a sudden major decline in 1969, from which it has not yet recovered. (From Pimm [1984b], using original data from the British Trust for Ornithology's Common Bird Census; see text.)

and how it may differ from place to place. Steele (1986) has done this, contrasting marine and terrestrial ecosystems in the process. Before I can explore what these results mean for population variability, I must devote the next section to some background material.

Background: White and Red Noise, Spectra, and Fractals

It is possible to express a time series—such as temperature records, water levels in rivers, or population densities over time—as a sum of sine waves (that is, cycles) of different amplitudes and periods. (Recall that frequency is the reciprocal of period.) The use of cycles is a purely descriptive device; it does not imply that the underlying causes are a neat set of oscillators. A long-term trend in the series, for example, can be viewed as just a piece of an even-longer-term cycle. Data for two model populations are shown in figure 5.3. The population densities are plotted against year. The model labeled "red" was formed by summing several sine waves that have different periods and amplitudes that increased proportionately with the periods. In other words, the longer the cycle, the lower the frequency and, in this model, the larger the cycle. The model labeled "white" was formed by summing sines waves that have different periods but equal amplitudes. Where the sine waves peak in relation to each other were chosen at random in both cases. (The sine waves

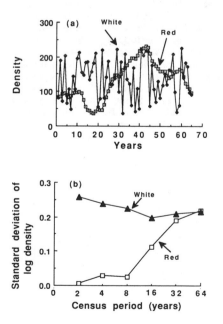

Figure 5.3. (a), Plots of "red" and "white" time series intended as models of population data. Both plots are the sum of seven sine waves of periods 2, 4, 8, 16, 32, 64, and 128; the waves have random phases. In the red model, the amplitudes of these sine waves are weighted in proportion to their periods. In the white model, all periods have equal amplitudes. The red population is relatively smooth; the white population is extremely rough, and its curve tends to fill the two-dimensional space. (b), Standard deviations of logarithms for increasing periods of time for the red and white models. For the former, the variability increases with length of time; for the latter, it does not. (Redrawn from Pimm and Redfearn 1988.)

could have all peaked together at the beginning of the model, for example.) Technically, the models have random phases.

These models are labeled "white" and "red" by analogy to light. Light is a composite of radiation of different periods and amplitudes. In white light, all frequencies are equally important; in colored light, some frequencies predominate. In red light, long wavelengths (low frequencies) predominate; for blue light, short wavelengths (high frequencies) predominate. Sound too can be expressed as the sum of cycles of different wavelengths, and so it can be described as "white noise" or "red noise," etc. The terminology of "colored" noise is also useful in describing population densities, even if, pardon the pun, the colorful language may seem remote from ecology.

Let us return to populations. One way in which we can explore population dynamics is to reverse this process of creating a model population from the sum of sine waves. We can decompose a time series into the sum of a number of sine waves; that is, we can investigate its spectrum. One way of graphing the results of this process is to plot a function of the amplitude of the cycles versus their frequency. This plot does not contain all the information, since the phase relationships (when the cycles peak) is lost; but it does show powerful cycles if they exist. A plot for the Canadian lynx densities shows a peak corresponding to a frequency of once in ten years (see chapter 6). There may not be an obvious cycle, or when there is we may want to look at the spectrum with the cycle removed. Obviously, analyses of temperature data show a marked peak at a frequency of once per year, but what does the rest of the spectrum look like? One way of summarizing the spectrum is to look at the slope of the amplitudes of a cycle versus their wavelengths, to see whether the time series is "blue," "white," or "red"—that is, which end of the spectrum predominates.

There is another term I wish to introduce: *fractal dimension*. Again, I must beg indulgence to develop a physical concept that initially may not seem to be relevant to population data—or, indeed, to my previous discussion of colored noise. Thanks to the pioneering work of Mandelbrot (a popular account is Gleick 1988), many picture books (my favorite is Peitgen et al. 1988), and even fractal planets in *Star Trek* movies, we are now aware of geometrical objects that resemble nature far more than they resemble squares, circles, and triangles. The idea of a fractal dimension is often introduced with a discussion of the length of the coastline of Britain. This length has a surprising property: the more detailed a map of the coastline we examine, the longer the coast appears to be. The more closely we examine the coastline, the more detail we see; and those details increase our measurements of the coastline's length. We start with a map of the planet, with Britain looking like a small triangle, and end up tracing out the spot where the water touches small pebbles on our favorite rocky shore. Most of the popular books have pictures which make this point better than any text can. Unlike a smooth boundary that we could measure with precision, Britain's coast is rough—it tends to be something intermediate, between a single dimension (a smooth line) and two dimensions (a surface), that the convoluted coastline tends to fill. The fractal dimension defines this intermediate dimension. Intuitively, fractal dimension defines how rough the coastline is.

These ideas have a relationship to time series, including those of population densities. After all, the idea of progressively finer detail seems at least reminiscent of the idea that time series are cycles on cycles on cycles. Indeed, with appropriate assumptions, there can be a relationship between the fractal dimension (the roughness) of a time series and the slope of the amplitude-frequency plot (Rothrock and Thorndike 1980). In figure 5.3, the model population with amplitudes increasing with decreasing frequency gives a relatively smooth curve that is close to being one-dimensional. Adding sine waves with different frequencies but with the same amplitude yields a curve that is very rough: it tends to fill the space and is close to being two-dimensional.

I hope these ideas may be helpful when I finally return to discussing populations. Notice that fractals have some odd properties—the length of the coastline is not defined, for example. The analogy most useful for ecology may be this. Suppose that we did *not* have a map of the entire coast of Britain but wanted to predict its appearance by just walking along the coast. We may only see the small features of rocks jutting into the sea and expect the coastline to continue like that. Or we might recognize the coast's fractal nature, with small features imposed on larger features and with those larger features imposed on yet larger features. There could, after all, be large bays, headlands, and even peninsulas beyond the part of the coastline we know. By analogy, populations may behave the same way. The analogies are useful if they make us think about how to best predict features on scales larger than we currently can measure.

Reddened Spectra

Steele (1986) demonstrates that physical variables in the ocean have a strongly reddened spectra over a wide range of periods, especially those of a year or longer. When he decomposes the time series of these variables into sums of sine waves, the amplitudes increase approximately as the period of the cycle. For periods less than about five years, Steele finds that the spectra of terrestrial physical variables appear white. Over longer periods, the terrestrial variables also show a reddened spectrum. Williamson (1981, 1983, 1987) has long argued that physical variables in terrestrial systems have reddened spectra too. If physical variables force variation in population densities over the time scales we are considering, then population densities would have reddened spectra too. A reddened spectrum suggests that longer-term changes become progressively more important over longer periods; population densities are likely to show the same pattern.

Finally, we return to ecology. Time series of population data are usually far too short to be broken into their constituent cycles. In any case, variability is a statistic that is more immediately compelling than any we can obtain from a spectral analysis. What does the red-noise idea say about population variabilities over time? By comparing the two models of figure 5.3, we can demonstrate the effect that a population's "color" has on how its variability changes through time. In the red model, variability increases through time; in the white model, variability remains approximately constant. In these terms, we can explain the result of increasing variance. What might appear as an average, normal population density over a few years is really a level which changes over a longer time period, and that change, in turn, is imposed on an even more powerful, less frequent trend. If we have few years of population data, we will have little confidence in our estimate of the average population density, because of our lack of data. As we accumulate more years of data, the average will change. And the more data we accumulate, the more important are those changes in density. The population's average will continue to change, and its variability will continue to increase.

If population densities have reddened spectra, we should obtain increasing variances of population abundances as we increase the period over which the variances are calculated. By implication, when we observe increasing variances, it *may* reflect the reddened spectra of physical variables forcing the variability of population densities. I shall return to various other possibilities later. First I should show you that the variability of most populations increases as we extend the interval of time over which we measure that variability.

Population Variability Increases the Longer We Count the Population

From the literature, Andrew Redfearn and I assembled twenty-six terrestrial population studies of species counted at least annually and for at least fifty years (Pimm and Redfearn 1988). Four of these species are insects; the rest are birds and mammals. To these we added the twenty-four years of data on breeding densities of ninety populations of birds in British farmlands and woodlands, data assembled by the British Trust for Ornithology's Common Bird Census. We calculated population variability over increasing lengths of time: two, four, eight, and sixteen years for the each of the birds species and, in addition, for thirty-two and fifty years for each of the twenty-six other species.

An example of a population—the skylark in English farmlands—and of how
these calculations were made is in figure 5.4. The average variability of these
populations increases continuously with the length of time over which we calculate that variability (figure 5.5).

It is possible that this average might be influenced by a small number of
populations that show large, continuous increases or declines in density, thus
causing large increases in variability. To test this possibility, we calculated the
proportion of the populations that show increases in variability with the length
of the census. A significant majority of populations show such an increase,
thus negating the possibility.

Only the four species of insect did not show an increase in variability, and
this was so only at the very longest time intervals, when the SDL levels off at

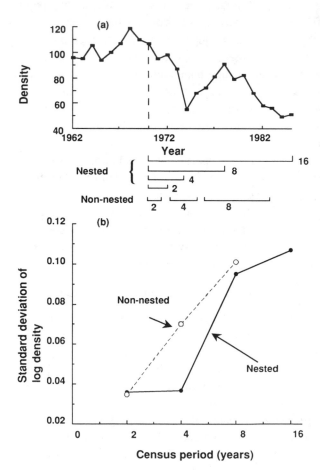

Figure 5.4. (a), Density
of the skylark in English
farmlands. Units of den-
sity are relative and are
set to 100 in 1966. Popu-
lation variabilities (standard deviations of the
logarithms of density
[SDL]) were calculated in
two ways. The nested de-
sign calculates variabilities in the first two years
of a set of years, the first
four years, and so on.
The non-nested design
calculates variabilities in
the first two years, the
next four years, and so
on. The first few years of
these data were not cho-
sen, because they con-
tained the unusually cold
winter of 1962–63. (b),
Population variability
(SDL) increasing with the
length of census period
for the skylark in farm-
land, for both nested and
non-nested designs. (Re-
drawn from Pimm and
Redfearn 1988.)

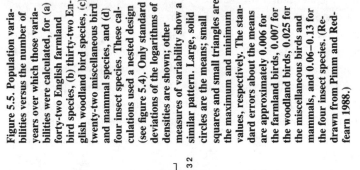

Figure 5.5. Population variabilities versus the number of
years over which those variabilities were calculated, for (a)
forty-two English farmland
bird species, (b) thirty-two English woodland bird species, (c)
twenty-two miscellaneous bird
and mammal species, and (d)
four insect species. These calculations used a nested design
(see figure 5.4). Only standard
deviations of the logarithms of
densities are shown; other
measures of variability show a
similar pattern. Large, solid
circles are the means; small
squares and small triangles are
the maximum and minimum
values, respectively. The standard errors about the means
are approximately 0.006 for
the farmland birds, 0.007 for
the woodland birds, 0.025 for
the miscellaneous birds and
mammals, and 0.06–0.13 for
the four insect species. (Redrawn from Pimm and Red-
fearn 1988.)

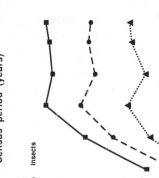

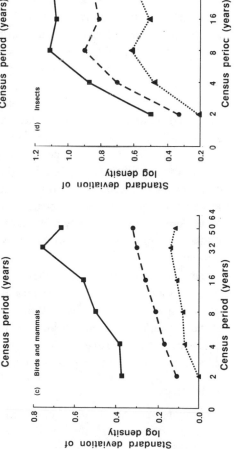

about 1.0 (figure 5.5). It is possible that an SDL of about unity is an upper limit to variability consistent with population survival: it is equivalent to approximately four orders of magnitude of density variation over thirty years. (There is certainly a lower limit of density below which a population cannot survive; see chapter 7.) It is equally possible that ecologists may be unable to estimate densities below a certain level—that these densities will be reported as zero and that the variability will be underestimated as a consequence (see the appendix to chapter 3). In a supplemental study, we found that seven moth species (data from Williams 1958) and three diatom species (data from Simola 1984) were more variable over thirty-two years than over sixteen years (Pimm and Redfearn 1989).

We might expect some populations to show long-term trends. Long-term trends in the data will produce an increasing variance as the population density reaches ever new levels. For example, all of the bird populations may have been recovering from some severe but infrequent disturbance. This would explain the prevalence of variability increasing with census duration. The British bird populations suffered such a major collapse during the winter of 1962–63. To be conservative, we analyzed data from 1970 onward: the pattern of increasing variability is evident in the postcollapse data. Alternatively, many of the populations may have been showing steady declines in abundance that were due to environmental degradation. Increasing variance in the bird populations, for example, might be due to degradation of, say, British farmlands, with all the populations decreasing for the same reason. But there is no overall trend in the species' abundances: the means of the total densities of all the woodland and farmland species only varied within a range of 6%. Of course, this could mean that an equal number of species benefit from the changes. The assorted species for which we have fifty or more years of data are from various countries in different centuries, so a *single* cause for the pattern of increasing variability seems unlikely.

The conclusion must be that the *majority* of populations show long-term trends in abundance, for *different* reasons and in *different* directions. These trends, moreover, continue over all the time scales across which we have calculated variability. This conclusion does not contradict the argument that populations have reddened spectra; it reinforces it. Recall that one way to describe a trend is as a piece of a longer-term cycle.

Implications

There are many implications of these results. Because the result of increasing variability appears for the majority of populations surveyed and over all the time intervals over which we calculate the variability, there are important implications for the debate over whether populations have an equilibrium. Alternative theories about the nature of equilibrium suggest—and are suggested by—different levels of variability. Some have argued that populations are best viewed as having a stable equilibrium toward which density-dependent processes return them after disturbances (Lack 1966). Such populations might not be highly variable. Alternatively, essentially random climatic processes may drive populations, leading to large fluctuations (Andrewartha and Birch 1954). But how can such populations persist, when they must eventually be driven to extremely low levels (Lack 1966)? There are alternatives consistent with both relatively high population variability and long-term persistence: populations may be attracted not to a fixed equilibrium but to either a limit cycle or some other, more complex attractor (chapter 6). Or populations may show wide fluctuations—and only at infrequent very low and very high levels encounter the processes that operate to keep the densities bounded (Strong 1986). Resolving these alternatives is not easy. In the natural world, where densities are buffeted by a variety of climatic factors, it seems unrealistic to expect the equilibrium to be absolutely constant; yet, how much inconstancy can we accept before rejecting an equilibrium view?

While all of these views provide important insights into the nature of population change, none captures the result of increasing variability. For some populations there is clearly an equilibrium in the sense of some approximate density to which densities will return after some major perturbation. (The bird populations and the hard winter are examples.) It is equally true that for some populations this equilibrium level is less than obvious. Even when that equilibrium is obvious, the results suggest it is constantly changing and, moreover, changing more the longer we measure the population. Older ideas of population change suggest that, the longer we measure a population, the better we will know the position of its equilibrium. I suggest that the converse is true. The notion that we know less about populations the longer we study them strikes a sympathetic chord with many field ecologists. All along they may have sensed the consequences implied by increasing population variance.

Explanations

Besides population densities being driven by the physical variables that are known to have reddened spectra, there are many processes that could lead increasing population variance. I have never claimed a single explanation of the phenomena, and in my previous discussion we have already glimpsed some of the alternative explanations. *Any* process that creates a trend in density will cause the population's variability to increase. Recall that Redfearn and I deliberately excluded bird densities recovering from a hard winter, because over the time course of the recovery we expected the variabilities to increase. Obviously, the consequence of density-dependent processes returning populations to normal values will be an increasing variance during the return. Densities of long-lived species will show high correlations between years (we are counting many of the same individuals), and so even random changes in the environment will be tracked by the densities quite slowly, leading to short-term trends in densities.

What I find curious is McArdle's (1989) argument that the increasing variance is solely a property of population features such as density dependence and year-to-year correlations in density caused by individual longevity. First, I cannot understand why a population-level explanation of increasing variance should be considered to enjoy a kind of logical supremacy. It is hard to dispute the evidence for long-term climatic changes—and the effects these changes have on species. One of the clearest examples is Clark's (1988) study of the frequency and intensity of forest fires. Forest fires are more severe the longer the interval between them (there is more wood to burn). The growth rate of the trees imposes a long-term trend on the community—as saplings become mature trees. Imposed on this population effect are the effects of climate changes, because the intervals between burns depend on how dry are the summers. Summer temperatures have changed over the past few centuries in ways one would expect. From decade to decade summer temperatures are roughly comparable, but there have been long periods when the climate was hotter or colder (mini–"ice ages"), and the frequency of forest fires shows this clearly.

Second, there are the technical details of McArdle's criticism. For the British birds, the increases in variability occur over much longer time periods (from eight to sixteen years, for example) than one would expect on the basis of their known recovery rates. (The song thrush, for example, has a return time of less than two years; chapter 2.) (McArdle estimates the duration of year-to-year correlations in the bird data, which would lead to trends—and so to increasing variance—but his analysis is circular. The issue is not that time delays cause increasing variance but whether the time delays arise within the population, the community, or the physical features of the environment. I shall return to this presently.) As already noted, many short-lived species (insects and diatoms, for example), for which demonstrations of density-dependent changes are rare, also increase in variability over long time periods.

The issue of what causes increasing variance comes down to our asking: What causes trends in population density? Trends in climate do. So do short-term trends in population recovery. It is equally obvious that interactions with other species also do so. Recall from chapter 2 that resilience—the rate of population recovery—depends not just on the natural history of the species but on the community in which it is embedded. When we look at population variability over the short term, we see the effects of population effects; over the longer term, the effects of interactions with other species; and, over even longer terms, the interaction with the climate. There are effects, imposed on effects, imposed on yet larger effects—rather like the coastline of Britain, with pebbles on rocks, on promontories, on headlands, and on peninsulas. It is not that increasing variability is hard to explain; we have more than enough explanations. What is surprising is that ecologists did not expect it, knowing, as they do from generations of fieldwork, to expect the unexpected. So, we ask, is the increasing variance best explained by processes at the population, community, or ecosystem level? By now, you should know that the answers to such a question depend on the time scale you have in mind. I think that the best reply is this: all populations belong to complex systems, and increasing variance is simply a consequence of that membership.

SUMMARY AND CONCLUSIONS

This chapter has dealt with environmental variability. Some environments are more variable than others, and, to a limited extent, these differences are reflected in differences in population variability. The hypothesized consequences that differences in environmental variability are alleged to have with regard to population variability are not always substantiated by careful analyses of population data.

The surprising result of this chapter involves how temporal variability changes with time. For the majority of animal populations surveyed, variability increases with the length of time over which it is calculated. Over short time scales, populations vary relatively slightly about some level. But this level is varying over a long time scale—and, moreover, varying much more than the variability about the shorter time scale; the longer we measure a population, the more we uncover long-term, large-amplitude changes in abundance.

Like resilience, the temporal variability of population density depends on factors at different spatial and temporal scales. Temporal variability will be greater in those places where critical environmental variables, such as rainfall in deserts, are themselves more variable. Food-web structure will modify temporal variability because it is affected by such factors as the diversity of a species' predators, prey, or competitors. Population parameters such as reproductive rate can modify variability because they determine both how fast populations recover from population crashes and how closely populations follow fluctuating environments. The relationships of these factors to variability are complicated, and they have complicated interactions with each other. The longer we count a population, the more we will notice effects external to the population itself. Increasing variance is a consequence of both community membership and ecosystem membership.

6

Nonlinear Dynamics, Strange Attractors, and Chaos

THE COMPLEXITY OF POPULATION CHANGE

In two previous chapters I postponed discussions of populations that show cyclical changes in density. In chapter 4, I noted that such populations are more variable than comparable populations that do not cycle, and, in chapter 3, I argued that high resilience may contribute to cycling and thus to high variability. As an example, I considered a rapidly growing population that overshoots its equilibrium density, because the increasing death rate lags behind birth rate. This hypothetical population would then decline to below its equilibrium density, and so the population could show regular cyclical changes in population density. Were population dynamics to be so simple as this example! If population densities fail to return to a fixed density, many other dynamical behaviors beside cycles are possible. The purpose of this chapter is to consider these possibilities.

The potential complexity of dynamical behaviors can be illustrated by this disarmingly simple example. In figure 6.1, I show three populations all modeled by the same equation that we previously encountered in chapter 2. The population size X_t is given by the equation

$$X_{t+1} = X_t + X_t \cdot r \cdot (1 - X_t/X^*) . \tag{6.1}$$

This is one of the simplest models of population growth: the population grows up to its equilibrium density X^* and declines above it, and the rate of approach to equilibrium, the resilience, depends on r. The three examples in figure 6.1 differ in the values of r. In the first example, the population returns almost directly to the equilibrium value, and, in the second example, there is a higher resilience, and the population cycles above and below equilibrium. The cycles are rather more complex than we might have imagined so far, for the population attains two distinct values of density above and two below equilibrium.

It is the behavior of the third example that is the most surprising. The resilience is only slightly higher than before, yet the population behaves most

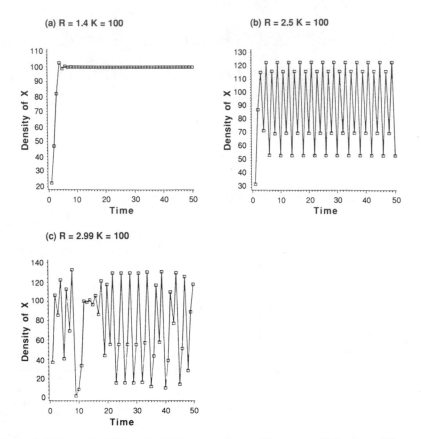

Figure 6.1. Three simulations of a simple population model, equation (6.1), which differ only in the population's resilience *r:* in the first simulation (a), *r* is relatively small, and the population returns directly to equilibrium; in the second simulation (b), *r* is larger, and the population cycles; and in the third simulation (c), *r* is even larger, and the population changes appear random.

erratically. It wanders unpredictably for about twenty generations, then cycles for about another fifteen generations, and then starts wandering again. No matter how long we simulate this model, this exact set of population densities never repeats itself. This behavior seems totally unpredictable and is technically called *chaos*. What is most surprising about the behavior is that it is driven by a strictly deterministic equation—there are no random elements in the equation. Even though the population looks random, it is not. Indeed, the population dynamics are driven by as simple a population model (eq. [6.1]) as one can imagine.

The ecological significance of this chaotic behavior is considerable. The chaotic behavior of the third example suggests that, when we observe highly variable, apparently unpredictable changes in population density, the populations may be obeying some quite simple population model. Moreover, the third example illustrates what Gleick (1988) in his excellent, popular account of the subject calls the "butterfly effect." Tiny changes in the initial conditions of a chaotic system lead to rapidly growing differences. Two systems started at almost identical initial states diverge rapidly and after a period no longer share any resemblance. Ecologically, this means that, unless we know the population parameters with infinite precision, our predictions of subsequent population changes will rapidly deviate from the actual changes. Finally, we might expect these chaotic behaviors to be characteristic of small-bodied species that are known both to have high values of population resilience and also to be highly variable.

The study of cycles, chaos, and related behaviors should be central to population ecology. The preceding ideas are from Robert May's work, recently summarized in his Croonian Lecture to Britain's Royal Society (May 1987). From ecology, the pioneering ideas of May and his colleagues have spread into other disciplines. Yet, as Schaffer and Kot (1985, 1986a, 1986b) have pointed out, the history of complex dynamics is extraordinary, for field ecologists have largely neglected the ideas. What work has been done was entirely theoretical until very recently.

Schaffer and Kot have brought studies of complex dynamics into the spotlight by asking three questions. First, what situations might generate complex dynamics? Second, what kinds of complex dynamics might we expect? Third and most important, how might we detect these dynamics *in nature?* Their approach has been like seeking rare animals or plants. First, we ought to know where to look for them, and, second, we ought to know roughly what they look like. Finally, we need to recognize specific features to make sure we have identified the rarities correctly. The reasons why ecologists have tended to ignore complex dynamics (and I was no exception) was that the dynamics seemed to be no more than mythical beasts that were fun to imagine and draw but that had unknown habits and were unlikely to ever be convincingly identified. What Schaffer and Kot have done is to find these beasts alive, very real, and living under our noses.

WHERE TO LOOK FOR THE BEAST

In finding any rare plant or animal, it helps to know where to look. Looking for complex dynamics is no different. The simple example of equation (6.1) suggests that such dynamics might be quite common, for the model is a very simple one. I shall now show that a broad range of population models can have complex behaviors under assumptions that seem eminently reasonable. Indeed, incorporating ecologically realistic features into population models often leads to complex dynamics. The "complex dynamics" beast could be common: its habitat is not severely restricted. Exploring the habitats in which complex dynamics might be found involves looking at mathematical models of various ecological processes—such as predation, competition, and disease transmission—and for varying numbers of species.

Single-Species Models

The simple population model, equation (6.1), has two parameters: the equilibrium density and the rate at which the population approaches that equilibrium. It has a feature that is not immediately obvious, however. This is the time delay of one time unit, so that the population density at one point in time depends on the population density one unit of time earlier. The population moves in finite jumps from X_t to X_{t+1}, and such models are called *finite-difference equations*. Difference equations are most appropriate to populations with distinct generations—for example, insects in temperate climates where diapause during winter months is common.

An equally reasonable alternative model format—known as *differential equations*—allows populations to grow continuously. No real population grows continuously, for births and deaths are discrete events. In populations in which generations overlap and births and deaths are relatively evenly distributed throughout the year, differential equations may be a reasonable approximation.

We write the continuously changing rate of population change as dX/dt. A simple single-species model is

$$dX/dt = r \cdot X \cdot (1 - X/X^*) \ . \tag{6.2}$$

The population increases when below X^*, decreases when above it, and approaches X^* at a rate dependent on r. In these ways it is very similar to its difference-equation analogue (eq. [6.1]). But, whatever the value of r, the population neither cycles nor shows anything other that a simple return to equilibrium. This is because the model has no time delays: death rates are expressed instantaneously and do not lag behind by one time unit. We can fix that!

$$dX/dt = r \cdot X_t \cdot [1 - (X_{t-q}/X^*)] \ . \tag{6.3}$$

The death-rate part of the equation is now dependent on the population density at some period q in the past. This parameter q could have a very simple meaning. The young of many animals and plants are much smaller than the adults. So the young may not consume as much resources as their parents and so will not depress those resources as much. In many marine, sessile organisms, the young are planktonic, and only when they settle do they begin to compete for space and food with the adults. So the death rate of a population may lag behind the birth rate by the length of time it takes for the individuals to become adults. Other interpretations of the lag q are possible, however. As the time lag q increases, the population dynamics change. For small q, the population density approaches equilibrium smoothly, and then, as q increases, the densities begin to cycle. Both finite-difference and differential equation models lead to the same conclusion: one place to look for complex dynamics is where there are time delays.

Generic Multispecies Models: Lotka-Volterra

No species exists in isolation, and we can fix that assumption too, by expanding equation (6.2) to include other species. We will need both an equation for the population density of each species and terms that reflect how each species interacts with every other species in the community. Obviously, when we do this, there will be a rapid proliferation of constants. So, first I want to introduce a new notation. Let me write equation (6.2) in a different form:

$$dX_1/dt = X_1(b_1 - a_{11} \cdot X_1). \qquad (6.4)$$

The equation is identical to equation (6.2), but I have labeled X as X_1 (in anticipation of there being other species) and have changed r to b_1 and r/X^* to a_{11}: the effect of X_1 on its own growth rate.

Now let's add another species. Suppose that X_2 reduces the growth rate of X_1, and vice versa: that is, the two species compete with each other. This gives us

$$dX_1/dt = X_1(b_1 - a_{11} \cdot X_1 - a_{12} \cdot X_2),$$
$$dX_2/dt = X_2(b_2 - a_{21} \cdot X_1 - a_{22} \cdot X_2). \qquad (6.5)$$

The constants, the a's, reflect the effect of each species on either its own or the other's growth rate. These terms are related to the familiar competition coefficients.

Of course, the interaction between the two species might not be competitive, but predatory. Species X_2 might feed on X_1—and not be limited by anything other than X_1. These assumptions give us

$$dX_1/dt = X_1(b_1 - a_{11} \cdot X_1 - a_{12} \cdot X_2),$$
$$dX_2/dt = X_2(-b_2 + a_{21} \cdot X_1). \qquad (6.6)$$

Equations (6.6) are models of a predator and its prey: we notice that X_2 declines when X_1 is absent (that is, the sign of b_2 is negative) and increases the more dense is X_1. X_1 suffers from both its predator which kills it and from itself, because high density reduces its own growth rate as more individuals starve.

These are two examples of two-species models. Now that I have established a notation, we can see that I could write much more complicated models involving n species, such as

$$dX_i/dt = X_i(b_i + \Sigma a_{ij} \cdot X_j) . \qquad (6.7)$$

The signs of the a_{ij} would reflect whether species interacted, and, if they did interact, whether that interaction involved competition, predation, or, indeed, mutualism. All these models are known generically as Lotka-Volterra models, after Lotka (1925) and Volterra (1926).

Both these two-species models (eqq. [6.5] and [6.6]) have simple behaviors: the densities return to equilibrium—or they do not, wherefore one of the species becomes extinct. The predator-prey model, however, is capable of cycling (though the cycles dampen over time) when the population densities are subject to a repeated external disturbance. Now consider the three-species system, created by Gilpin (1979) and shown in figure 6.2, in which one predator feeds on two competitors. This is hardly an implausible model, but the dynamics seem totally unpredictable. The population densities cycle, but the same cycle never repeats itself. Gilpin's conclusion was that very simple multispecies models could also yield chaotic dynamics.

Before I leave Lotka-Volterra models, I would like to alter the predator-prey equation in two simple ways. First, equation (6.6) assumes that the prey's reproductive rate b_1 is a constant. Yet most organisms have seasonal variation in reproduction. To make the equations more realistic, we might replace the constant birth rate b_1 with a value that changes cyclically, thereby reflecting seasonal changes. This is called *seasonally forcing* the equation. Seasonal forcing can produce population densities that cycle perpetually in the absence of repeated disturbances.

The second simple modification involves the term that represents the interaction between the predator and prey species: in the Lotka-Volterra models it is proportional to the product of the predator and prey densities. This assumes that the predators can consume prey at a rate proportional to the prey density no matter how abundant the prey become. The way in which the rate of prey killed by the predator increases with the increasing density of its prey is called the predator's *functional response*. Lotka-Volterra models assume a linear functional response. A more realistic assumption is embodied in the idea of the *type II* functional response, as defined by Holling (1959). With this kind of functional response, the predators kill relatively fewer prey as the prey become

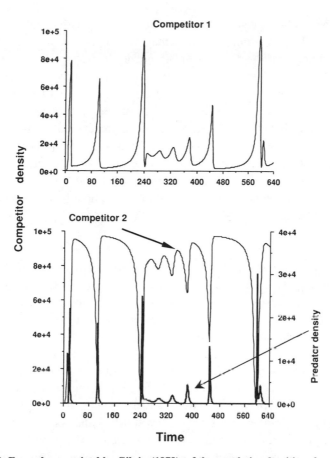

Figure 6.2. Example, conceived by Gilpin (1979), of the population densities of a predator and its two prey which compete with each other. The population densities of all three species appear to change randomly. With thanks to Michael Gilpin for supplying data for this figure.

more abundant, because the predators eventually are killing all the prey they can physically handle (or need). This simple modification of the Lotka-Volterra dynamics can also lead to cycles which are generated by the system itself and are called *limit cycles*. (So, if seasonal forcing leads to cycles and if a type II functional response leads to limit cycles, what happens when we seasonally force a model that would have limit cycles even without seasonal forcing? Intuitively we might get cycles on cycles, and, indeed, this is what happens. I shall present an example of this latter.)

So far we have seen three ways of generating cycles. The first was where episodic, external perturbations (perhaps a hard winter) moved a predator-prey

system away from its equilibrium and where the populations subsequently returned toward equilibrium, cycling as they did so. The environment starts the
system cycling, but the environment is not itself cycling. In the second example, cycles arise from the external seasonal cycling of the environment, and,
in the third example, the cycles are generated internally, but the interaction of
the predator and its prey. So just because we see cycles does not mean we
know anything about their causes. An identical point could be made for all the
complicated dynamics we shall encounter in this chapter. A population may
show the most unusual, yet deterministic changes in density, but these changes
may only reflect how the population tracks some feature of the physical environment and have nothing else to do with the ecology of either the species or
its community.

Diseases

Models of diseases often include separate equations for the different parts
of the host population they infect: susceptible individuals (of which there are
S), exposed individuals (of which there are E), infectious individuals (of which
there are I), and individuals that have recovered from the disease (of which
there are R) (hence the name "*SEIR* models"). The critical process is that of
infection: susceptible individuals become exposed individuals by their encounters with infectious individuals. If this happens at a rate simply dependent on
the product of S and I (which assumes random encounters between the two
groups), then the dynamics can be relatively simple. As we all know from
experience, this is often not the case for some of the common diseases we
suffer. During winter, we are more likely to catch colds, perhaps because we
are more likely to be inside and so close to larger numbers of potentially infectious people. For this and other, similar reasons, disease transmission may be
seasonally forced, which can lead to very complicated dynamics.

Summary

Time delays, multispecies interactions, between-species interactions that are
nonlinear (such as the type II functional response), and seasonally varying
parameters can lead to very complicated dynamics. These are hardly minor
features of the environment. It is hard to image any realistic population model
that would not include at least one of these features. These features do not
inevitably lead to complicated dynamics. So I am not arguing that all populations *must* show complicated dynamics, but I do suggest that such dynamics
may be found almost anywhere. So, if the beast is common in a variety of
habitats, how do we recognize it?

WHAT TO EXPECT: A BESTIARY

How to Describe the Beast

Inexperienced naturalists often fail to identify birds solely on the basis of their color patterns. More experienced observers record less striking features—such as shape, relative size, and habitat—and are usually more successful. Likewise, identifying complex dynamics involves knowing what to look for. As the examples in figure 6.1 show, looking at the population densities over time may not be very helpful in distinguishing between complex yet deterministic population changes and those that are essentially random. In one example we can see that the population returns to a fixed point, in another that it cycles; but what do we learn from looking at the chaotic population? Let us consider several methods that *can* detect complex dynamics.

Method 1: Spectral Analysis.   Spectral analysis decomposes a time series into its constituent cycles (chapter 5). For the data in figure 6.1, this would easily identify the cycle in the second example. If we had only the middle section of the time series in the third example, spectral analyses would show a prominent cycle. This cycle, which develops midway through the time series, is subsequently lost, so if we had another part of the series we would detect no cycles at all. Spectral analyses of one of the three species in Gilpin's one-predator-plus-two-competitor model show little in the way of recognizable periodic behavior. (The spectrum of this model can be reddened for very long periods [Schaffer and Kot 1986b]. So it is possible that complex dynamics alone are responsible for the increasing variability of densities that is so prevalent in animal populations; chapter 5.) In short, spectral analyses may often help to detect—but by themselves may not always be powerful enough to detect—the most complex changes. Do any other ways of describing these dynamics show their underlying unity? The problem posed by the data in figure 6.1 is particularly vexing, for, after all, all three examples are based on the same equation.

Method 2: Successive Time Plots.   In figure 6.3, I show the same data as in figure 6.1 but plotted differently. The three parts of the figure plot consecutive pairs of density values: X_{t+1} versus X_t. Figure 6.3a corresponds to figure 6.1a, and we see the path of the populations moving to a fixed point—the equilibrium density of 100. In figure 6.3b, we see the population moves toward a cyclical attractor, moving between the points labeled 1–4, back to 1 again, and so on. None of this tells you anything that you did not know from looking at

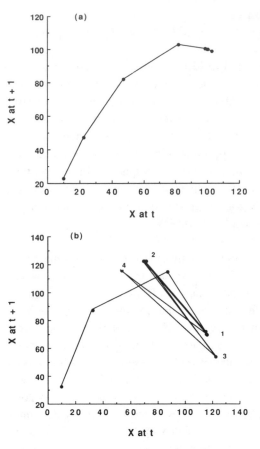

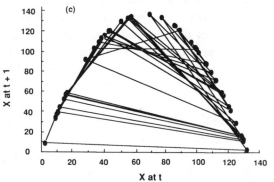

Figure 6.3. Plots of successive pairs of population densities for the three simulations of figure 6.1. The densities converge to the equilibrium (a) and to a repeated sequence of points labeled, 1, 2, 3, 4, 1, etc. (b), and what appeared to be random population changes in figure 6.1c can be seen to be highly predictable as the populations move over a simple parabolic attractor (c).

the plot's density versus time. Look at figure 6.3c; the plot of density against time is apparently random and chaotic, yet we see that the densities are moving backward and forward on a curved line that is actually a parabola.

What is not apparent in these figures is how the models behave if we start the model densities at different values. In fact, for this model, it almost does not matter from where we start the model simulations: the three models converge to the same point, the same cycle, and the same curved line, respectively. (I say "almost" because, if the initial population is too high, the population becomes immediately extinct.) The curved line, for obvious reasons, is called the *attractor*. While I have not shown figures of this convergence, it is extremely easy to demonstrate empirically. Equation (6.1) can be programmed into either a handheld calculator or any computer. (May [1987], who provided this example, strongly urges everyone to try to do this. The example shows how chaos emerges from an extremely simple system, how the underlying attractor is nonetheless simple. Finally, one can also show by simulation that from any two initial densities, however similar, we will eventually obtain highly divergent sequences of population densities, although the sequences of the densities will still be on the curved attractor.)

So far we have encountered three beasts in our dynamical zoo: (1) a point attractor, (2) a cyclical attractor, and (3) something rather more complex and strange. All three were detected by plotting successive pairs of population measurements.

Method 3: Density versus Density Plots. The next method can be applied to multispecies models. Rather than plot successive pairs of densities for one species, we can plot the densities of each species against each other at each period of time and connect the points. These plots would be two-dimensional for two species, three-dimensional for three species, and so on. For a predator-prey model with a type II functional response, this plot shows the existence of a closed loop—the limit cycle (figure 6.4). This analysis applied to Gilpin's one-predator-plus-two competitor system (figure 6.5) provides a remarkable result! The plot reveals that the population densities are not random but are instead moving on a beautiful—and very strange—surface. The various spirals etched out by the densities plotted against each other led to Gilpin calling his model "spiral chaos." The population trajectories never repeat themselves even though they continue to spiral over the strange surface. Moreover, simulations show that points arbitrarily close on this surface eventually lend to widely divergent trajectories across it. This surface is called a *strange attractor.*

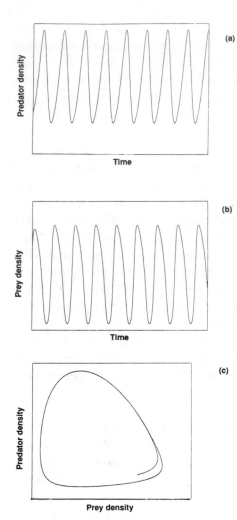

(a)

(b)

(c)

Figure 6.4. Predator densities (a) and prey densities (b) plotted against time for a model in which the predator shows a type II functional response (see text). In (c), successive pairs of predator and prey values are plotted. The trajectory is attracted to a limit cycle, from its initial position off the cycle.

Method 4: Other Plots. For a model of a prey species with a seasonally changing birth rate and exploited by a predator with a type II functional response, the plot of predator density versus prey density produces a rather messy-looking limit cycle (figures 6.6–6.8). With some imagination, we might recognize that there are some cycles imposed on the limit cycle (figure 6.6); but how might we display this? Since it is the prey's birth rate that is changing seasonally, I have plotted the birth rate versus the prey's density versus the predator's density, in figures 6.7 and 6.8. We see that the motion of successive values of predator density, prey density, and prey birth rate is on the surface of a torus (a "doughnut").

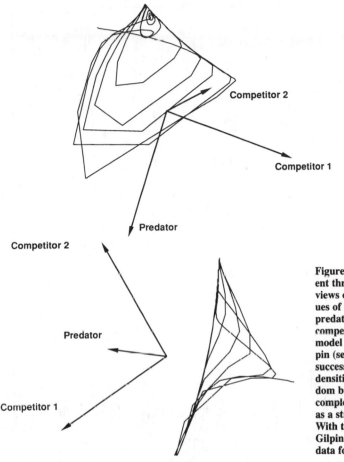

Figure 6.5. Two different three-dimensional views of successive values of the density of the predator and of the two competitors, for the model conceived by Gilpin (see figure 6.2). The successive population densities are not random but move over a complex surface known as a strange attractor. With thanks to Michael Gilpin for supplying data for this figure.

A Bestiary

We now have seen various ways in which we can describe complicated dynamics. Here is a simple taxonomy of the various types of dynamics we might encounter:

Point attractors are the most simple: population densities return to a set value (see figures 6.1a and 6.3a). In the real world, population densities may constantly be nudged away from this point and may cycle as they try to return to the point attractor.

In *limit cycles* the population densities are attracted to a cyclically changing set of densities (figures 6.1b, 6.3b, and 6.4). Even in the absence of disturbance, the population densities would cycle.

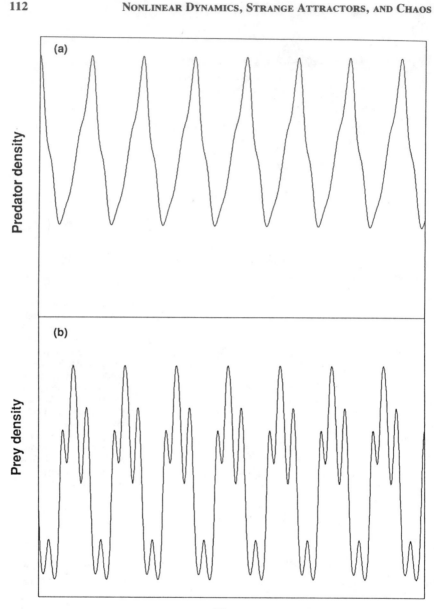

Figure 6.6. Densities of predator (a) and prey (b) from a model that is seasonally forced (see text) and that would show limit cycles in the absence of the seasonal forcing because of the predator's type II functional response. As might be expected, there are obvious cycles on cycles, which are most easily seen in the prey population.

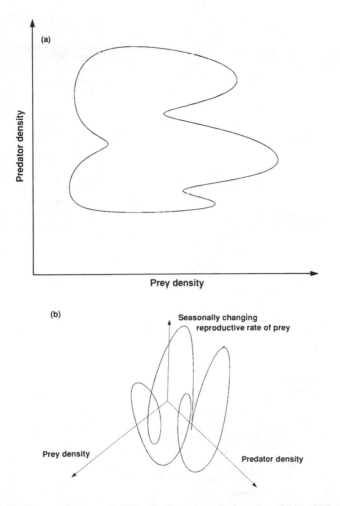

Figure 6.7. Predator versus prey densities for the seasonally forced model in which the
predator has a type II functional response. The plot of predator versus prey densities (a) is
complicated and relatively uninformative. With some imagination, one can see that, in the
bottom plot (b), in which the two axes toward the viewer are the predator and prey densities and in which the vertical axis plots the seasonally cycling reproductive rate of the prey,
the trajectory is moving around a torus. The trajectory makes exactly four seasonally
forced orbits as it moves around one complete predator-prey cycle. If you have trouble
seeing this, look at figure 6.8.

In *toroidal flow,* the motion is on the surface of a torus (figures 6.6, 6.7,
and 6.8). The motion can be periodic, when the trajectories wind around the
torus an exact number of times before the motion repeats itself (figure 6.7).
The motion can be more complex than this, for it may never exactly repeat
itself and, in time, may cover the torus's entire surface (figure 6.8). This be-

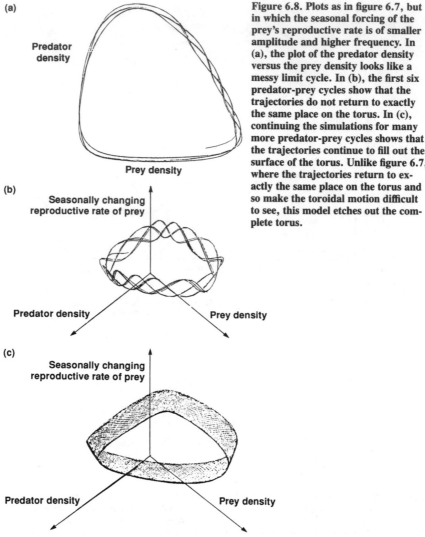

(a)

Predator
density

Prey density

(b)

Seasonally changing
reproductive rate of prey

Predator density   Prey density

(c)

Seasonally changing
reproductive rate of prey

Predator density   Prey density

Figure 6.8. Plots as in figure 6.7, but in which the seasonal forcing of the prey's reproductive rate is of smaller amplitude and higher frequency. In (a), the plot of the predator density versus the prey density looks like a messy limit cycle. In (b), the first six predator-prey cycles show that the trajectories do not return to exactly the same place on the torus. In (c), continuing the simulations for many more predator-prey cycles shows that the trajectories continue to fill out the surface of the torus. Unlike figure 6.7, where the trajectories return to exactly the same place on the torus and so make the toroidal motion difficult to see, this model etches out the complete torus.

havior is called *quasi-periodic*. For, although quasi-periodic population densities never exactly repeat themselves, they come close to doing so. Moreover, small differences in initial densities do *not* lead to widely divergent sequences of densities.

Strange attractors are neither periodic nor quasi-periodic, but the motion is confined to a low-dimensional attractor. The motion is chaotic: it never exactly repeats itself. Moreover, even the most minute differences in initial densities will eventually lead the population densities to wander over entirely different parts of the attractor.

Finally, it is possible that the motion may be very much more complex than even strange attractors: the attractors may be highly dimensional, and, even though the motion may be entirely deterministic, predicting its behavior in any detail may be impossible. Turbulence in fluids is a physical example of this.

MEANWHILE, IN THE REAL WORLD . . .

There is no doubt either that the mathematical objects I have introduced in the previous section are exquisite or that they appear from ecological models so simple that they are found in even the introductory ecology texts. In fact, the model dynamics I have discussed only touch the surface of those possible. It is as if I had just described the animal kingdom by introducing you to a worm, a mollusc, a shark, and a white rat. For example, some of the possible attractors have fractal structures (see chapter 5) becoming more complex the closer we examine them. Other models show *phase locking:* the motions cycle at the same frequency even though the parameter values are changing. One wonders whether this phase locking explains the behavior of, say, vole cycles, which occur at approximately four-year intervals. Why do not voles cycle over a whole range of different frequencies? It is easy to be seduced by the computer programs, however. I have spent many afternoons fascinated by population trajectories being etched on the screen of Mark Kot's computer, and I have spent many evenings looking at beautiful photographs of attractors in popular books on chaos. But there comes a time when one asks: What does all this have to do with real populations? Can we detect these complicated dynamics in the real world?

There are several clues to the detection of complicated dynamics. Some populations—voles and lynx, for example—show regular cycles, and these are readily detected either by plots of density versus time or by spectral analyses (chapter 4). Complicated dynamics lead to high population variability, and so high variability is one clue to their existence. How might we detect motions that are not simply periodic? How might we decide whether these complicated behaviors are widespread in nature?

The first attempt to answer these questions was by Hassell et al. in 1976 soon after May recognized that simple population models could lead to complicated dynamics. The paper is clear, its scope is broad, and its three authors' influence on ecology is indisputable. Yet, I think this paper probably seriously misled ecologists about the importance of complicated dynamics in ecology, despite the authors' carefully worded caveats. So what did Hassell et al. do, why did it cause most field ecologists to neglect complex dynamics for over a decade, and what has lead to my belief that the existence of complex dynamics has been seriously underestimated?

Hassell et al. on the Dynamics of Real Populations

Hassell et al. surveyed four laboratory and twenty-four field populations of insects. These populations might be expected to have approximately discrete generations, and the relatively high reproductive rates of insects makes them a likely group in which to detect dynamics more complicated than point attractors. To these data the authors fit the parameters of a simple population model. The model parameters from this fit were then examined, in light of the kind of dynamical behavior a model with those parameters might be expected to show.

For example, we could estimate, on the basis of successive population estimates, the two parameters (resilience and equilibrium density) in the model of equation (6.1). Because we know what values of resilience would lead to a stable point, limit cycles, or chaos, we could then classify the populations' behavior accordingly. (We have already seen how to do this in chapters 2 and 3 [see figure 2.2], where I introduced a regression approach to calculating the parameters in the eq. (6.1). Recall that, of nearly one hundred bird populations, I found only two that had high enough values of resilience to generate cyclical changes in density [Pimm 1984b].)

Hassell et al. used a more sophisticated model, but the recipe was much the same. We have seen how plots of X_{t+1} versus X_t can be important in elucidating population dynamics. The solutions to equation (6.1) become clear when we perform these plots (figure 6.3). Equation (6.1) describes the relationship between X_{t+1} and X_t that is a simple parabola. As the parabola goes from being relatively flat to being more pointed, the dynamics go from a point attractor to a limit cycle to chaos. Hassell et al. argued that, for real populations, a curve more flexible than a parabola might better describe the relationship between X_{t+1} and X_t. This flexibility is achieved with an additional parameter, for a total of three. Two of these parameters are ones we encounter in the simpler model of equation (6.1), and they can be thought of as measures both of equilibrium density and of fecundity or resilience. The third parameter describes how the individuals compete for resources; call it the *competition parameter,* if you like. In the simpler model of equation (6.1), mortality from the individuals competing for limited resources increases modestly with increasing density. The maximum growth rate of the population occurs at a density that is half of the species' equilibrium density. Such a situation has been described as a "contest" for resources. Under other circumstances, competition may be more of a "scramble," leading to mortality that increases dramatically once some threshold density has been reached. Scramble competition causes the graph of X_{t+1} versus X_t to become much less symmetrical than a parabola.

The dynamical properties of the three-parameter model can be explored theoretically. The model dynamics may show a point attractor either with the densities returning straight to equilibrium or with dampening oscillations, limit cycles, or chaos. The parameter boundaries that lead to these different outcomes are shown in figure 6.9. As with the model of equation (6.1), the dynamics of the model depend on the species' fecundity: the higher the fecundity, the more likely the population densities will cycle or be chaotic. The dynamics also depend on the competition parameter. This is also intuitively reasonable. The more the competition is a scramble, the greater the crash in population density at high population densities—and the lower the population will be the next generation.

Hassell et al. estimated the parameters of the model by using data on the life history of the species, data such as how mortality changed with density

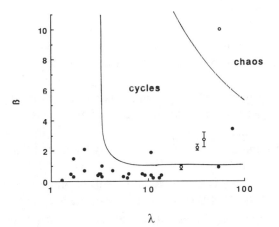

Figure 6.9. Boundaries between parameters that describe the population growth rate and the shape of competition, which determine the nature of the population dynamics for the single-species model investigated by Hassell et al. (1976) (see text). For parameter values near the origin, the population densities return directly to equilibrium, and, for high values, the population densities are chaotic. The points on the figure represent the estimated values for twenty-four field populations (solid circles) and four laboratory populations (open circles) of insects. The positions of these points suggest that most field populations should have densities that return to equilibrium, that do not cycle, and that certainly do not show chaotic changes. Notice, however, that the population represented by the point nearest the origin (the moth *Zeiraphera*) shows clear population cycles (see table 6.2), yets its parameter estimates suggest that it should be the *least* likely of all these populations to cycle. (Redrawn from Hassell et al. 1976.)

and how fecund were the species. They placed these parameter estimates on the graph (figure 6.9) and drew some simple conclusions. Of the field populations, one had parameters consistent with limit cycles, and one was consistent with dampened oscillations. Of the four laboratory populations, one was consistent with chaos, two more were consistent with at least some kind of cycling, and the fourth was close to the boundary of dampened oscillations. Their conclusions are worth quoting directly: "We find that the majority of populations show a monotonic return to a stable equilibrium following a disturbance, with relatively few examples of oscillatory damping or low order limit cycles. Examples of stable cyclic behaviour, or of chaotic fluctuations are mostly found in laboratory populations."

The differences between laboratory and field populations are easily understood, they argued. In the laboratory, the absence of natural mortality factors such as predators will inflate the reproductive rates, and the lack of dispersal will tend to make the competition for resources more of a scramble.

The authors were careful to add several caveats, the last of which is also worth repeating: "There are grounds for thinking that chaotic dynamics may arise more readily in multi-species situations. Thus, morals drawn from single-species studies must be treated with caution."

There was one uncomfortable result in their paper. It involves a moth, *Zeiraphera diniana*—and the parameters they estimate for it place it in the most extreme bottom left of figure 6.9. That is, its densities should be the *least* likely to cycle of all the species surveyed. Unfortunately, this species shows some of the most extraordinarily regular cycles (with a period of about nine years) in the animal kingdom (see below and table 6.2). Despite the elegant approach, Hassell et al. found more evidence for complex dynamics by looking at the population densities through time than they did by statistically analyzing the data.

Despite the caveats, the conclusions of Hassell et al. were emphatic. A follow-up study by Bellows (1981), using the same data but a different model, came to the same conclusion. Similar studies on laboratory populations of fruit flies, *Drosophila,* also found an overwhelming preponderance of parameters consistent with simple equilibrial behaviors (Muellar and Ayala 1981; Phillipi et al. 1987). Ecologists in general (including me) stopped looking for complex dynamics in real populations.

Schaffer and Kot's Use of Takens's Solution

The theory developed in this chapter suggests that an obvious way to detect complex dynamics is to plot the corresponding densities of several species. We can quickly reject this method. It is unlikely we will simultaneously have data

on more than one species, and, if we do, they may not be the two or more
species that we need to construct the appropriate portrait. For the periodically
forced predator-prey model, the needs are even more constraining, for we need
predator *and* prey densities *and* simultaneous knowledge of the prey's reproductive rate.

A most ingenious solution to this problem of inadequate data has been suggested by Takens (1981), and Schaffer and Kot were quick to appreciate the
significance of his solution (Schaffer 1984; Schaffer and Kot 1985, 1986a,
1986b). Takens considered a group of several interesting populations for which
only repeated observations of *one* population are available. Then, for almost
every population and almost every time lag T, the multidimensional portrait
constructed by plotting $x(t)$ versus $x(t-T)$ versus $x(t-2T)$ versus $x(t-3T)$ and
so on out of m dimensions will have the same dynamical properties as does the
dynamical portrait constructed for the multidimensional group of populations.

This is an extraordinary result. So often in analyzing ecological data what
ecologists need and what they have are very different from each other. What
ecologists might seem to need to analyze population data are the densities of
many species. Moreover, they might expect to need data collected at time intervals that are determined not by the convenience of the ecologist but by the
nature of the species' dynamics. Takens suggests this is not necessary: one
species will do. This is fortunate, because generally only data for one species
in any system are available—and that species was not chosen because it was
the "right" one for understanding the dynamics of the system. Moreover, censuses are conducted at intervals (often years but sometimes months or weeks)
that reflect observer convenience—and not at some interval, such as every
generation, determined by the dynamics themselves.

Takens's ideas arose in the context of physics and chemistry. Schaffer and
Kot asked whether the procedure could be applied to ecological systems, for
these would likely be far more variable than those studied in the physics laboratory. They showed not only that Takens's idea could be applied but that it
could demonstrate the existence of some surprising patterns. Here are three of
their examples.

The Classic Case of a Variable Insect Population.   Inspection of nearly any
modern introductory ecology text shows how influential has been Andrewartha
and Birch's discussion of a fourteen-year study of the insect *Thrips imaginis* in
flowers in an Australian garden (Davidson and Andrewartha 1948a, 1948b;
Andrewartha and Birch 1954, 568–83). The numbers of thrips were always
well below those set by food and space, and the numbers were closely correlated with a number of weather factors. Indeed, the study is generally held to

be the classic example of how population numbers are determined by changes in the weather.

Schaffer and Kot (1986a) reexamined the data. Spectral analyses showed a peak at a frequency of close to one a year, which one might expect from a population driven by annual weather patterns. Applying Takens's solution, they obtained plots that are far more interesting. Rather than being the tangled knot one would expect from random motion, the trajectories move over a thin, curved surface. Now, two-dimensional drawings of three-dimensional objects, be they on a computer screen or on a printed page, are not easy to interpret. For, while the portrait of the thrips data in three dimensions looks like "something," it equally clearly does not look as elegant as those attractors drawn out by entirely deterministic equations. Physical intuition is useful at this stage of the argument. Just how "thick" the "thing" is is obviously quite important in separating deterministic yet complex dynamics from dynamics that are essentially random. For an attractor, the trajectories are moving over a paper-thin surface, while, for random motion, the three-dimensional portrait is of a thickly filled blob.

So how do we decide how thick or thin are the real data? The answer is, just as we would for any other physical object. If you want to know how thick a pie is, you take a slice of it. In the context of these portraits, that slice is called the *Poincaré section*. It can be placed anywhere sensible in the three-dimensional space, and it yields a two-dimensional picture of where the trajectories pierce it. For random population changes, we would observe a circular scatter. But, for the thrips data, the points look more like a line, indicating the thinness of the attractor on which the population trajectories are moving. Poincaré sections, therefore, are an essential tool in analyzing the lag portraits formed by using Takens's analysis.

Schaffer and Kot conclude that, while Davidson and Andrewartha were probably correct in denying the importance of food as a limiting factor, "their conclusion that the magnitude of thrips outbreaks is essential random turns out to be wholly in error" (Schaffer and Kot 1986a). What drives the complex, if nonrandom changes in the thrip's densities, is not obvious. There are no data on the species' predators and parasites. Whatever the cause of the population variation, when viewed appropriately, the population densities seem remarkably deterministic.

Canadian Lynx Cycles.   For their second example, Schaffer and Kot examined data on the Canadian lynx. Here again, they found a surprising result. The lynx have a well-known cycle of approximately ten years in duration. Two-dimensional plots of density versus density the previous year show the fuzzy circles we might expect from a noisy limit cycle. But three-dimensional plots of density versus density the previous year versus density two years previously suggest that the motion is more complicated than a simple limit cycle. In three dimensions, the orbit shows successive stretching and folding, with large-amplitude cycles alternating with small-amplitude cycles. Schaffer (1984) suggests that the lynx data may be moving on a low-dimensional strange attractor. Although he concedes the data are few, he does create doubt that the lynx dynamics are a simple cycle.

Childhood Diseases.   One of Schaffer and Kot's most interesting analyses involves three childhood diseases: measles, mumps, and chickenpox. Data for mumps and chicken pox suggest nothing more than a yearly cycle, probably driven by the greater tendency for children to be in close contact with each other during the wintertime. The data for measles, however, suggest a strange attractor. The data for New York City and Baltimore, for example consist of monthly physicians reports from 1928 until 1963. Prior to the introduction of vaccines, major outbreaks occurred every second or third year in New York and less frequently in Baltimore. There is a strong annual cycle in the data, with the maximum number of cases reported during the winter. Superimposed on this annual cycle is substantial year-to-year variation. In New York, for example, yearly totals ranged from a low of about 2,000 cases in 1945 to a high of over 79,000 in 1941. What initially appears as a complex, unpredictable sequence of population data can be shown to be remarkably predictable when the three-dimensional lag portraits, with the lags set to three and six months, are examined. Once the shape of the strange attractor has been appreciated, it is a simple matter to use the information to make predictions about the number of infections to be expected. Another interesting feature of this system is that it is what we might expect. Schaffer and Kot subjected a simple SEIR model (see above) to seasonal forcing. Then they plotted the model data in exactly the same way as they plotted the real data. They found that the model populations also showed highly erratic outbreaks of infections, but that, when plotted by using Takens's approach, the strange attractor was clearly identifiable.

Caveats.   What Schaffer and Kot have shown is that deterministic yet complicated or even chaotic population changes may have remained undetected because of the lack of a suitable way of analyzing the data. They suggest that lynx data may be more complex than the simple ten-year cycle they were supposed to be. Their analyses of thrips and of childhood disease suggest that these changes may be complex yet inherently predictable. There are, however, a number of difficulties that have led to a good deal of controversy.

The data are less than compelling. One wonders whether New York physicians faced with an unprecedented number of cases of measles during a period of global warfare really had time to report their experiences accurately. The lynx data also have some special sampling problems, and their collection spanned intervals of political upheaval in Canada (Williamson 1987).

The approach outlined does not provide statistical tests, and, though such tests exist, how well they perform on the very short time series available to ecologists is far from clear. An important development is the work of Sugihara and May (1990). They develop a technique of nonlinear forcasting based on the idea that there may be an attractor in a small number of dimensions which, like the parabolic attractor in figure 6.3, serves to predict subsequent population densities. Moreover, the errors associated with such forcasting may also be diagnostic. With chaotic systems, the difference between predicted and observed values increases particularly rapidly, more so than in random processes. Among other things, Sugihara and May agree that a low-dimensional attractor underlies the measles data, and they claim another low-dimensional attractor for plankton data collected off San Diego, California.

Finally, even if we concede that these examples represent complex dynamics, this says only that complex population dynamics have been overlooked. They may be more common than we had thought, but they are not necessarily common. Only if complex dynamics are shown to be common will the ways in which ecologists view ecological processes be changed radically.

How Prevalent are Complex Dynamics in Natural Populations?

There is the need to conduct a survey of long-term population data and ask what proportion of the data are consistent with complex dynamics. There are obvious difficulties with this. Long-term data in ecology may still involve very few years, relative to the requirements of these methods. Moreover, as the length of the study increases, the quality of the data decreases. A professional ecologist may study a population for a decade, but a century of data will have been collected for purposes other than the ones we plan for them. Worse, the statistical techniques are so complicated and so new that any results must be considered tentative until the scientific community has had time to fully digest them. Nonetheless, we should still ask what long term data can tell us about population changes.

There are two surveys that fit the need outlined in the previous paragraph. Two colleagues and I surveyed all the terrestrial population data we could find that spanned fifty years or more (Witteman et al. 1990). These data included birds, mammals, and insects. Turchin (1990) analyzed, with considerably more sophistication, fourteen species of insects counted over ten years or more, including four of the insect species we analyzed. In both surveys, the emphasis was on finding populations that cycled. Recall that the existence of significant cycles or indeed other suggestive patterns will not tell us anything about what causes them, a point to which I shall return later. I shall deal with the vertebrates first, then the insects.

Vertebrates. Witteman et al.'s analyses of long-term data proceeded in two ways. The first was to look for cycles directly by using studies of autocorrelation (Chatfield 1984; Finerty 1980 discusses the methods in an ecological setting). Each density is correlated with the density one year in the past, two years in the past, and so on. When there are cycles of x years in duration, the correlations will be significant and positive for x, $2x$, $3x$, . . . years and significant and negative for $x/2$, $3x/2$, $5x/2$, . . . years. The second way was to use plots of a single successively lagged variable (densities $x[t]$, $x[t-T]$, . . . $x[t-nT]$), as suggested by Schaffer and Kot (above). This procedure was used simply in a descriptive fashion: we made no inferences about the patterns generated but suggested likely candidates for more detailed analyses.

We assembled seventy-one populations counted for at least fifty years in temperate and arctic habitats. There were twenty-seven bird populations of seven species and twenty-seven mammal populations of ten species (table 6.1). Wherever possible, we excluded segments of data at the beginning or end of censuses if these segments contained many zeros. None of the population censuses included more than two consecutive zeros during the years that we selected for analysis. Of the bird and mammal populations, only one—the grey heron (population 1 in table 6.1)—was censused directly. Abundances for the remaining populations were estimated from historical records of annual harvests, records based either on game bag records from various estates in Britain (populations 2–32), on the number of animals exported from Iceland (population 50), or on the number of fur returns, shipments, or sales from Canada and Alaska (populations 51–71).

Twenty-six of the bird populations (seven species) and six of the mammal populations (three species) were censused in temperate regions. The remaining populations (twenty-three mammal populations [seven species] and one bird population) are from arctic regions. There are no tropical populations. We have not yet compiled long-term studies on aquatic and marine communities.

We classified our results into three categories: significant, likely significant, and not significant. We considered an autocorrelation *significant* when a population density was both significantly negatively correlated with a population density at the half-cycle period and positively correlated with the population density at the full-cycle period. The significance level chosen was 5%. An autocorrelation was considered *likely to be significant* when the half- or full-cycle correlation was significant *and* when the full- or half-cycle (respectively) correlations were significant at the 10% level.

This is undoubtedly an unsophisticated analysis reflecting our lack of experience of time-series analyses. There were, however, reasons for our choices. First, the densities often showed significant autocorrelations over very short time lags—particularly those of one year. This is not surprising. For many of the species, the densities in one year would correlate with those of the previous year because the same individuals may be present in both years. Our choice excludes such significant autocorrelations, for they say nothing about the presence of cycles. Second, our significance levels were calculated to correct for the fact that many different correlations were being performed on each population. Obviously, if we did not correct for this, then we would typically find one "significant" correlation per population, because we always looked at correlations out to lags of at least twenty years. However, even with this correction, we sometimes found significant autocorrelations that seemed meaningless in the context of cycles. For example, we might find a twenty-year significant positive autocorrelation in an insect species—when the ten-year autocorrelation was neither negative nor significant. Our choices would reject such an autocorrelation as being either significant or likely significant.

We felt that our choices were conservative. We rejected, as even likely significant, those populations in which there were marginally significant positive and negative correlations at matching integer and half-integer periods of cycling. These were all in species for which other populations of the species did show significant autocorrelations and did so, moreover, at the same periods. These correspondences suggests that we ought to have included these as being significant. (Indeed, Turchin [1990] analyzed four of the same insect species as we did, and, while he came to very similar conclusions, he found statistical significance where we did not; see below.)

The data and the results of our analyses are summarized in table 6.1. In the data from arctic regions, the one bird population showed a likely significant autocorrelation. Significant autocorrelations were found in sixteen of the twenty-one mammal populations, and a further three populations showed autocorrelations that were likely significant. In short, in *every* species we found at least likely cycles.

TABLE 6.1   Cycling Characteristics of Vertebrate Species

Name	Population number	Cycles? (period)	Dates of Census[a]	N[b]	Location[c]
Temperate birds:					
Ardea cinerea (grey heron)	1	Likely (16)	1928–1977	50	England and Wales
Tetrao tetrix (black grouse)	2[d]	No	1866–1938	73	Central Scotland
	3[d]	No	1850–1938	89	South Scotland
T. urogallus (capercailie)	4[d]	No	1866–1938	73	Central Scotland
Lagopus mutus (ptarmigan)	5[d]	No	1876–1939	64	Central Scotland
L. scoticus (red grouse)	6	Likely (10)	1866–1938	73	Blair Atholl, Scotland (moor 1)
	7	Likely (10)	1866–1938	73	Blair Atholl, Scotland (moor 2)
	8	No	1866–1938	73	Blair Atholl, Scotland (moor 3)
	9	No	1866–1983	73	Blair Atholl, Scotland (moor 4)
	10[d]	No	1859–1938	80	Central Scotland
	11[d]	No	1854–1938	85	Wales
	12[d]	No	1886–1938	53	Hebrides, Scotland
	13[d]	No	1834–1938	105	South Scotland
	14[d]	No	1870–1938	69	East and central England
	15[d]	No	1855–1938	84	Ireland
	16[d]	No	1887–1938	52	Northwest Scotland
	17[d]	No	1879–1938	60	West Scotland
	18[d]		1869–1938	70	Northeast Scotland
	19[e]	Likely (5)	1848–1904	57	Cumberland, England
	20[e]	Yes (6)	1849–1904	56	Aberdeenshire, Scotland
	21[e]	Yes (6)	1858–1909	52	Lanarkshire, Scotland
Scolopax rusticola (woodcock):	22[e]	No	1846–1932	87	Sussex, England
Perdix perdix (partridge):	23[e]	No	1843–1933	91	Norfolk (estate a), England
	24[e]	Likely (5)	1862–1932	71	Norfolk (estate b), England
	25[e]	No	1843–1933	71	Yorkshire (estate b), England
	26[e]	No	1879–1932	54	Hertfordshire, England
Temperate mammals:					
Lepus europaeus (European hare):	27[e]	No	1862–1932	71	Norfolk (estate b), England
	28[e]	No	1843–1932	90	Yorkshire (Estate a), England
Oryctolagus cuniculus (European rabbit)	29[e]	No	1867–1928	62	Yorkshire (estate b), England
	30[e]	Likely (7)	1862–1932	71	Norfolk (estate b), England
	31[e]	No	1869–1933	65	Norfolk (estate a) England
Mustela vulgaris (weasel)	32[e]	No	1879–1930	52	Perthshire, Scotland
Arctic Birds:					
Lagopus mutus (ptarmigan)	50[f]	Likely (12)	1864–1914	51	Iceland

TABLE 6.1   (continued)

Name	Population number	Cycles? (period)	Dates of Census[a]	N[b]	Location[c]
Arctic mammals:					
Alopex lagopus (Arctic fox)	51	No	1834–1925	92	Hebron, MM, Canada
	52	Yes (4)	1834–1925	92	Hopedale, MM, Canada
	53	Yes (4)	1834–1910	77	Nain, MM, Canada
	54	Yes (4)	1834–1906	73	Okak, MM, Canada
	55	Yes (4)	1868–1924	57	Ungava District, HBC, Canada
Vulpes fulva (colored fox)	56	Yes (4)	1834–1918	85	Hopedale, MM, Canada
	57	Yes (4)	1854–1918	65	Nain, MM, Canada
	58	Yes (4)	1854–1918	65	Okak, MM, Canada
	59	Yes (4)	1854–1918	65	Hebron, MM, Canada
Martes americana (marten)	60	No	1846–1916	71	Hopedale, MM, Canada
	61	No	1853–1908	56	Nain, MM, Canada
	62	Likely (8)	1853–1909	57	Okak, MM, Canada
Ondatra zibethicus (muskrat)	63	Likely (10)	1849–1927	79	MacKenzie River, HBC, Canada
Lynx canadensis (lynx)	64	Yes (10)	1821–91	71	Athabasca Basin, HBC, Canada
	65	Yes (10)	1821–91	71	West central, HBC, Canada
	66	Yes (10)	1821–1934	114	MacKenzie River HBC, Canada
	67	Yes (10)	1821–91	71	Upper Saskatchewan, HBC, Canada
	68	Yes (10)	1821–91	71	Winnipeg Basin, HBC, Canada
	69	Yes (10)	1821–91	71	North central, HBC, Canada
Lepus americanus (snowshoe hare)	70	Yes (11)	1849–1904	56	HBC
Gulo gulo (wolverine)	71	Likely (10)	1910–64	55	Alaska

Note.—See Witteman et al. (1990) for the original sources of data.

[a]Dates refer to the years of the census that we included in our analyses.

[b] *N* refers to the number of observations included in our analyses.

[c]MM = Moravian Missions; HBC = Hudson's Bay Company.

[d]Annual abundances were estimated as the total of game bags from a variable number of moors within the region shown. For each population, we corrected data according to the number of moors included in the annual counts.

[e]Annual abundances are based on the number of animals killed on one individual estate in each of the locations shown.

[f]Abundances were estimated as the total number of birds exported from Iceland each year. We corrected these totals by the duration of the open season, which changed throughout the census period.

In the data from temperate regions, two of the twenty-six bird populations (both of the same species) showed significant autocorrelations. Moreover, populations of two additional bird species had a likely significant autocorrelation according to our definition. There were no significant autocorrelations in the six populations of the three species of mammals; however, one population had a likely significant autocorrelation.

For the populations with significant autocorrelations, Takens's method should reproduce a closed curve, and, indeed, this is what we find. For two additional populations that did not show even likely autocorrelations, we found patterns that *to us* suggest cycles for one temperate mammal (rabbit; population 30 in table 6.1) and one temperate bird population (red grouse; population 9 in table 6.1). These species were those in which autocorrelations had already been found, and, indeed, both populations were physically close to populations for which significant autocorrelations had been found.

In short, cycles are common in the arctic; indeed, they appear to be the major pattern of population changes there. Two of seven species in temperate birds and one of three species in temperate mammals were at least suggestive of complex dynamics.

Insects.   Witteman et al. (1990) also assembled fourteen insect populations counted for at least fifty years; ten were of migratory moths and butterflies recorded by amateur naturalists in Britain (populations 33–42 in table 6.2). The other four were of German forest pest Lepidoptera that were directly counted (populations 43–46 in table 6.2). Turchin (1990) assembled fourteen populations of forest insects; four of these were the same four German forest Lepidoptera, and the other ten were different species (table 6.2).

Witteman et al. applied to the insects the same methods that they applied to the vertebrates discussed in the previous section and used the same classification: significant, likely to be significant, and nonsignificant. Turchin looked for delayed density-dependent changes; these are the kind of time lags that are indicative of cycles. Now, there are often correlations between population densities in consecutive years (see chapter 3), and, as a direct consequence, there are progressively weaker correlations between densities two and more years apart. This lag of one year can be factored out statistically, and any partial correlation that remains between density changes and the density two years in the past indicates delayed density dependence. Turchin's approach, therefore, involved formal statistical tests of this delayed density dependence.

In the ten migratory insect species, Witteman et al. found that two showed "likely" autocorrelations (two hawk moths; populations 41 and 42 in table 6.2). More interesting, in two more of these migrant populations (populations

TABLE 6.2   Cycling Characteristics of Insect Species

Name	Cycles? (period)	Dates of [a]Census	N[a]	Location
Species only in Witteman et al.'s (1990) analysis:				
Colias croceus (clouded yellow)	No	1855–1955	101	Britain
H. convolvuli (convolvulus hawk moth)	No	1851–1955	105	Britain
A. atropos (death's head moth)	No	1856–1955	100	Britain
M. stellatarum (hummingbird hawk moth)	No	1881–1955	75	Britain
Nymphalis antiopa (Camberwell beauty)	No	1855–1955	101	Britain
Vanessa cardui (painted lady)	No	1877–1955	79	Britain
V. atalanta (red admiral)	No	1875–1955	81	Britain
Plusia gamma (silver Y)	No	1877–1955	79	Britain
H. celerio (silver-striped hawk moth)	Likely (3)	1877–1926	50	Britain
C. livornica (striped hawk moth)	Likely (10)	1856–1906	51	Britain
Species in both Turchin (1990) and Witteman et al. (1990):				
Bupalus piniarius (bordered white moth)	No/yes[b]	1881–1940	60	Letzlingen, Germany
Dendrolimus pini (pine spinner moth)	No/yes[b]	1881–1940	60	Letzlingen, Germany
Hyloicus pinastri (pine hawk moth)	Uncertain[c]	1881–1930	50	Letzlingen, Germany
Panolis griseovariegata (pine beauty moth)	No/no[b]	1881–1940	60	Letzlingen, Germany
Species in Turchin (1990) only:				
Acleris variana	Yes		12	
Dendroctonus frontalis	Yes		30	
Zeiraphera diniana	Yes		38	
Lymantria dispar	Yes		26	
Orgiya pseudotsugata	Yes		15	
Operophtera burmata	Yes		19	
Diprion hercyniae	No		24	
Hyphantria cunea	No		22	
Lymantria monacha	No		42	
Choristoneura fumiferana	No		28	

[a]Defined as in table 6.1.

[b]Conclusion of Witteman/conclusion of Turchin. Notice that Turchin found evidence for effects, in cases Witteman et al. were unable to make a definite decision.

[c]Witteman et al. concluded that a nine-year cycle was likely. Turchin did not consider oscillations significant in this species. However, examination of his figure 1.j shows that this species comes very close to being significant.

33 and 36 in table 6.2), we found patterns that *to us* suggest attractors; I show these in figure 6.10. The hummingbird hawk moth (population 36 in figure 6.10a) has a pattern suggestive of a cycle. The clouded yellow butterfly (population 33 in figure 6.10b) shows a pattern that I find particularly intriguing, and I know of no model that might generate such a pattern. (It is clear that such a model would have to be firmly based in the detailed natural history of this species. If such a model predicted the patterns a priori, then we would have more confidence that the patterns are not spurious.)

In one species (the pine beauty moth) of the four analyzed by both Turchin and Witteman et al., both studies found no evidence of cycling. In another species (the pine spinner moth), Turchin found a significant result; we did not. For the bordered white moth, our analyses showed marginally significant positive and marginally significant negative correlations at compatible times, and we felt that our exclusion of it from the "likely significant" category was probably a reflection of a too strict criterion. Turchin would agree; he found a significant result. Finally, on the basis of the lag plots, we suspected that the pine hawk moth might show cycles. Turchin did not find significance, though inspection of his figures shows that the analysis came very close to being significant. Of the ten forest insect species analyzed only by Turchin, he found significant results for six of them. Thus, in a total of twenty-four species of

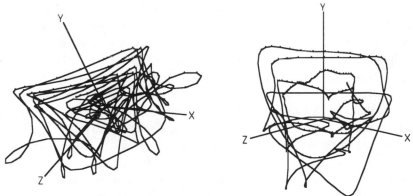

Clouded Yellow Butterfly

Hummingbird Hawk Moth

Figure 6.10. Plots of population densities at current time, at one time interval in the past and at two time intervals in the past for two species: hummingbird hawk moth and clouded yellow butterly. The resultant patterns are suggestive of complex yet possibly predictable dynamics. Original census data were logarithmically transformed and were smoothed by using a three-point moving average. The time intervals chosen for plotting were approximately 25% of the most powerful peak in the spectral analysis of these data. For the technical details, see the work of Witteman et al. (1990).

insects, there is evidence of cycling in eight, of probable cycling in another three, and of possible cycling in another two.

Caveats and Conclusions.   There are many limitations to these analyses, as both studies went to some length to explain. The data often were collected haphazardly. The migratory Lepidoptera were counted by several generations of amateurs. The game bag records span two world wars, when sportsmen and game keepers were occupied in other activities. Furthermore, these and other harvesting records may reflect changing management strategies as well as the intrinsic dynamics of the populations. However, such problems might be expected to obscure patterns in the data rather than to create them. The relatively few data points (114 years—the maximum) should make it difficult to detect anything other than the most pronounced of patterns. The data in Turchin's survey were collected more carefully—but over much shorter periods of time. So, these surveys may have underestimated the prevalence of complex dynamics.

Conversely, all the insect species should be expected to be unusually variable in density, because migratory species and species considered pests are found to be more variable than are species classified otherwise (chapter 5). Highly variable populations may be more likely to demonstrate complex density changes than are a random selection of populations (chapter 4). In any case, game birds and mammals, as well as migratory and pestilential insects, are hardly a random selection of species. I hope that these analyses will prompt others to report and analyze other long-term studies that we missed.

I have three very tentative conclusions. First, it appears that cycles in insects are common and—outside the arctic at least—more common than in vertebrates. Second, these patterns seem sufficiently compelling to refute the hypothesis that complex dynamics are special cases in geographically restricted habitats. Cycles are more prevalent in the arctic, but they also occur elsewhere. When changes in density do show predictable cycles, the mechanisms still need to be understood. There is no reason to think that cycles will be caused by ecological interactions between species and their enemies, for example. As discussed above, there are several explanations for cycles. This prompts me to ask what situations conducive to complex dynamics are more common in the arctic than elsewhere—rather than to ask what situations are restricted to the arctic. Studies of vole cycles have suggested that it is the relative simplicity of arctic communities that leads to cycles (Hansson and Hentonnen 1985; Hansson 1987; Hentonnen et al. 1987; also see chapter 4 above). Yet it is model complexity—not model simplicity—that leads to cycles. This raises the possibility that complex dynamics may be prevalent in temperate areas too but that only in the arctic are the dynamics routinely complex yet simple enough (that is, they are cyclical) for us to appreciate them.

Third, the potential ubiquity of complex dynamics suggests alternative approaches to population and community theory are required.

Summary and Conclusions

Even the most simple models of population change show a variety of behaviors that include a simple return to equilibrium, cycles, cycles on cycles, and chaos: aperiodic, seemingly random behavior that develops when initially small differences between models with identical parameters lead to differences in density that widen rapidly. Dynamical behavior more complex than a simple return to equilibrium is a characteristic of both (a) models with high resilience, time lags, realistic features of predator-prey encounters, and seasonal changes in the degree of interactions between species and (b) models with many species. It is hard to find realistic population models that do not have complex dynamics for some ranges of their parameters. Moreover, many real populations have highly variable and seemingly unpredictable changes in density, the kind of changes one might expect.

Despite all the good reasons for ecologists to appreciate the importance of complicated dynamics, investigations into these dynamics have been of interest to theoretical ecologists but not to field ecologists. Why has this been the case? There are many reasons—perhaps including field ecologists' antipathy toward theory but *certainly* including the failure to view population changes over periods of time long enough for complex dynamics to become apparent. Perhaps the most important reason for dismissing complex dynamics were attempts to fit parameter-estimated models from field data and so deduce the dynamical behaviors of the populations under consideration. Such attempts routinely concluded that the parameters of most field populations and of many in the laboratory were consistent with a simple approach to equilibrium. Ecologists concluded that complex dynamics were not important and that the unpredictable and highly variable nature of many populations was due to genuinely unpredictable factors such as climate.

The most influential paper of this genre was that of Hassell et al. (1976). Their own theoretical work on complex dynamics plus their choice of highly resilient insect populations made their finding that complex dynamics are rare all the more compelling. Yet a species of moth that, in their parameter-estimation approach, ought to have been the least likely to cycle showed a pronounced cycle. So they found as much evidence for complex dynamics by simply examining the population data by eye as they did by analyzing their complex statistics.

The difficulties in detecting complex dynamics in the field data are not trivial. Certainly, cycles may be obvious by inspection or by a spectral analysis; but neither technique will work for chaos. For chaotic systems driven by multispecies interactions, we can illuminate the underlying simplicity of the dynamics by plotting successive densities of the species involved in generating to chaos. Such a graph would seem a vain hope for the field ecologist—who rarely has long-term data for one species, let alone simultaneous data from several. Schaffer and Kot were the first ecologists to appreciate Takens's solution to this problem. The existence of an attractor can be made clear by using repeated observations of just one species, by plotting successive values of density against each other in two, three, or more dimensions. Doing this suggests complex dynamics in a number of cases. Analyses of long-term population data sets for natural populations suggest that dynamics as complex as cycles are a feature far more common than had previously been expected.

In what ways would the ubiquity of complex dynamics alter ecology? The most obvious way is in improved prediction. Schaffer and Kot (1986a) and Sugihara and May (1990) show that their knowledge of the measles attractor could have been used to predict both the timing and the severity of disease outbreaks. The uncertainty of changes in a population seem to be more apparent than real. The less obvious way in which the existence of complex dynamics would alter ecology seems to me to be fundamentally more important, for it involves how we construct ecological theories, be these simple word images of how ecologists think populations behave or more complex computer simulations of ecological processes.

The Underlying Nature of Population Change

There are at least four different ideas about how population densities change, and all have advantages and limitations. First, there is the "simple equilibrium plus noise" model, which posits an equilibrium density to which densities return after inevitable environmental disturbances (noise). The great majority of populations analyzed do show a tendency to increase when unusually rare and to decrease when unusually abundant (Tanner 1966; Pimm 1982). More detailed evidence for density dependence is compelling for many species—and it may get better for others, the longer we study them (Hassell et al. 1989). This idea does not capture the complexity that can arise from delays in density dependence, although these complications might be considered friendly amendments. Nor does the idea anticipate the great range of year-to-year density variations that populations experience. For some species, modest recoveries from obvious environmental disasters are obvious; in others, populations fluctuate spectacularly from year to year.

The second idea is the "all is noise" model, which emphasizes the unpredictability of population change and its dependence on essentially random changes in environmental conditions. It is a model different from all the others in not considering the internal, ecological interactions as playing a role; in this model, it is the external physical environment that drives populations. Perhaps as another friendly amendment, this model may recognize that at extremely low densities processes will prevent the species' extinction and that at extreme high densities processes will lead to density reductions (Strong 1986). This model misses the mark for many populations in which the role of ecological interactions with resources and enemies is clear. Yet the model does focus on the differences between species' variabilities.

Third, there are the ideas of this chapter, which at least raise the possibility that a substantial proportion of populations may be driven by deterministic processes that yield complex population changes. Environmental noise adds a further component of variation to these populations. In the past, this idea may have been viewed as applicable only to special cases. This view is surely no longer tenable, but just how prevalent are complex changes is likely to be a matter of intense debate for some time to come. Some populations do not show obvious complex changes, but that may be because the changes are so complex we cannot detect them. To argue that all populations show complex dynamics, however, is to simply retreat beyond the limits of the data, for the the argument cannot be proved until we have very long time series for many species.

Finally, there is the "red noise" model, which recognizes short-term noise imposed on longer-term trends, themselves imposed on even longer and more powerful trends. Neither of the three previous models anticipates the empirical result that population variabilities should increase over time; yet it is interesting that all are consistent with it. The "equilibrium plus noise" model stresses the return to equilibrium, but for multispecies systems that return will be a very complex one—with some species densities changing rapidly, others changing more slowly, and every species dependent on the slowest-moving species. Even with a simple return, there will be a nested pattern of trends and a consequent increase in variance while the entire system recovers. The "all is noise" model recognizes the influence of climatic variables, and these are known to have a reddened spectrum. Finally, at least one strange attractor, Gilpin's one-predator-plus-two-competitor model, can generate a reddened spectrum for very long time periods.

The differences between the four ideas may be neither profound nor important, for the ideas are only strict alternatives when we describe them by using the caricatures that we often use to teach complex ideas to others. The first three ideas place emphasis on different facets of population change. The difference between the first three ideas and the red noise alternative is that the latter is an empirical description of population changes, while the others postulate mechanisms. In some ways the interesting difference is between the first two and last two ideas. The first two ideas were motivated by experiences that were short term; the latter two ideas only emerge from long-term studies. The challenge of the last two ideas is in their common feature of emphasizing the importance of the long-term trends in population densities. Most of our ecological ideas were developed with the assumption of an underlying constancy. So how would the recognition of long-term cycles, complex attractors, or progressively more important trends affect our ecological theories?

Some ecological theory has been developed which incidentally accepts the uncertainty of the values of the parameters that describe the interactions between species (and consequently their equilibrium values). Models with uncertain equilibrium values may be preadapted to a circumstance of constantly varying equilibrium values. We shall encounter such models soon. On the other hand, some ecological theory may be incorrect or need to be rethought. The hypotheses embodied by the last two ideas are new one, and their implications have not been fully thought out.

7

Extinctions

From Population Variability to Community Persistence

This chapter considers the topic of extinctions, and so it is a transition. How long communities persist is the topic of the next set of chapters. Communities change when they gain or lose species, so one reason to look at extinctions is to look at why losses occur. This chapter is also part of the discussion about population variability: why do some species become so rare that they become extinct? Many extinctions occur because the populations are driven inexorably to ever lower numbers. The existence of too many predators, parasites, or competitors may eliminate a species. Such extinctions involve the interactions between species, but in this chapter I will consider populations that are not being driven to extinction by such external factors. Populations that can potentially increase in numbers or that at least have the potential to remain unchanged are not obvious candidates for extinction. Yet, under the appropriate circumstances—small population sizes, for example—we may readily observe at least their local extinction. Many species now occupy only a small fraction of their former range, and the length of time they persist is an important question for conservation biologists. How small a population size is needed to doom a species, and what factors make species more or less prone to extinction?

The Effect of Population Size on Extinction

Demographic Accidents and Environmental Disturbances

The ideas to be presented next are condensed from several early models (MacArthur and Wilson 1967; Richter-Dyn and Goel 1972; Leigh 1975, 1981) and from the chapters in Michael Soulé's (1987) book on the viability of small populations. Of these chapters, Goodman's (1987) is a particularly valuable source of ideas. I will first explore the factors that cause the extinction of small

135

populations and then examine how species with different life histories will respond to these factors.

Even in a perfectly constant environment, small populations may face an appreciable risk from demographic accidents. These are vagaries of birth and death schedules—all the individuals dying in the same year, for example— and of sex-ratio fluctuations—all the young of a year being of the same sex, for example. In practice, the environment does fluctuate, and this superimposes an additional risk of extinction on that arising from demographic accidents alone. Finally, even large populations may be destroyed by some extraordinary perturbation. Clearly, there is a continuum, with different populations facing from little to extreme environmental perturbations.

Consider two extreme cases. In the first case, extinction is caused solely by demographic accidents in an unvarying environment. Some simple models of this case assume constant per-capita birth and death rates up to some population ceiling, K, above which the birth rates are assumed to be zero. These models predict that the length of time a population lasts until it comes extinct will increase rapidly as K increases. While for small populations extinctions may be quite common, a small increase in population ceiling will make the time a populations lasts so long that its extinction will be observable only on geological time scales. Goodman (1987) finds that MacArthur's (1972) model has T (time until extinction) increasing approximately as a constant to the power K (i.e., $T = a^K$; figure 7.1a). (Quite what K has to do with real populations is far from obvious. I have no idea whether it exists or how to calculate it. However, in simulating a large number of different models of species extinction, I find that K is closely related to the average population size. Typically, average population size is about half of K. Average population size is easy to calculate for real populations, and the data I shall present will explore the relationship of T to average population size and not that of T to K.)

For the opposite extreme case, consider the ultimate form of external environmental disturbance—total destruction of the habitat, such as might result from logging of a forest, or an asteroid collision, or a nuclear holocaust. The time until extinction will be the same for all species, irrespective of population size (figure 7.1c). (This is true provided that this time is less than the time to be expected in the absence of the disturbance.)

The first case is not realistic, and the second case is, one hopes, avoidable; but the cases demonstrate an important point. For demographic accidents alone, time to extinction increases rapidly with population size. For a given population size, environmental variation reduces the time to extinction below that expected from demographic accidents alone in an unvarying environment. As environmental variability increases, the time until extinction increases more

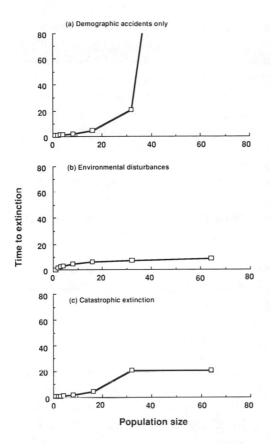

Figure 7.1. Time that a population is likely to last (*T*) before it becomes extinct, versus population size. For extinctions resulting only from demographic accidents in an unvarying environment (a), *T* increases approximately as a constant to the power of maximum population size. Leigh's (1981) model (b) incorporates moderate environmental variation, and *T* increases approximately as the logarithm of maximum population size. For environmental catastrophes that strike after a time T_{max} (20 in this example) and destroy all populations (c), no population survives beyond T_{max}. Some small populations might die out before the catastrophe; hence the line is kinked. These graphs illustrate that *T* increases much more steeply with population size if demographic effects predominate than if effects of environmental variation predominate.

slowly with increasing population size. When environmental disasters are extreme, we reach the extreme case of *no* increase at all in time before extinction. Leigh (1981) develops a model in which the time until extinction may increase as the logarithm of *K* (figure 7.1b), and one of Goodman's (1987) models has time until extinction increasing as "less than the square of the logarithm of *K*." Quite which relationship we should expect will depend on the severity of the environmental disturbances, of course. (None of these models considers the effect of the increasing variance in population densities over time [chapter 5]. As Lawton [1988] realized, the effect of this increasing variance on time until extinction should be to considerably reduce the time a population survives, even if the population is now abundant.)

Empirical Tests

This overwhelming importance of population size in determining how long a population will survive means that I should discuss this factor first and look at the other effects only after the effect of population size has been taken into account. Diamond (1984a, 1984b) provides an extensive compilation of previous studies on extinctions and the role of population size. He includes three kinds of case histories. First, there are extinctions of bird populations on islands, as documented by usually yearly censuses. Such studies typically examine data spanning about a decade or two. To these studies, I have added a comparable study of spider populations. Second, there are extinctions of bird populations in habitat patches created by human fragmentation of previously continuous habitat. Such studies typically have data spanning from a couple of decades to half a century or more. Third, there are extinctions of mammal and bird populations due either to late-Pleistocene fragmentation of land masses or isolation of habitats. Such studies consider extinctions that occur over several millenia.

Short-Term Population Censuses I: Birds.   Jared Diamond, Lee Jones, and I have analyzed studies of birds on sixteen small islands that are off the coast of Britain and Ireland (Pimm et al. 1988) and are 0.07–7.65 km². These islands are popular with bird-watchers; they are places where rare birds often turn up. The popularity of these islands has led to their having bird-watchers counting the nesting bird populations at yearly intervals, for several decades. The results of these extraordinary efforts are published in local reports; complete censuses for some islands, as well as other analyses, have been published elsewhere (Lack 1969; Diamond and May 1977; Williamson 1983; Diamond 1984a, 1984b). The data sets for all the islands include 100 bird species and 355 populations (that is, island/species combinations).

Typical of these islands is Bardsey, off the coast of Wales. A total of thirty-seven species nested on the island between 1954 and 1969, although not simultaneously. Only twenty-three species nested in 1956, and thirty species nested in 1965. For example, the cuckoo nested in 1955, 1961, 1962, 1964, 1965, and 1969. We do not know the fate of the population after 1969, so we excluded that year. Of the remaining five years, the cuckoo became extinct on Bardsey after three of them: 1955, 1962, and 1965. Therefore, the average duration of the population was 1.67 years. The average number was 1.4 females: a single female was present in each of the years 1955, 1962, and 1964, and two females were present in each of the years 1961 and 1965. (In general, we analyzed the number of mating pairs. Cuckoos have a rather odd mating system and also do not build their own nests, so for this species we used number of females seen during the breeding season.)

For each population we calculated the mean number of nesting pairs over those years when that species actually bred on that island. For each species we then calculated a mean (averaged over all the islands on which the species bred) of those mean numbers of nesting pairs on each island. We also recorded the number of years each population lasted—that is, the consecutive number of years of breeding, from immigration to extinction. Some populations repeatedly became extinct and reimmigrated on a given island and thus yielded several values for number of years lasted. We calculated the reciprocal of the number of years that each population lasted, and called this reciprocal the *risk of extinction*. This reciprocal was taken as zero for the many populations that survived continuously for the entire sequence of census years.

There are two obvious features in the data (figure 7.2). First, the risk of extinction falls very steeply with increasing population size. Second, notice that some species are much more prone to extinction than others. We would expect the species to differ considerably in their sensitivity to demographic accidents or environmental disturbances, and this may explain the scatter in figure 7.2. I will return later to the question of how and why species differ in their risk of extinction; first I must present other data that document the effects of population size on risk of extinction.

Short-Term Population Censuses II: Spiders. Toft and Schoener (1983) counted spiders on a number of tiny islands in the Bahamas. The islands ranged in size from 10 to nearly 9,000 m². Figure 7.3 shows the sizes of populations for each of the four common species, these populations being grouped according to whether they did or did not become extinct during the one-year duration of the study. The data show that small populations were more susceptible to extinction.

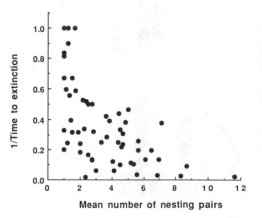

Figure 7.2. Risk per year of extinction (the reciprocal of time to extinction, in years) for sixty-two bird species, plotted against mean population size (number of nesting pairs). Note that risk decreases steeply with population size but that there is a good deal of variation, with some species much more prone to extinction than others. (After Pimm et al. 1988.)

Medium-Term Habitat-Fragmentation Studies of Birds.   In a second group
of studies, Diamond (1984b) examined the fate of bird populations when once-
continuous forest was largely removed, thereby leaving only isolated forest
patches. The first of two studies was of the forest birds in southeastern Brazil
that were counted by Willis (1974, 1980). Over the past 150 years, the for-
merly unbroken forest has been cleared for agriculture, leaving isolated forest
islands. Willis surveyed two tracks of 250 and 21 hectares that supported 119
and 76 breeding bird species, respectively. The original forest housed about
203 bird species. Terborgh and Winter (1980) analyzed these data and grouped
species in four categories according to abundance estimated from the numbers
of individuals encountered in one hundred hours of observation. They took the
percentage of these species missing from the two smaller forest patches as their

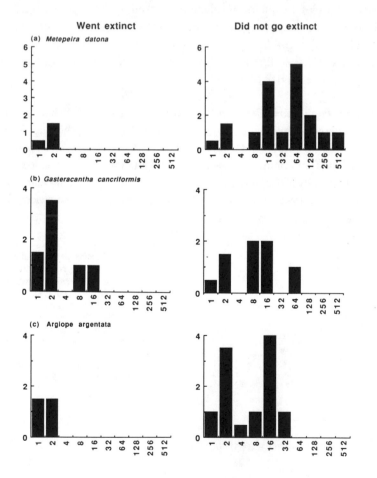

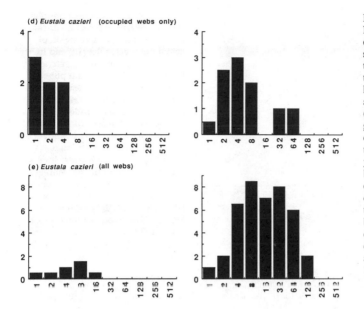

(d) *Eustala cazieri* (occupied webs only)

(e) *Eustala cazieri* (all webs)

Figure 7.3. Numbers of spiders on very small islands in the Bahamas that did (left-hand graphs) and did not (right-hand graphs) become extinct during the course of a year. The numbers that became extinct were generally smaller than the numbers of those that did not. (Redrawn from Toft and Schoener 1983.)

estimate of the percentage extinctions (Figure 7.4). The extinction rates are greater for the species presumed to be rare in the original forest and are greater for species in the smaller forest tract.

The second study involves the island of Barro Colorado, which was formed when the Chagras River was dammed to form Gatun Lake during the creation of the Panama Canal early in this century. Of approximately 108 species thought to have bred on the island, 45 have disappeared. Many of these extinctions were due to habitat changes. Forest has now covered that part of the island that was cleared at the time of flooding. Thirteen of the former residents species, however, are forest species. Karr (1982a) argues that 45 is probably an underestimate of the number of species lost. Intensive studies of the island's birds did not begin until well after the island was formed. Lovejoy et al. (1984) found in studies of habitat fragmentation in Brazil that some extinctions happen quickly. Rapid extinctions would necessarily be missed in the Panamanian study. In comparably sized plots of Panamanian forest off the island, there are many more species than those recorded for the island soon after its flooding. Karr (1982b) found that the abundance of birds, as estimated by numbers caught in nets, was a poor predictor of extinction. However, Diamond records that abundances estimated by annual Christmas counts on the Panamanian mainland do predict which species were lost from the island, with the rarer species being those more prone to extinction.

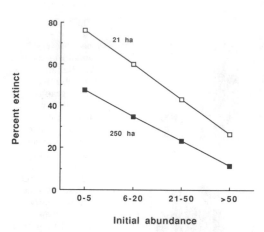

Figure 7.4. Percentage of populations of Brazilian forest birds that become extinct, as a function of initial population density and habitat area. For two forest patches (21 ha and 250 ha) initial population densities were estimated and expressed as bird individuals encountered per one hundred hours of observation. Populations were then grouped into four abundance classes, as indicated on the horizontal axis. For each class, the vertical axis gives the percentages of populations that become extinct over the period of a few decades. (Redrawn from Terborgh and Winter 1980, from original data in Willis 1974.)

Long-Term Studies of Habitat Fragmentation.   At the end of the Pleistocene, climatic changes fragmented populations in a number of ways. The melting glaciers caused sea level to rise more than 100 m, thus converting former lowland areas into shallow seas. Britain was separated from Europe, Tasmania from Australia, Java from the mainland of Asia. The presence of a rhinoceros on Java was recognized by Wallace (1876) a century ago as evidence for a landbridge that had once connected the island to the Asian mainland. This species did not swim but walked to Java; the island was cut off later. Changes in climate have also dramatically altered vegetation worldwide. Moreau (1952), for example, appreciated the similarities and differences in the isolated bird communities in African montane forest. Before detailed studies of forest changes were available, he had predicted that montane forest had been continuous over larger areas than at present. (Incidentally, he also predicted where the gaps were in that more extensive ancient forest.) In short, Pleistocene changes have left a clear impression on modern species distributions.

The area of Tasmania is about 70,000 km² and in the Bass Straight between it and Australia are four islands 100–1,500 km² size: Flinders, King, Cape Barren, and Clarke. There are many smaller islands; all were connected during the Pleistocene. Not only are the modern mammal faunas known, but Hope (1973) has reported their recent fossil species. The water gaps are probably too wide to be crossed by mammals other than bats and man. At present, there are eleven species on the largest island (Tasmania), and islands smaller than 1.4 km² have no species. The distributional pattern is nested with each species present on all islands greater than a particlular size. This size is 1.4 km² for a species of wallaby (*Thylogale billardierii*); in contrast, a mouse (*Pseudomys novahollandiae*) and a rat-kangaroo (*Aepyprymus rufescens*) have survived only on the Australian mainland.

Diamond argues that these patterns of extinctions suggest that the species most prone to extinction were likely to have been the rarest species. Carnivores are usually rarer than herbivores, and the herbivores survived on smaller islands. Smaller carnivores survived on islands smaller than those on which the larger carnivores survived. Habitat specialists were also more likely to be lost: the platypus is confined to freshwater, and it survived only on Tasmania and two of the larger islands.

A similar study—albeit one involving habitat islands—is Brown's (1971) survey of the mammals on montane "islands" above the Great Basin desert of Nevada (figure 7.5). In continuous tracts of forest, the number of species increases with increasing area. In the isolated forest islands, there are fewer species than in an area of continuous forest of the same size. Moreover, there are *proportionately* fewer species in the smaller forest fragments. About 7% of the expected number of species are present in islands of about 10 square miles, but this increases to about 40% for islands of about 1,000 square miles. As in the Tasmania study, the pattern in which species disappear correlates roughly with the species' expected rarity as based on their habitat specialization.

Belovsky (1987) has analyzed these data in more detail, along with the comparable data collected by Patterson (1984) for mammals in the southern Rocky Mountains. For both cases, Belovsky found that extinction rates were higher on smaller mountaintops, where one would expect to find smaller populations. He also found that species with smaller reproductive rates were more likely to become extinct than those with larger reproductive rates, a result that anticipates the topic of my next section.

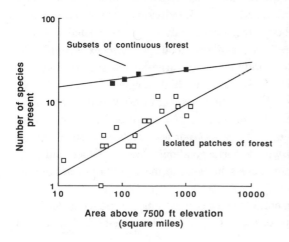

Figure 7.5. Number of small mammal species in areas of different sizes. Open squares denote the number of species present on montane forest "islands" (isolated areas of forest) in Nevada, Utah, and California. The areas were those above 7,500 feet (approximately 2,280 m). Closed squares denote the number of species in subsets of large, continuously forested areas in the same region. Small, isolated forest areas have *relatively* fewer species than might be expected on the basis of their size alone. (Redrawn from Brown 1971.)

Diamond reviews many other comparable studies, involving birds, mammals, and lizards. Isolation is a potent factor in reducing population sizes, and the effect of population size on extinction is obvious at all time scales. There is nothing really surprising here. Many factors doom small populations to extinction, and small populations become extinct much more rapidly than larger ones.

THE EFFECTS OF INDIVIDUAL LONGEVITY AND POPULATION RESILIENCE

Ideas

How should times before extinction vary among species having the same average population size? Other things being equal (which they are not!), a population of a long-lived species would have a lower risk of extinction from demographic accidents *per year* than would a short-lived species. A population of no more than ten aphids may well become extinct in a year. Each individual lives only a few weeks, and, even if each individual tends to be replaced, there is still a good chance that during the year enough individuals may die within a short period for the population to become extinct. In contrast, a population of no more than ten elephants will survive much longer than the aphids. (Although, in terms of generations, they may last only as long as the aphids.) It is also obvious that populations with low resilience should be at increased risk of extinction. A population that recovers slowly from a severe reduction in density will remain longer at risk of extinction from demographic accidents.

In sum, extinction rates should be lower for species whose population densities are highly resilient and whose individuals live a long time. The difficulty is that the two parameters are linked, for both individual longevity and population resilience depend on body size. Large body size is associated with low population resilience (figure 2.5a), and this tends to *raise* extinction rates (Leigh 1981; Goodman 1987). Individuals of large-bodied species also live longer (Bonner 1965; Peters 1983), and this high longevity tends to *lower* yearly extinction rates. What, then, is the net effect of body size on how long a population should survive? Should we expect a small-bodied species with high resilience and short lifetimes to be more or less likely to become extinct than a large-bodied species with a correspondingly low resilience and long lifetimes? (I assume that both species have the same equilibrium density, of course).

For a simple verbal argument, consider two extremes. First, consider identically sized large populations of the two species. The large-bodied species is at a disadvantage because, after some severe reduction in numbers, its lower resilience makes it take longer to climb from low numbers. For a species to remain at low levels greatly increases the risk of extinction—a large-bodied, slowly growing species may still be at low levels when the next severe reduction in numbers occurs. In most models of extinction it is this effect that causes times to extinction to decrease so rapidly with increases in resilience, across species with the same equilibrium numbers but different population resilience. Second, consider a single individual of a large-bodied species and a single individual of a small-bodied species. Both are doomed to die, but the former will probably live longer; its yearly extinction rate is lower, even if the per-lifetime rates of extinction for the two species are the same. Naturally, you may object to my calling hypothetical single individuals "a population." It is certainly not one that can reproduce. Suppose we have a large-bodied, long-lived breeding pair. Their long life gives them several opportunities to breed in the event that they experience one or more bad years. They may not breed until they are several years old, leaving the young in a possible position to take over from their parents should their parents die. In short, small populations of long-lived species may get several attempts to compensate for bad years. This will not be true, of course, if their numbers fluctuate considerably.

These two extreme cases suggest that, at low numbers, large-bodied species could have an advantage over small-bodied species; at high numbers the reverse is true. Is there a population size at which the advantage is reversed? If we look at time to extinction measured in lifetimes (not years), small-bodied species are always at an advantage because of their high resilience. Measuring extinction times in generations may be sensible for some applications, but conservationists' plans for managing species will be measured in years, not in the different generation times of different species. Nor is the year a biologically arbitrary time unit. For many species, dispersal—and hence movement between isolated populations—will occur on a yearly cycle. Thus, years—not lifetimes—may be the appropriate measure of the time until a foundering population may be rescued by immigrants.

Empirical Tests

There are potentially complex relationships between the time until extinction, individual longevity, and population growth rates. Certainly, long lifetimes and fast population growth rates reduce the chance of extinction, but this combination is unlikely. The relationships lead to an unusual theoretical prediction about the extinctions observed over a fixed number of years. At low population densities where demographic accidents are a powerful though not the only cause of extinction, large, long-lived, slow-growing species are less likely to become extinct than small, short-lived, fast-growing species. At high populations, only environmental disasters can doom species, giving the small, short-lived, fast-growing species a considerable advantage.

The British Island Birds and Their Natural History. To test this theory, I return to the study of birds on small islands off the coast of Britain (Pimm et al. 1988). Because the theory involves the way species differ in their chances of extinction, we must first discuss the species and their natural history—that is, factors which could also affect this chance. The species are diverse taxonomically and trophically. They include insectivores (swallows, martins, pipits, wagtails, warblers, and, while on the breeding grounds, many of the shore birds), granivores (doves, finches, and buntings), and carnivores (hawks, buzzards, and falcons). Some of the species are residents, but others are migratory. There are several reasons why migratory and resident species might differ in their chance of extinction. First, some individuals of migratory species may return one year to a breeding area that differs from their breeding area of the previous year. If these individuals were from an island, their wandering might cause a local extinction. Alternatively, by wandering from the mainland they might establish a short-lived island population. In either case, the populations of migrants in any location would appear more prone to extinction than residents. Second, migrants and residents might also be expected to differ in the variability of their densities. On the one hand, migrants avoid the unusually hard winters that depress resident bird populations. On the other hand, migration itself is risky. The net result is that migratory species are more variable in their densities than are resident species (Pimm 1984b; see chapter 5 above). Because of these considerations, Pimm et al. analyzed migrants and residents separately.

Analysis. There is much scatter in Figure 7.2, suggesting that population size is only one factor determining the risk of extinction. One way to clarify species differences is to statistically remove, as much as possible, this dependence of risk of extinction on population size, in order to observe how the risk of extinction depends on factors other than population size. We calculated a function of risk of extinction which we called the *corrected risk of extinction.* Corrected risk of extinction will be constant (and not depend on average population size, N) if time until extinction increases as a constant raised to the power N—the relationship expected from demographic accidents alone. In other words, corrected risk of extinction represents a species-specific risk of extinction partly corrected for the large effect that mean population size has on risk of extinction.

As a dichotomous expression of body size, we categorized species as either large bodied (nonpasseriform birds, plus the large-bodied passerine family Corvidae [crows and ravens]) or small-bodied (passeriform birds other than the Corvidae). The two groups are divided by weight: 100–1,000 grams (with only one exception: ringed plover, 65 grams) for large species and 5–100 grams (also with only one exception: mistle thrush, 120 grams) for small species.

The analyses for large-bodied and small-bodied species differ significantly. For small-bodied birds, the corrected risk of extinction does not depend on population size. Therefore, for small-bodied species the times until extinction increase as a constant raised to the power N (as in figure 7.1a). For large-bodied birds, however, the corrected risk of extinction increases significantly with N. This means that large-bodied species at progressively higher population sizes are at greater risk of extinction (have shorter times until extinction) than one would expect if demographic accidents were the sole cause of extinction. Below average numbers of about seven pairs, the large-bodied species have significantly *lower* corrected risks of extinction than do small-bodied species. However, while the corrected risks start low for large-bodied species, they increase faster. At the average number of seven pairs, the large-body advantage disappears, and the results, if extrapolated, would predict that, at higher numbers, large-bodied species would become extinct faster than small-bodied species. (The statistical justification for the crossover is that, for large-bodied species, the intercept of the regression line is significantly *less* than that for small-bodied species—and that the slope of the regression line is significantly *greater*.) The fact that the data only clearly show an equality of extinction risks at about the seven-pairs level—and not a clear advantage to small-bodied species above the seven-pairs level has troubled some critics. There are not enough data above the seven-pairs level (such extinctions are rare over the time spans of this study) to allow us to see the advantage. I am less bothered by the criticism, for the following reason: For large- and small-bodied species, the data show an equality of *risk calculated on a yearly basis*. Yet, on a *per-generation basis*, the large-bodied (and hence long-lived) species would be reaching extinction at a much faster rate. Therefore, whether one views the advantage of small versus large body size as actually reversing depends entirely on how one measures time. I conclude that these results support the theory developed earlier—a satisfying conclusion, given the rather complex nature of the theory's predictions.

Effects of Migratory Status.   The status of a species as a migrant or a resident complicates these results but does not alter them. Given the effects of body size and population size, we found that migrants tended to have a significantly greater corrected risk of extinction. Given this important difference between migrants and residents, I have plotted their corrected risks of extinction separately in figure 7.6.

For resident species at low population densities, large-bodied species are less prone to extinction than small-bodied species, but, with increasing numbers, the corrected risk of extinction increases much more rapidly in large-bodied species. This result is the same as for the entire data set and confirms the theoretical predictions.

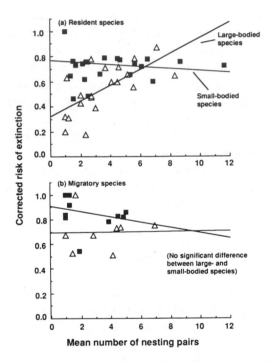

Figure 7.6. Risk of extinction for bird species on British islands, partly corrected for population size in order to show species differences plotted against mean population size (mean number of nesting pairs). Large-bodied (open triangles) and small-bodied (solid squares) species are defined in the text. Resident species (a) and migratory species (b) are plotted separately. For resident species, large-bodied species are less prone to extinction at low population densities (fewer than seven pairs) but are equally prone to extinction at higher population densities (at about seven pairs) than are small-bodied species. (The lines differ significantly in slope as well as in intercept. The slope for large-bodied species differs significantly from zero; that for small-bodied species does not. A line of zero slope for this corrected risk of extinction means that the time before extinction increases as a constant raised to the power of the mean population size.) For migratory species, there are no significant differences between large-bodied and small-bodied species, and there is no significant effect of population size. Migratory species, however, have higher corrected risks of extinction than do resident species of the same body size and population size. (From Pimm et al. 1988.)

For migrant species, there is no significant difference between large-bodied and small-bodied species in—and no significant effect of population size on— corrected risk of extinction. Of course, there are rather few migrant species (fewer than one-third of the sixty-two species analyzed), and the data span a small range of population sizes. The theoretical predictions are still supported if we combine the migrant and resident species.

TEMPORAL VARIABILITY IN DENSITY

Temporal variability differs greatly between species and has many causes (chapters 3–6). Whatever the causes, the greater is the variability of a population's density, the more likely it is that population will become extinct, for a highly variable population will more often encounter those low population levels from which recovery will be impossible. Other things being equal, then, populations varying greatly in their densities should be more prone to extinc tions than populations which are not so variable. Here again, the complication is that other things are not equal. The first complication is that highly variable populations tend to be small-bodied species, and such species have highly resilient populations (chapter 5). Indeed, I have argued that populations may be highly variable *because* they are highly resilient. Are small-bodied species, then, more likely to become extinct than large-bodied species because they vary more—or less likely to become extinct because their populations are more resilient?

If one considers the entire range of animal species, then I do not yet know how to answer this question. Some simple calculations suggest that, for some ranges of population density, small-bodied and hence highly variable species must be doomed to extinction while large-bodied species are not. There are many insect species which have densities that vary over four orders of magnitude over twenty to thirty years (chapter 3 and figure VAR.1c). Such populations must average in the thousands if they are to survive this length of time. Yet bird populations, with their much less variable populations, can survive decades if their average density is measured in tens (figure 7.2).

Within a group such as birds, the answer to whether small- or large-bodied species are more prone to become extinct should be more simple. Birds with highly resilient populations are *less* variable than those with lower resilience (figure 3.7). For birds, then, less variable and more resilient species should be less likely to become extinct than more variable and less resilient ones. This theoretical prediction was also tested using the data on birds on small British islands. Having calculated the effects that population size, migratory status, and body size had on time before extinction, my colleagues and I looked for an additional effect of population variability. We calculated the coefficient of variation of the number of nesting pairs for each population for which there were five or more years of continuous breeding. (Hence we did not calculate coefficients for many populations that bred for only short periods.) For each species, we then calculated a mean coefficient of variation by averaging over all islands for which the coefficient had been calculated. Reliable values of these coefficients were available for thirty-nine species.

The deviations from our regression lines measure how far a species differs, in its corrected risk of extinction, from what we expect on the basis of the factors discussed so far: the species' population size, body size, and migratory status. A positive deviation, for example, means that a species is more prone to extinction than one would expect on the basis of these factors. Figure 7.7 plots the deviations from the regression lines of our statistical analyses against the temporal coefficients of variation. Temporally variable populations are more prone to extinction than populations with low temporal variability. This result also confirms the theoretical prediction.

The second complication in evaluating the effect of body size on risk of extinction is that small-bodied species are often more abundant than large bodied species (Peters 1983). Perhaps they have to be, in order to survive. High

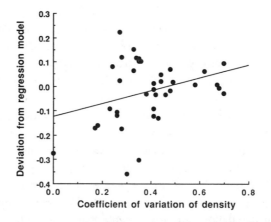

Figure 7.7. **More temporal variable populations are more likely to become extinct. Shown are deviations from the regression lines through the data of figure 7.6, plotted against temporal coefficient of variation of density. Positive or negative deviations mean that a species is, respectively, more or less prone to extinction than one would expect on the basis of its population size, body size, and migratory status. The slope of a regression line through the points is significantly greater than zero. (From Pimm et al. 1988.)**

variability and rarity are not compatible with persistence! Schoener and Spiller (1992) examined patterns of the extinction of spiders on very small islands. They found that the populations which varied the least were the very rare ones, yet it was these populations that were more prone to extinction; in contrast, the more abundant species were less prone to extinction even though they were more variable populations. Thus, this conclusion is in contrast to the result found for the island birds. Interestingly, their data also support results for the island birds. This paradox is apparent when one compares the extinction rates of the birds and spiders. The latter are more variable and become extinct more quickly even when their average densities are higher.

The third complication is the skewness of the population (chapter 4). Variable species may be so variable because of excursions to very high densities. Less variable species may encounter smaller—yet fatal—crashes in density. This possibility has yet to be explored empirically.

THE EFFECTS OF SPATIAL AND DEMOGRAPHIC STRUCTURE

Ideas

So far, the populations I have considered have been (a) single, isolated populations and (b) treated as if their individuals were all identical. Of course, neither of these assumptions may be correct. The individuals of a species may differ in age, and age affects the chance of death. A population may be spatially structured too, with a network of local populations. After all, when a species of bird on small British islands beome extinct, it was the local population that was lost, not the entire species.

The relationship between a population's structure and its risk of extinction has been explored mathematically by Gilpin (1987) and Quinn and Hastings (1987). The ideas follow simply from what we have discussed so far. If a population suffers only demographic accidents, we are not likely to observe populations of forty individuals going extinct very often. In reality, a population of forty individuals may be highly prone to extinction. In the first place, the sex ratio may differ from what is ideal for maximum population growth. Those forty individuals may contain many fewer than twenty females. Similarly, the population may have an age structure that is less than ideal—too many old or young individuals. And there may be difficulties with both the sex ratio and the age structure, so that, despite its relatively large size, the future of the population of forty really depends on a handful of mature females. That handful may be very vulnerable to the simultaneous chance deaths that will cause the species to become extinct.

The spatial structure of a population may also increase a population's chance of extinction. The forty individuals may be divided into a number of relatively isolated subpopulations. Dividing the population into eight subpopulations of five, for example, means that each subpopulation will last only a very short time. Of course, immigration from one subpopulation to another could rescue a subpopulation on the verge of extinction.

How should different demographic and spatial population structures modify a population's chance of extinction from environmental disturbances? Interestingly, the existence of partially isolated subpopulations could also reduce, rather than increase, that chance. Environmental disturbances may not fall equally on the isolated subpopulations. Only parts of a population may be affected by a bad storm or a disease epidemic, though other catastrophes may fall over the entire range of the species. Consequently, a widely distributed population may be less vulnerable to environmental catastrophes than a locally concentrated population. Similarly, one might expect that mortality would fall unevenly in different sexes and different age classes. If so, a population with an uneven distribution of age classes may be less vulnerable than one in which most individuals are of similar age.

In short, the best chance for a population may occur when there is some migration between the spatially isolated subpopulations. This leads to a "rescue effect" (Brown and Kodric-Brown 1977)—local extinction should be less likely when there is substantial migration between the groups of a population. The worst that may happen to a population is for it to be fragmented into spatially isolated subpopulations, with little migration between them, each so small that demographic accidents rapidly drive the population to extinction.

Demonstrating these various effects will not be easy, since the factors interact in such complex ways. I shall begin with some compelling examples of how population structure protects a species and then provide examples of relatively abundant species endangered by their fragmentation.

Case Histories

Farne Island Shags.   My first example involves age and spatial structure in a colonially nesting seabird, the shag, studied by Coulson et al. (1968). Shags on the Farne islands off the English coast compete for nest sites on steep cliffs above the ocean. Cliff nest sites are limited because a site has to be flat enough to support the eggs and young while also being far enough above the waves for protection from the water during storms. Male shags do not breed until they are at least two years old, and females do not join the breeding population until they are three or four years old or even older. These young birds do not remain around the islands but disperse along the coasts. Shags eat small fish, particularly sandeels; these small fish feed on zooplankton, and the zooplankton feed on phytoplankton.

The average annual adult mortality is known, from banding studies, to be about 17%. In mid-May of 1968 the Farne Island shags suffered an exceptionally heavy mortality. About 90% of the nests were deserted, and the young were left to die from predation and starvation. Eighty-five percent of the adult females and 79% of the adult males died within two weeks. An abundance of a particular single-celled phytoplankter, a dinoflagellate, is known to have been the cause. These dinoflagellates are the cause of "red tides," and they contain a powerful nerve toxin that works its way through shellfish (thereby also resulting in human poisonings) and through zooplankton and sandeels to the shags and various other birds.

The Farne islands shags were hit very hard by this disaster, for it came at almost the peak of the nesting season. The disaster killed off nearly all the young of that year and nearly all the adults, which take several years to reach breeding age. Despite this, the population structure protected the species in two ways. First, many young birds were not breeding when the disaster struck and so were away from the colony. Two-year-old shags returned to nest in unusual numbers during the following year (Potts et al. 1980), presumably in response to the greater availability of nest sites. Second, the abundance of the toxic dinoflagellate was locally restricted; other colonies of seabirds off the English coast were not affected. That the colony recovered so quickly reflects both the demographic structure of the population and the spatial structure of many different colonies with some migration between them.

Not all environmental disasters are as geographically restricted as this one. England's unusually cold winter of 1962–63 depressed bird populations over a very wide area and killed young and old birds alike (chapters 2–5). For such a widespread catastrophe, spatial and demographic population structures will provide little protection against extinction.

Edith's Checkerspot Butterfly.   My second example is a detailed study of the fate of three spatially fragmented populations, the subject of a study that Paul Ehrlich and his colleagues conducted on Edith's checkerspot butterfly. (Reviews of this work are in Ehrlich 1965; Ehrlich et al., 1975, 1980). This butterfly has been studied intensively both at a grassland at Jasper Ridge, south of San Francisco, California, for more than twenty-five years and at other locations in northern California and Oregon. The food plants in the open grassy areas grow in the cool, wet winters and dry up and die in early summer. At

Jasper Ridge, the butterfly population occurs in three areas of serpentine soil that are a few hundred meters apart and are not separated by any physical barriers. Nonetheless, butterflies rarely move from area to area, and the three populations act largely independently. The two larger populations average about 1,000 butterflies each season, but the variations are wide, the approximate range being 100–5,000 butterflies. The population sizes in these two larger units changed in parallel in the second decade of the study, but not in the first decade. The smallest population averaged about 30 individuals; its numbers were uncorrelated with those of the other two populations, and it also fluctuated considerably. The numbers of individuals in this population have reached nearly 200. As we might expect for a small, variable population, it has experienced extinction twice: in 1965 and from 1975 to the present.

At Jasper Ridge, the butterfly larvae feed on a native annual plaintain, but these plants dry up during the time the larvae are feeding on them, and most of the larvae die also. Those larvae that survive do so by one of three routes. Some come from eggs laid very early in the season, and these have enough time to mature before the host plants die. Others survive because the eggs are laid on plaintains growing on gopher-tilled soil. In such soil the plants' roots go deeper and the plants remain green longer. The most important route involves the individual larvae finding a second host plant: the owl's clover. This plant is restricted to only some of the grassy areas, and the butterfly's distribution matches it—and not that of the more widespread plaintain. Moreover, the abundance of the owl's clover was an accurate predictor of the butterfly's numbers the following year until about the mid-1970s. (Since then, the weather seems to have become the more important factor.) In view of the uncertainty of finding food resources, it is not surprising that the population fluctuates so wildly.

At other sites in California and Oregon, the butterfly feeds on different plant species. At some sites, only one food plant species is used; at others there are unimportant alternative food plants; and at yet others there are essential secondary food plants.

The years 1975–77 were unusually dry ones in California. In 1975, rainfall in central California was reduced to 50% of normal and down to 10% in some areas. In 1977, the critical January-April rainfall was 30%–40% of normal. It is not surprising that some butterfly populations were severely reduced. The small population at Jasper Ridge became extinct—it was down to fewer than 10 individuals in 1974. Other small local populations suffered a similar fate. The two larger populations at Jasper Ridge dropped to 125 and 500 by 1978—a severe reduction but proportionately less than at some other sites. The greater sunlight that accompanied the lower rainfall enabled the adult butterflies to lay their eggs earlier. At one inland site the population was reduced from about 1,000 individuals to about 20; at this site, the species feeds on one plant, and there is intense competition for food. At other sites, the local populations actually increased; at high elevations, one population increased so much that it defoliated its food plant.

There are a number of lessons from this story. The diversity of food plants at a given location may make the population more vulnerable, not less, because, as at Jasper Ridge, two consecutive food plants may both be essential for survival. (Contrast this with the arguments of chapter 4.) However, the species' use of different food plants in different places means that the species as a whole was buffered against the unusual drought. Indeed, some populations benefited from the drought. Yet the spatial structure of the population must be viewed in the context of the species' very poor dispersal characteristics. The breeding colony of Farne Island shags was recolonized quickly by individuals from other areas and other age groups. The butterfly, however, has only one generation per year, and all the adults die each autumn. Equally important, the butterfly is reluctant to cross even narrow barriers or to fly even very short distances over unsuitable habitat, though it has clearly done this and so recolonized the smallest population at Jasper Ridge. Given the reluctance to cross such barriers, the difficulty of crossing barriers of many kilometers, compounded by the change in food plants, must make the time before some populations are recolonized on the order of decades or perhaps centuries. Even when a local population is recolonized, the fate of such a population is far from secure, as frequent extinction of the small population at Jasper Ridge demonstrates. We might predict that, among similar species, the species might be absent from what would seem to be perfectly suitable habitats. Extinctions may happen far more often than recolonizations. Indeed, Ehrlich and his colleagues have used transplant experiments to demonstrate that some butterfly distributions are limited by just such infrequent colonization (Holdren and Ehrlich 1981).

Too Many Other Species.   It is depressingly easy to find examples that demonstrate how spatial structure can increase a species' chances of becoming extinct. Many species are threatened as their once continuous ranges are broken into smaller fragments. Indeed, earlier in this chapter I used the extinctions that follow such fragmentation to demonstrate the effects of small population size. Within the United States of America, there are many well-known examples of vertebrates that are threatened not because their total numbers are particularly low but because the populations are scattered across many locations. The grizzly bears in the states outside Alaska, the northern spotted owl, and the red-cockaded woodpecker are all endangered, despite their relatively large populations (Norse et al. 1986; Wilcove 1987). In each case, the species require relatively large areas of relatively old forest, and human intervention has reduced the area of the forest and replaced old forest with younger trees. In the case of the birds, the total population numbers in the thousands (Norse et al. 1986), while the subpopulations often may be numbered in tens.

What these and other examples tell us is more than meets the eye. Spatial structure can greatly increase the chance of extinction, for, if the populations are not isolated, the species are abundant enough that their extinction would not be so imminent. Together, the examples suggest the importance of migration between subpopulations. The approximately one million hectares of the Yellowstone and Grand Teton National Parks may only be able to sustain a breeding population of about fifty grizzly bears (Norse et al. 1986). The fate of this population is uncertain; it would help greatly if bears annually wandered down from the larger populations in Alaska. Goshawks are large predatory birds which also prefer large tracts of old forest; however, unlike the bears, they occasionally migrate long distances, and so their future seems more certain as a consequence. The point is that, while a widely distributed, vagile species may not be so vulnerable as a locally restricted one of the same abundance, a sedentary, widely distributed and rare species may be the most vulnerable of all.

Another feature of the three species I have discussed is their relatively large body size among mammals or birds. Perhaps, it is because large-bodied species tend to be rare—or charismatic—that these examples come to mind. Large-bodied species tend to be long-lived as well, however, and their population densities tend to vary little. Possibly the real reason we have the opportunity to contemplate the fate of these species is that, with rare small-bodied species, demographic accidents and high population variability would have caused them to become extinct long ago.

CHANGES IN REPRODUCTIVE RATE AT LOW DENSITY

So far, I have assumed that populations have constant birth and death rates. More likely, reproductive rate will change as a function of population size. There are many factors that cause reproductive rate to be greatest and death rate to be lowest at low densities: for example, there will be more resources per individual and less chance of disease, and predators may pay less attention to a rare species. Yet at low densities there are also a number of factors that may cause net growth rates to decrease and so make extinction more probable.

If predator numbers remain constant, then the number of the prey species taken by predators may either also remain constant or fall only slowly as the species' numbers are reduced. The proportion of prey taken, therefore, will increase as the prey species becomes rarer. Many species nest colonially, thereby simultaneously swamping predators, but such a tactic falls at low population sizes. Another important reason for the decrease in reproductive rate with decreasing population size comes from genetic effects.

Genetic Effects

Small populations often lose genetic variability—they become inbred. In small populations not all the genes of the parents may be present in, or "sampled by," their offspring, and so genes are lost. Now, many genetic mutations are both deleterious and recessive—their phenotypic effects appear rarely in large populations because they are masked by the overwhelming numerical superiority of the other alleles. Inbreeding greatly enhances the likelihood of homozygous recessives and, thereby, the increased frequency of individuals bearing traits that may reduce the chances of their survival.

European royalty seem not to have appreciated the effects of inbreeding, and the distribution of their genetic ailments, such as hemophilia, is well known. Royalty have perceived the population of potential spouses as being very small. To join this select club, it helps to have Queen Victoria as an ancestor—she carried the gene for hemophilia. Similarly, in a wild population that has been greatly reduced in numbers, the individuals may also share many of the same ancestors—and those ancestors' genetic defects. Thus, reduced population size leads to inbreeding, which may lower reproductive rate, further decreasing the species' chance of survival.

Royalty apart, the reason that animals should avoid inbreeding is best illustrated by Ballou and Ralls's (1982) study on ungulates kept in zoos. They surveyed breeding records of twelve species of captive ungulates. Animals founding a herd were assumed to be unrelated to each other, and individuals subsequently added to the herds were also assumed to be unrelated unless the exact relationship of these individuals was known. In all but one species (Pere David's deer) there was a considerable decrease in survivorship with inbreeding. In a later study, Ralls et al. (1988) looked at inbreeding in forty zoo populations of thirty-eight species of mammals including marsupials, insectivores, primates, rodents, carnivores, and ungulates. They found that inbreeding depressed the survivorship of the young in thirty-six of these populations. In this study, Pere David's deer showed a significant decrease in survivorship with increased levels of inbreeding, in contrast to the results of their earlier study.

A genetic effect on small populations that is quite separate from inbreeding depression is that, with the loss of genetic variability, the species will be more vulnerable to external disasters. Disasters are also likely to affect different phenotypes in different ways (Ralls et al. 1986). The more genetically uniform a population, the proportionately greater will be the effect of a given disaster. Moreover, the species may be less able to capitalize on new opportunities if it lacks the genes (and the phenotype) to exploit them.

Reproductive-capacity depression through inbreeding seems easy to demonstrate in captive populations. But is it an effect likely to doom wild populations? Can wild populations suffer from genetic difficulties at population sizes high enough that they are unlikely to be doomed by the simple population consequences I already have discussed (Lande 1988)? Simple models show that the amount of genetic variability retained per generation is $(1 - [1/2N])$, where N is the *effective population size* of the species. Effective population size equals population size when each individual contributes genes equally to the next and subsequent generations—and is less than population size when this is not true. Thus a population of ten individuals loses only 5% of its genetic variability per generation. A population of ten individuals (that is, five pairs), however, has very little chance of survival in any case. Look at the data for birds in figure 7.2, and recall that birds are among the least temporally variable species. In sum, the threat of losing genetic variability should, for many species, be a minor, if additional, problem of being rare. The emphasis that some scientists place on inbreeding as a cause of extinction needs to be examined carefully.

Can Inbreeding Doom a Species in the Wild?

Perhaps the most suggestive examples that genetic difficulties can be important in the wild comes from studies of lions and cheetahs. The effects of inbreeding are often manifested in physiological problems, including reduced sperm counts and increased proportions of abnormal sperm. In an ingenious study, three populations of lions were compared (Wildt et al. 1987). Two populations were from Tanzania—one from the Serengeti National Park where inbreeding was minimal and the other, an isolated population of about 3,000 lions descended from between 6 and 15 individuals that survived an epidemic in 1962, from inside the Ngorongoro Crater. The third population was of Asiatic lions from the Gir forest in India and consisted of fewer than 250 animals descended from the fewer than 20 animals constituting the population in the early part of this century. The Indian and Ngorongoro Crater lions had less genetic variability than those from the Serengeti. Data on the sperm of the male lions are consistent with the known population histories, with the Serengeti population showing the most healthy sperm and with the Indian population the least healthy sperm. This study does not show that these physiological problems translate into a loss of reproductive fitness.

Estimates of the numbers of cheetahs in the wild range from 7,000 to 23,000; the species that once ranged over most of Africa and southwest Asia to India is now virtually extinct north and east of the Sahara (Ashwood and Gittleman 1991). O'Brien et al. (1986) have studied this species and its genetic problems in some detail. They report that, since the first litter in 1956, no more than 15% of wild-caught cheetahs have reproduced in captivity and that about 30% of these cubs die before they reach six months of age. In the wild, infant mortality is estimated to be as high as 70%. The physiological problems the cheetahs experience are considerable. Sperm concentration is only a tenth of that in domestic cats, sperm motility is low, and 71% of the sperm have abnormal shapes (the corresponding figure in the domestic cat is 29%). In addition, cheetahs seem unusually vulnerable to disease. As O'Brien et al. put it, "cheetahs throughout Africa are virtual genetic twins." This lack of genetic variability is not uniformly accepted, however. Hedrick (1987) analyzed O'Brien et al.'s data and found that only 26% of the young born to parents known to be unrelated died but that 44% of the young born to parents known (on the basis of their common ancestry) to be inbred died. This was a statistically significant difference, strongly reminiscent of the studies I discussed earlier, and one that seems inconsistent with the idea that even noninbred cheetahs are "virtual genetic twins."

Population Histories and Genetic Difficulties

O'Brien et al. argue that the physiological defects and the low variability of cheetahs is a product of "an ancient population bottleneck." They argue that "at least once and perhaps several times in the past, the cheetah [population] must have dropped to a very few individuals" and, moreover, that "the surviving cheetahs never managed to expand their numbers rapidly." I have already noted that this simple argument is possible but extremely improbable. For a once genetically variable population to lose all its variation (or even most of it), population levels must remain at very low levels for a long time. After all, the inbred lions of the Serengeti and those of the Gir forest appear to have considerably more genetic variability than cheetahs, yet both populations have been reduced to very low population sizes.

Basically, O'Brien asks us to accept that a population consisting of at least several thousand individuals is now in grave genetic trouble because at least once (and possibly several times) it was reduced to a handful of individuals and then later recovered and spread over most of one continent and a good piece of another. The species managed this feat with a reproductive rate that is very low even at the relatively high numbers that now survive. If the cheetah population is in trouble now, how much more serious would be the situation if the cheetah were a hundred or a thousand times rarer?

Not content with simply attributing the cheetah's problems to a population bottleneck, O'Brien and Evermann (1988) attempt to make a case that reduced genetic variability is generally a consequence of "repeated bottlenecks." Their data are not convincing. Of their thirteen populations, only four provide data simultaneously showing both how small a population became and the loss of genetic variability. Another population, the United States population of the Swiss mouse, was founded with as few as nine individuals and did not lose genetic diversity. They omit from their discussion sea otters, which apparently have not lost genetic variability (Ralls et al. 1983), and the Indian rhinoceros, which has genetic diversity "at the high end [of the diversity] observed in over 140 mammalian species" (Dinerstein and McCracken 1990). Both these species were hunted to very low levels; and I shall discuss their histories presently.

For a species lacking genetic variability, are there alternative explanations consistent with its numbers not going through the extremely low levels that would surely lead to its extinction? Can a relatively large population lose genetic variability? And, in the same vein, why do some rare species *not* lose genetic variability?

Social Structure and Inbreeding

Consider first the elephant seals and sea otters of the Pacific coast of North America. Both were hunted to virtual extinction in the late 1800s, though it later became apparent that perhaps twenty to thirty individuals of both species persisted locally for a couple of decades (Bonnell and Selander 1974; Ralls et al. 1983). The elephant seal is genetically uniform (Bonnell and Selander 1974), and the sea otter is not (Ralls et al. 1983). The population histories of the species are similar, so why the differences in genetic variability?

Seals, in general, have low levels of genetic variability (Testa 1986). The social structures of the elephant seals (and perhaps of seals in general) and the sea otters are dramatically different. In particular, each year most of the young of a seal colony are likely to be fathered by the one dominant bull guarding the harem. A seal population well above the level threatened by demographic accidents may be genetically uniform because very few males father a given generation of young. This same circumstance may apply to the lions I discussed earlier. Sea otters are largely monomagous, so a group of them will have a greater number of fathers than will a comparably sized group of seals. As a consequence, Ralls et al. (1983) estimated that the sea otters retained nearly 80% of their genetic variability.

The Indian rhinoceros was reduced to isolated populations, one numbering 12–100 and the other numbering 60–80. Moreover, not all of the males were likely to have bred, although the females produced similar numbers of young. Dinerstein and McCracken argue that this species has probably lost little of its genetic variability, both because the individuals are long-lived and because, in terms of generations, the bottleneck is recent. For a population to lose a good deal of its genetic variability, it must not only be reduced to low levels but must *stay* at low levels over a long period of time.

Demographic Structure and Inbreeding

Another scenario consistent with low genetic variability yet low risk of extinction from demographic accidents or environmental disturbances was suggested to me by Michael Gilpin (see Pimm et al. 1989). Consider one of two subpopulations that grow geometrically—2, 4, 8, 16, etc.—to some maximum limit, say, 1,024. As one subpopulation reaches higher densities, the pressure of population density causes that subpopulation to disperse and so colonize the other, which had become locally extinct. After the first subpopulation has reached its upper limit, it becomes extinct but is later recolonized by the second, surviving subpopulation. This hypothetical population never falls below a total density of 1,026 (= 1,024 + 2) individuals, and its spatial structure in conjunction with its limited migration minimizes the chance of extinction from environmental disturbances. Yet the local extinction of subpopulations can reduce the population's genetic variability (Ewens et al. 1987). Indeed, Gilpin's hypothetical population rapidly loses genetic variability with individuals always being recent descendants of each founding pair of colonists.

This model is highly contrived: plausible models retaining any of the basic ideas of O'Brien et al. appear to be lacking. Clearly, much work is needed to explore models that combine population change, chances of extinction, and the degree of inbreeding. Moreover, the model begs the question of why each population should be so completely eliminated after it reaches some high density. There is a possible ecological mechanism for such a catastrophe. Diseases may only invade a population when it reaches a critical size (Dobson and May 1986). Do cheetahs have a series of partly isolated subpopulations each periodically exterminated by disease and periodically recolonized by a few individuals? This I do not know.

Conclusions

Consider the enormous natural variability of population densities and its likely effect on extinction rate. Contrast this with the tenuous scenario of variability leading to bottlenecks which species survive several times only to lose genetic variability, suffer reduction of litter sizes as a consequence, and so, eventually, succumb. The conclusion must be that inbreeding effects *may* doom relatively large populations but that the conditions required for this—certain social or demographic structures—are special and, therefore, likely to be rare. There is a moral here as well, one which I shall draw in my final conclusions.

FEEDBACK CYCLES

Gilpin and Soulé (1986) recognized that the factors I have discussed interact in a number of quite complex ways. The factors are not independent of each other; they feed back on each other in ways that can amplify their effects. The extinction of a small population may occur with great speed because of the positive feedback of the various processes I have discussed.

For example, a population may reach a low level because of some catastrophe. At this low level, it may simultaneously (1) have less than optimal age structure, (2) have a less than optimal sex ratio, (3) be spatially fragmented, and (4) have lost genetic variability—and so be suffering from inbreeding depression. Each of these effects increases the chance of population decline. A population decrease makes each of the four effects more likely to cause a further population decrease. There may be a minimum viable population (MVP) size (Shaffer 1981) for a species, below which these four effects (and the positive feedback among them) will result in rapid extinction. Each of the effects may be the critical one because it may be the effect that causes the population to be sucked into a downward spiral.

Estimating MVP is likely to be very difficult. I have argued that spatial and demographic structures affect the chance of extinction in complex ways. Partially isolated subpopulations, for example, increase the possibility of extinction from chance demographic changes but make a population less vulnerable to external disasters. A population's demographic structure is likely to be closely related to its genetic structure. Obviously, a highly fragmented population with little migration between areas is also going to have isolated gene pools. However, the relationship between structure and genetics is much more complicated than this. The patterns of social behavior, for example, might lead to small inbred groups, but this result might be offset by dispersal between groups. Dispersal of individuals is important in rescuing isolated populations which may have reached the brink of extinction, because the dispersing individuals carry with them genes which may have been lost through inbreeding.

SUMMARY AND CONCLUSIONS

Small populations are prone to extinction. The time a population lasts before becoming extinct should increase very rapidly as population size increases, when demographic accidents are the only cause of extinctions. When environmental disturbances are the principal cause of extinction, the time until extinction will increase only slowly with increasing population size. Indeed, for the most extreme environmental catastrophes, large population size may offer no protection against extinction. There is abundant evidence that small populations persist for shorter periods of time than large populations. Are there circumstances where larger populations are more prone to extinction that small ones? For this to occur, it would mean that the extinction process must be catastrophic for large populations, while small populations must be unaffected. Diseases may act in this way.

Slowly growing populations should be more likely to become extinct than more resilient populations, because they can recover less quickly from environmental disturbances. Species in which the individuals live a long time should be less likely to become extinct, because, in any given interval, demographic accidents are less likely to exterminate a population of long-lived individuals. Generally, small-bodied species have fast-growing populations but short-lived individuals, and these features affect the chance of extinction in opposite ways. At relatively high densities, the cause of extinction will be environmental accidents, and so small-bodied species with their fast-growing populations will persist longer than large-bodied species. At very low densities, demographic accidents are the more likely cause of extinction, and large-bodied species with their longer-lived individuals will most likely persist longer.

These differences in the susceptibilities of small-bodied species versus large-bodied species enable us to test in the field the contributions that demographic accidents and environmental disturbances make to the process of extinction. For birds on small islands off the coast of Britain, the theoretical predictions are supported. Below the crossover number of about seven pairs, large-bodied species persist longer than small-bodied species with the same average population size. At this crossover density, the small-bodied species and large-bodied species have the same risk of extinction *per year*, which means that the large-bodied species (which are longer lived) have a higher risk of extinction *per lifetime*.

Species whose population sizes fluctuate greatly should be more likely to become extinct than those that vary little from year to year. This supposes that

the mean abundances of the populations are the same. Rare species may fluctuate less, but they still become extinct more often than common, more variable species; this pattern is found in spiders on small islands. An additional
supposition is that the population densities must be similarly skewed: a very
variable population may be so variable just because it reaches unusually high
densities, and these obviously do not shorten its time until extinction. For the
final complication, recall from chapters 2–5 that small-bodied species have
more variable densities than do large-bodied species (making their extinction
more likely) but that small-bodied species also have more resilient populations
(making their extinction less likely). For some species of birds, at least, more
resilient populations are less variable. So, for these birds, more variable populations should always be more likely to become extinct; and analyses confirm
this.

Spatial and demographic structure can influence the chance of extinction in
either direction. Spatially partially isolated populations may be more likely to
become extinct from demographic accidents: a population of forty is in little
danger from demographic accidents, but eight isolated populations each of
which comprises five individuals will likely persist for only a short period of
time. In contrast, a spatially partially isolated population will be better protected against environmental disasters, which typically may be geographically
restricted or affect only some age classes of the population. Not all environmental disasters are geographically restricted, of course. And, if the populations are isolated from one another, a local extinction may not be quickly followed by a recolonization. The final complication with regard to spatial
structure involves the crossover population size discussed above. A fragmented
population of a large-bodied species may be less vulnerable than a comparably
structured population of a small-bodied species, because the latter is more vulnerable to demographic accidents.

The different contributions of spatial and demographic structure, of environmental disasters, and of migration between local populations can be illustrated
with a number of examples. The spatial and demographic structure of the shag
was responsible for its rapid recovery from a catastrophic poisoning of its food
supply. The spatial structure of the Edith's checkerspot butterfly may give it
some protection from severe droughts, but the low rate of dispersal means that
some populations may take decades to be recolonized. Finally, many species,
such as grizzly bears and spotted owls, are threatened because, although their
total numbers are not small, the populations are scattered over many areas—
and in each the population may be doomed by demographic accidents. Migration between local populations will minimize the effects of demographic acci-

dents while reducing the risk of a single, catastrophic event affecting the entire population.

Small populations become inbred, inbreeding tends to reduce growth rate, and a reduced growth rate in turn may be expected to increase the chance of extinction. Moreover, species that have lost genetic variability may be less able to exploit new ecological opportunities. Inbred species may be particularly vulnerable to disasters such as epidemics, since different genotypes often differ in their susceptibility to diseases. Both the inability of inbred species to exploit new opportunities and their increased susceptibility to disease may be very important processes in the long term, but they are difficult features to demonstrate in field populations.

Is inbreeding important in natural populations—or, by the time they are inbred, are species doomed for the dynamical reasons I have already discussed? If inbreeding is important in the wild, then species must lose genetic variability even at quite large population sizes. Sea otters and Indian rhinos have been reduced to very low numbers but have retained their genetic variability. Populations of cheetahs, lions, and elephant seals have appeared to lose variability while at relatively high numbers. Special social systems, such as harems in elephant seals, can lead to common yet highly inbred species. And there are some contrived population models in which populations may remain abundant but in which the individuals are highly inbred.

The various factors of demographic accidents and genetics may interact to give positive feedback whereby, below a minimum viable population size, the population may be dragged inexorably downward to extinction. A decrease in population size may fragment a population, or lead to an unbalanced age or sex distribution, or reduce genetic variability with attendant reduction of reproductive rate, inability to respond to new environments, or increased susceptibility to disease. Any of these effects will then make the population even more likely to decrease, and the effects will increase in severity. Estimating the minimum population size is likely to be difficult because of the conflicting effects of demographic and spatial structure, the sensitivity of a population's growth rate to inbreeding, and the relationship of this sensitivity to demographic and social structure.

Issues of Scales

A recurrent theme of this book has been the demonstration that the various factors explaining the various aspects of "stability" predominate on different spatial and temporal scales. This has been a diffuse theme of this chapter too, and now I wish to bring the various issues together. Consider the smallest scale on which we might think about extinction. What immediately comes to my mind are a number of species effectively extinct in the wild and with captive populations each comprising half a dozen or fewer individuals. What determines whether these populations will survive will often be aspects of the individuals themselves—their longevity, their reproductive rate, and, in particular, their genetic relatedness. Now consider slightly larger populations: the bird populations on the British islands. Below the crossover population size of about seven pairs, demographic accidents appear to be the dominant cause of extinction, because, below seven pairs, species with long-lived individuals survive longer than species with short-lived individuals. It is a feature of the species' natural history that determines the extinction rate. Above seven pairs it appears to be environmental disturbances that are the dominant cause, for now it is the species with the faster-growing populations that survive longer, because these species recover more quickly from crashes in numbers. I have discussed in some detail what determines population resilience of a species; while it depends on features of the species itself, it also depends on interactions between other species in the community. What determines the fate of a much larger population—say, one numbering in the hundreds or thousands—will be largely the magnitude of the populations' year-to-year changes. From what we know about the variability of bird populations, a bird population averaging a thousand might be expected to persist for a long time. Most insect populations vary over several orders of magnitude over a decade or two, and so an insect population averaging a thousand may not persist very long. Many factors in addition to the features of a species' natural history affect variability, as chapters 4 and 5 attest.

Now consider spatial scales. How long a large population will last before becoming extinct will likely depend on how it is structured spatially. Quite large populations may be vulnerable if composed of isolated small subpopulations; smaller populations may be unusually persistent if scattered across areas which suffer independent environmental disasters and where the species has the ability to colonize areas where the population became locally extinct.

Earlier I promised a moral from my criticisms of those who claim that loss of genetic variability will doom populations that are currently numerous. While I agree that this is possible in some very special cases, it seems an improbable explanation for the great majority of species. The moral is that the importance of genetic loss is certainly correct at one scale—say, five individuals in a zoo— but that at other spatial and temporal scales other factors predominate.

Why I Did Not Talk about Island Biogeography

Two results of this chapter are that the chance of extinction increases as population sizes decline and as populations become increasingly isolated. So, small islands should have fewer species than large islands, because extinctions happen faster on smaller islands. Distant islands should have fewer species than islands near a source of colonists, because populations on distant islands are less likely (*a*) to be rescued when their numbers decline and (*b*) to receive immigrants in the first place. Of course, these are the features of MacArthur and Wilson's (1967) theory to explain the patterns of species numbers found on islands.

I have not discussed large-scale patterns of species diversity on islands, for two reasons. First, while the empirical patterns of more species on larger islands and on nearer islands are well documented, there are many explanations for the patterns. There are several reasons why larger areas should have more species than smaller ones. For example, as we increase area, the number of types of rocks and soils increases (Williamson 1987); this could be why the number of species of plants increases. Increases in plant species cause an increase in the number of animal species: more habitats, more species. Even if small islands have fewer species mainly because extinction rates are higher there, this tells us little about exactly why rare species are more prone to extinction. Similarly, distant islands are not replicas of the mainland: distant islands often have distinct geologies and climates (Lack 1976). In short, island biogeographic patterns are less helpful in an exploration of the fine details of extinction than they are in suggesting how rapidly species composition changes. MacArthur and Wilson imagined that very small islands near a mainland might constantly be changing their species. In contrast, on very remote islands immigration may be so slow that the islands gain most of their species through the slow process of speciation. These are issues of community persistence, and I must next introduce the varied factors that determine it.

A Note on Persistence

How long will a community retain its current species composition? Communities change their compositions through extinctions and invasions. Over the shortest time scales, communities may appear persistent, with invasions and extinctions rare enough for us to consider them as individual events and to seek their individual causes. Over long time scales, no community is ever persistent. Still, some communities change more rapidly than others, and we may very well ask why.

Consider a two-dimensional graph defined by increasing rates of extinction and increasing rates of successful invasions (figure PER. 1). Along the diagonal of this graph are communities whose species richness is constant. Away from the diagonal in one direction are communities gradually losing species and, in the other direction, are communities gradually gaining species. Near the bottom left-hand corner of this graph are communities having an approximately constant species richness and a slowly changing community composition. Finally, at the top right are those communities having approximately constant species richness but in which community membership is changing rapidly.

To understand persistence we need to discuss the rates at which extinctions take place (last chapter) and at which communities gain species. This second topic is covered by this next set of chapters. The complication is that extinctions and invasions will not be independent events. The extinction of one species may precipitate the extinction of others, and extinctions may cascade through the community. Similarly, an invasion may cause a cascade of extinctions. Alternatively, the extinction or invasion of one species may facilitate the invasion of others. These are topics of resistance: the degree to which changes in the density of one species affects the density of other species. Resistance will be the subject of chapters 12–15.

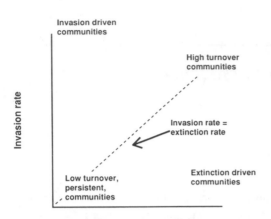

Figure PER.1. Extinction rates and invasion rates, combine to give persistence. Communities with high extinction rates relative to their invasion rates will gradually lose species. Communities where extinction rates and invasion rates are similar may be persistent (and so have low turnover) or may not be persistent (and so have much the same number of species through time but with the species constantly changing).

Extinction-Driven Communities

The length of time over which some communities retain their current species composition is determined principally by the rate at which their species become extinct. Extinctions were the topic in the previous chapter; what follows is a consideration of the scales over which they occur. Extinctions need to be understood at a variety of organizational levels. If the rate of extinctions depends on factors at different organizational levels, then obviously so does persistence. The physical features of the environment play a major role in extinctions. Communities that are fragmented will progressively lose species because the number of individuals of many species in each fragment may be too small to persist. Likewise, communities isolated by changing sea levels during the Pleistocene or by forests recently fragmented by human activities will not persist. The smaller the islands or the fragments, the faster they will lose species and the less persistent they will be. How isolated the fragments are will determine whether locally rare populations may be rescued when their numbers fall below critical levels.

The temporal variability of the environment affects population variability (chapter 5), which in turn affects the rate of extinction (chapter 7). There are community-level effects on extinction rate; population variability, for example, also depends on community structure (chapter 4). In addition, there are population-level effects; species differences in birth rate and longevity affect the chances of a particular species being prone to extinction (chapter 7).

Invasion-Driven Communities

For other communities, the rate at which species composition changes will be determined by how fast the community gains species. There are the familiar examples of species colonizing new ground—lava flows; entire islands of lava and ash, such as Krakatoa; or areas freed from ice cover by retreating glaciers. There are also less obvious examples. As fragmented communities begin to lose species, they may also begin to gain species from the human-modified communities that surround them. Again, we should expect population-, community-, and ecosystem-level factors to determine the rate of invasion.

Turnover

Yet other communities may have essentially similar rates of extinction and invasion, and so it will be sensible to combine them into a single measure, *turnover,* that measures how rapidly a community changes in its composition. MacArthur and Wilson's (1967) theory of island biogeography emphasized the fact that floras and faunas (particularly those on islands) could be constantly changing. Moreover, the theory showed that even the simplest of conditions would tend to equalize invasion and extinction rates, leading to a dynamic equilibrium.

As the number of species in a community increases, invasion rates must decrease. Inevitably, the more species the community has, the less likely that an invading individual will be of a species that it lacks. As the number of species in a community increases, extinction rates will increase because, the more species there are, the more there are to go extinct. (The question of whether the per-species extinction rate will increase as the number of species increases is a much more controversial issue.) Certainly, there are many other processes that will alter extinction and invasion rates, and these processes will alter the position of the equilibrium. Nevertheless, relative to the equilibrium number of species, species-poor communities will gain more species than they lose, and species-rich communities will lose more than they gain. Although there is a tendency toward equilibrium, the process of invasion and extinction can be so slow that we will often even observe islands progressively gaining or losing species.

The MacArthur and Wilson theory has stimulated theoretical studies of turnover and empirical tests of these theories. Within any taxonomic group, the theory suggests that the greatest turnover should be on small islands (or habitat patches) close to sources of invading species. Extinction rates will be greatest on small islands, and invasions rates will be highest where species can readily reach the islands. Large remote islands should have low turnover rates. There are studies that support this prediction: Jones and Diamond's (1976) study of the bird communities on islands off California is an example.

The Rates of Species Invasion

The controversy over turnover rates centers on those "other factors" which I noticed earlier. There are two lines of reasoning to explain the phenomenon of a community which gains no new species. First, there may be no species capable of reaching the community in numbers sufficient to overcome the difficulties encountered by small populations (see the previous chapter), or the species may be unsuited for the physical and chemical environment which that community affords. Second, the community may be constantly bombarded with species seeking admission, each capable of flourishing there. The community composition persists because the invading species finds either insufficient prey or mutualists or too many competitors or predators. Some species are excluded by the presence of others.

The contrast between these two lines of reasoning introduces a long-running and most contentious subject. To begin with, I will consider the characteristics which render a particular species unable to invade a community. Following this, I will introduce a more extended discussion of the community's role in determining the success of invasions.

8

Species Differences and Community Structure as Explanations of Why Introductions Fail

THE EFFECTS OF INDIVIDUAL ADAPTATIONS

Obviously, any community has only a tiny fraction of the Earth's species because those it lacks could not walk, swim, or fly there—or could not survive in the environment if they could. Even when species are deliberately introduced into a community, many will fail because they lack the physiological or genetic adaptations necessary to survive in the new environment. Because some introduced species become pests, the topic of whether an introduced species succeeds has generated a large literature. In the past five years SCOPE (The Scientific Council on Problems of the Environment) has organized a worldwide synthesis of work on introduced animals and plants. The various national and international volumes and comparable other studies are replete with case studies and reviews of the factors that enable or prevent species from being successfully introduced. These edited volumes include those by Brown et al. (1985), MacDonald and Jarman (1985), and MacDonald et al. all three considering plants in southern Africa; Groves and Burdon (1986) and Kitching (1986) both on Australia; Kornberg and Williamson (1986) on Britain; Mooney and Drake (1986) on the United States; Joemje et al. (1987) on The Netherlands; and Drake et al. (1989) in a worldwide synthesis. The collective wisdom of these volumes is that there are many reasons why species fail, including the species having inappropriate adaptations to the physical environments of their new homes.

Nonetheless, I urge caution in always attributing failure to a lack of the appropriate adaptations. In the next chapters I will document many examples of situations in which community composition and structure play a role in determining whether a species will invade. In addition to these examples, there are a few studies specifically addressing the question of how well arguments about adaptations to the physical environment predict which species will successfully invade.

The Hawaiian Bird Introductions

Michael Moulton and I have analyzed the introductions of birds to Hawai'i, to test a number of hypotheses about community persistence (see also chapters 9, 11, and 15). Among other things, Moulton and I asked whether prior experience of a habitat predicted whether an invasion would be a success (Moulton and Pimm 1986). At least fifty species of passerine birds have been introduced, some of them several times, to seven of the main Hawaiian Islands (Hawai'i, Maui, Moloka'i, Lana'i, O'ahu, Kaua'i, and Ni'ihau). Of these introductions, most were *intentional,* made by private citizens or agencies whose desire was to establish populations of these species. The Hawaiian Islands are tropical, and the birds were introduced into lowland habitats which range from rain forest on the windward sides of the islands to dry savannas on the leeward sides. The islands afford a wide range of tropical habitats. In these lowlands, all of the native passerine birds are either absent or, at best, very rare.

We grouped the introduced species according to whether their native ranges were tropical or temperate; three species could not be classified. If adaptations to tropical habitats were an important feature in determining success, tropical species might be expected to stand a better chance of being introduced than temperate species. This was not the case. Of twenty-four tropical species, seven failed; of twenty-three temperate species, ten failed. This difference was not significant; bird-watchers can walk through hot, humid forests in Hawai'i and see cardinals from North America and white-eyes from temperate Asia.

Crawley (1987) addressed the same topic by considering which introduced insects species could be successfully used to control weeds. Noting that it is usually assumed that "climate matching" between the home and introduced ranges would likely improve success, he concluded that there seemed little evidence to support the idea. What Crawley *did* find was that widespread species were more successful than species with small geographical ranges. Moulton and I found exactly the same result with the introduced Hawaiian birds— a greater percentage of widespread species were successful, compared with those species of limited range. Now, it may be that widespread species, in tolerating a wide range of environmental conditions, have an advantage over restricted species which may not find suitable conditions in their new homes; but there are many other reasons why widespread species may be more successful. In Hawai'i, several of the widespread introduced species were introduced earlier than species with more restricted ranges, with the result that they were able to build substantial populations before the later introductions arrived. As we shall see in the next chapter, these early introductions had the added advantage of encountering less competition. Widespread species may be so because they are better able to cope with competitors and enemies than are more geographically restricted species.

Finally, there is a wealth of anecdotal evidence—especially from gardeners and, to a lesser extent, those who keep animals outside in zoological parks—suggesting that many species are able to eke out an existence in climatic regions far from their native haunts, as long as they are protected from competitors, predators, and other enemies.

THE ROLE OF POPULATION PARAMETERS

Even when a species has the necessary adaptations to the environment and might reasonably expect to flourish despite the prey, mutualists, competitors, parasites, and predators it finds in the community, it may still fail to invade. Natural invasions and many species introduced by humans will be at low numbers at first and very rare species are prone to extinction (chapter 7). Many introductions may fail because the populations cannot quickly reach the high numbers which make unlikely their extinction from demographic accidents or environmental disturbances.

Game-bird introductions

G. J. Witteman and I (in preparation) explored what factors were important in the fate of small introduced populations by using data on the introductions of various species of game bird. Between 1960 and 1970, the United States Fish and Wildlife Service's Foreign Game Importation Program (FGIP) released over 130,000 game birds into forty-four states and three territories (Bohl and Bump 1970). After the termination of this initial program, game-bird introductions continued until 1978 in twenty-three of these states (Banks 1981). We based our analyses on a compilation of these two sources and include 790 separate releases of twenty species. In these data, the number of individuals involved in each release (henceforth, *propagule size*) ranged from 6 to over 10,000.

The relationship between propagule size and the probability that an introduction will be successful reflects a variety of factors. As population size increases, the time a population persists should rapidly accelerate as it overcomes demographic accidents (chapter 7) and the problems stemming from rare individuals finding mates (Dennis 1989). Yet, even in populations large enough for these effects to be unimportant, environmental disturbances may cause considerable fluctuations in density. In general, small populations may still be more vulnerable to environmental disturbances than large ones, but the probability of surviving may rise only slowly with increasing population density, if that density is very variable.

Combined, these theories suggest the shape for the relationship between the probability of success of an introduction and the propagule size; relationship should have a logistic form. The smallest propagules (one, for example!) have no chance of success. At small population sizes, the probability of success should accelerate, but, above a given propagule size, the chance of success will increase more slowly because population size is less important in determining success. Eventually, an upper bound is reached where increase propagule size has no effect on the chance of success, since that success is determined by other factors.

Of the total of 790 documented game-bird releases, we only included the seven species that in some place and at some time were successful. Of the 424 releases of these seven species, 360 failed. There were no differences between these seven species in their proportions of successful releases, so in the remaining analysis we treated all seven species together. To obtain a continuous measure of the relative success of introduction, we sorted the data for all releases, on the basis of propagule size. We then grouped these releases into sets with an equal number of releases in each. This set size was arbitrary, but we found that its choice did not alter our results. For each set we then calculated the mean propagule size and calculated the probability of success as the proportion of the releases that were successful.

We then fit a logistic curve to these data by using statistical techniques for nonlinear models. The results of one such analysis are shown in figure 8.1: the fit of the data to the logistic curve is both statistically significant and visually compelling. The inflection of the logistic curve represents the point at which the per-individual success of adding more individuals to the release begins to decrease. We estimate the inflection point as being approximately seventy-five birds. (Estimating the confidence interval of this parameter is a nontrivial matter. My guess is that it is likely to be wide.)

The data show that 85% of introductions fail, even when sufficiently large numbers are introduced. Yet we only include in our analyses species known to succeed at some place at some time. It is almost certain that most individual introductions will fail even when we know that the species is going to succeed somewhere and when sufficient numbers are introduced. This suggests that the conditions needed for a successful introduction are scarce in time and space. It would seem that, at least for these data, different places and times often afford very different chances of success. If this result is general, then the best way to introduce game birds would be to split a large number of individuals into smaller propagules—but not so small that the numbers fall to below seventy-

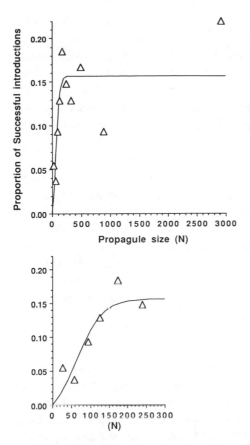

Figure 8.1. Proportions of game-bird introductions that succeed, increasing in a sigmoidal fashion with the numbers of the individuals introduced (propagule size). The proportions are the number of successful introductions divided by the total number of introductions in each set of introductions. These sets are the first, second, etc. groups of about fifty introductions sorted by the number of individuals introduced. The propagule size is the mean number of individuals introduced for each set of introductions. The lower graph is an enlargement of the first few points of the upper graph, in order to show more clearly how well the sigmoidal curve fits these data. (From G. J. Witteman and S. L. Pimm, in preparation.)

five—when the chance of failure is very high. The more releases attempted, the better the chance of hitting the right time and place. For species not deliberately introduced, the implication is obvious. Species that eventually succeed may usually fail. The more times they try, the more likely they will be successful. Conversely, a few failed invasion attempts does not mean that a species will not eventually become established.

Other factors affected the outcome of the releases but do not alter the conclusions drawn above. We were surprised that releases in the wintertime were more likely to be successful than those during the rest of the year. The birds released were either raised in captivity or caught in the wild. The birds caught in the wild were significantly more likely to be introduced successfully than were birds raised in captivity. (Unfortunately, we know nothing about the way the releases were structured spatially. It would have been fascinating to see whether several partially isolated subpopulations would have been more successful than a single population of the same total size.)

A Comparison of Game Birds and Insects

The maximum chance of success for the game birds is about 15% (see figure 8.1). Only about two hundred individuals need to be released for the probability of success to come within 5% of this maximum value. How quickly the maximum chance of success is reached with increasing numbers of released individuals is a measure of a species' risk from being at small numbers. The year-to-year variability of a population must be expected to play a major role in determining this approach to the maximum success. Species that have very variable densities will often be driven to those low levels from which recovery will be impossible. In comparable manner, highly variable species are likely to require the introductions of many individuals if they are to be successful. In contrast, populations that do not vary greatly may succeed with quite small propagule sizes, because fluctuations do not drive them to the low levels that ensure failure. Recall that, while passerine birds have standard deviations of the logarithms of density (SDLs) of about 0.3, some insects have SDLs of 1.0 or more (chapter 3). Thus, simple and approximate calculations suggest that small populations of insects may not persist whereas similiar-sized populations of birds will.

For introduced insect populations, in contrast to the game birds, such an increasing chance of success should be obvious over much larger ranges of propagule sizes. *Some* introductions of very small numbers of insects may still succeed, for fast-growing insect populations may quickly reach such high densities that population fluctuations will not cause the introduction to fail. Even so, many introductions of large numbers of insects may still fail as chance fluctuations drive them to extinction. This predicted contrast between birds and insects is supported. Witteman and I detected no increase in the success of an introduction when more than two hundred individuals were introduced (figure 8.1). In reviewing biological control attempts against pest insects in Canada, Beirne (1975) found that only 10% of introductions of fewer than 5,000 individuals were successful, that this rose to 40% of introductions of 5,000–31,200 individuals, and that, above this number of individuals, 78% of the introduction attempts were successful. Of course, it is possible that the studies using larger releases of insects were those with larger financial budgets and so were carried out more carefully than releases with small numbers of insects.

The Effects of Body Size

Using data for which the propagule size was not usually known, Lawton and Brown (1986) examined the introductions of various invertebrates and vertebrates into Britain. They considered how the success of an introduction might vary with the body size of the species introduced. Small-bodied species have higher population resiliences than large-bodied species (chapter 2), thereby reducing their chance of failure (chapter 7). However, the densities of small-bodied species also vary more than those of large-bodied species (chapter 3), and this will increase their chance of failure. Indeed, the comparison of insects and game birds shows that insects can be more or less likely to succeed than birds. Lawton and Brown's results (figure 8.2) perhaps demonstrate this conflict. A greater proportion of (large-bodied) vertebrate introductions than of (small-bodied) invertebrate introductions succeed; within insects, however, a greater proportion of the small-bodied taxa than of the large-bodied taxa succeed. Crawley (1987) analyzed the success of insects introduced to control particularly widely introduced weeds (*Lantana camara* and *Opuntia* species). He also found that small-bodied insects species were more likely to be successful.

A similar conflict should arise in studies of turnover, for turnover is driven in part by the rate of extinction. Small-bodied species could turn over faster than large-bodied species, because they either are more prone to extinction through their high variability or turn over more slowly because they are less prone to extinction because of their high resilience. Schoener (1983b) found that community turnover is slower for large-bodied than for small-bodied organisms (figure 8.3).

Small-Mammal Introductions

There are at least two experimental tests of the idea that species with relatively high birth rates should be less likely to become extinct at low population densities—and, as a consequence, should be more likely to become intro-

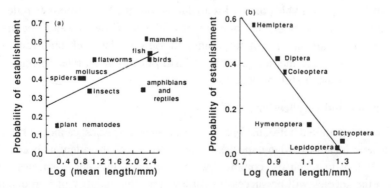

Figure 8.2. Proportion of successful introductions, increasing with increasing body size over a wide range of taxa (a) but decreasing with body size for different families of insects (b). (Redrawn from Lawton and Brown 1986.)

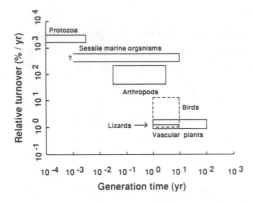

Figure 8.3. Turnover rates of species on islands. Turnover rates are greater for species with shorter generation times (and, therefore, for small-bodied species). (From Schoener 1983b.)

duced. Crowell (1973) introduced three species of rodent to small islands off Maine, in the northeastern United States. He calculated the ratio of their birth rates to death rates and found that the meadow vole had the highest, followed by the deer mouse, followed by the red-backed vole. Over a ten-year period, the meadow vole was found on almost every island but not always on each island each year, a pattern suggesting frequent recolonizations. Their ubiquity meant that there were few available islands onto which to release them. Deer mice were released twelve times, and five of these introductions failed. The red-backed voles were released eleven times, and eight of these introductions failed. In short, the higher was the species' birth rate, the more successful was the species.

Ebenhard (1987) performed a comparable experiment by introducing bank voles from northern and southern Sweden to islands in the Aland archipelago in the Baltic sea. The northern voles had higher birth rates than the southern voles, but they had much higher death rates too. Ebenhard successfully introduced the southern voles onto all ten islands, but the northern voles failed on three of the ten islands. The sample sizes for both Crowell's and Ebenhard's studies are small; this is not surprising, in view of the work involved. Still, both are suggestive that a high ratio of birth rates to death rates increases the chance for a successful introduction.

Summary and Implications

Many attempted introductions fail because the species cannot survive in their new locations. Many species cannot even make the attempt, because of their failure to reach the community. In repeatedly introduced species, even when the species will be successful somewhere, the majority of introductions fail because of the difficulties faced by very small populations. I have reviewed these ideas only briefly because my emphasis is on community-level processes.

The significance of these results comes when we try to test the community-level theories about persistence. In order to conclude that a particular community or structure cannot be invaded by a given species, we must rule out other explanations. We must be sure that the species reached the community, survived in the community, and arrived in sufficient numbers that the factors dooming small populations were not responsible for its failure. These conclusions are difficult to attain with certainty, and doubts about them underlie some of the more vigorous ecological debates. Ecologists resort to community-level ideas about persistence not just because they reject population-level explanations. The community-level processes make many novel predictions as to which communities should and should not be readily invaded. To introduce these ideas, I begin with the models of how communities might be assembled from simple, few-species systems to more complex ones.

COMMUNITY STRUCTURE AND SPECIES ASSEMBLY

Models

Models of community assembly utilize one of two approaches. The first takes a "snapshot" view of a community and examines how readily a given community might be invaded. The other approach models communities as they assemble continuously from relatively simple communities—it is a "movie," if you like. Both approaches have three essential steps. The models create an initial number of species by choosing parameters such as their intrinsic growth rates. Then the interactions (predation, competition, etc.) with the other species present in the community are specified, and parameters are chosen to quantify them. Next, the models must create, by specifying its growth rate and its interactions with the species already present, a new species potentially capable of invading the community. Finally, each new species must be tested to see whether it will be able to invade the community. This test involves calculating whether the new species can increase in density when it is rare.

For models that consider consecutive invasions, there is another step required. If the new species can invade, its population will grow. The densities of some, and probably all, of the species in the community will be changed as a consequence of this growth. The new species may either add on to the community or cause the extinction of one or more of the species already present. The models must find where the new equilibrium densities of the species will be. The assembly process then creates another potential invader, and the last step is repeated.

An example of this process is given in figure 8.4. The community starts

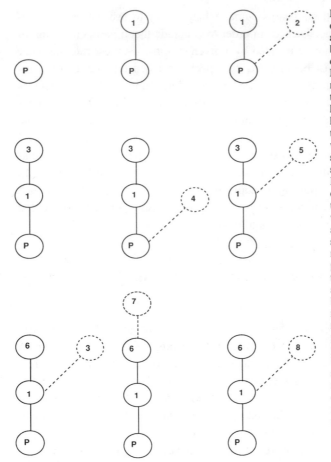

Figure 8.4. Model of community assembly, as might be generated by the approach discussed in the text. Figures should be read like a book— that is, starting at top left, going along the line, and then on to the next line. Starting with a single plant species, herbivore 1 successfully invades. Herbivore 2, however, fails, but predator 3 succeeds. Herbivore 4 and predator 5 also fail. Predator 6 succeeds and eliminates predator 3. Top-predator 7 and predator 8 also fail. This assembly process can be continued indefinitely and with larger numbers of initial plant species. The process generates both the number of species in the community and the number of times an invasion must be attempted before it is successful.

with a single species of plant, and the model tries to add an herbivore (species 1); it succeeds. Next, the program tries to add another species (species 2). It, too, is an herbivore, and it fails to invade the community because it is a poorer competitor than species 1. Species 3 is a carnivore, and it succeeds. Species 4 and 5 also fail. Species 6 succeeds and is a better competitor (for species 1) than species 3, causing the extinction of species 3. Species 7 and 8 also fail.

In its general features, the above recipe is one that takes a static food-web structure (or a sequence of food-web structures), adds population dynamics to the structure, and so creates a dynamic model of a community. Each species has a separate equation detailing its growth rate and interactions with other species. The recipe requires the choice of the equation form, rules for choosing which species interact, and ways of generating interaction coefficients. This process underlies theories that are central to the remainder of this book, but, because the details of the models are detailed and make many assumptions, I will postpone an extended discussion of them until I reach chapter 16.

Simulations of Food-Web Assembly

The simulations for assembling community models used the basic recipe that I have described already (Post and Pimm 1983). Our models had at least two idiosyncrasies. First, species were continuously created, and the assembly process stopped after a fixed number of invasion attempts. This is a realistic assumption because the pool of species available to colonize any community must be limited. However, species were allowed only one chance to invade, and this is not very realistic. The simulations started with six species of plant and finished when a total of two hundred invasions had been attempted.

Following this, we established a rule for the number of interactions between the invading species and both its predators and prey. This rule was motivated by an empirical result. The number of species with which a species interacts is approximately independent of the number of species in the community (Cohen 1977; Pimm 1982). For example, a species interacts with about the same number of species when there are twenty species in a community as it does when there are only five. Post and I chose interactions so that a species invading a species-rich community would not feed on more species of prey than would a species invading a species-poor community. Nor would an invading species suffer the attentions of more predatory species in the species-rich community. We varied the average number of interactions per species across different sets of simulations.

For a given number of interactions per species, the interactions between the invading species and those already in the community can still be distributed in different ways. The invading species must always be able to feed on at least one prey species. Given this restriction, the invading species may have many prey species and few predatory species, or vice versa. This distribution of the invading species' interactions with predators and prey was also varied across different sets of simulations.

Preliminary Results.   Post and I performed two hundred simulations for each of the many different combinations of selecting parameters and of selecting how species would interact. Figure 8.5 shows the averages, over two hundred simulations, from just two selections, but they are quite typical. The number of species in the communities rise quickly and then remains approximately constant (figure 8.5a). After the number of species has stopped increasing, each successful invasion causes the replacement of a species, on average. This

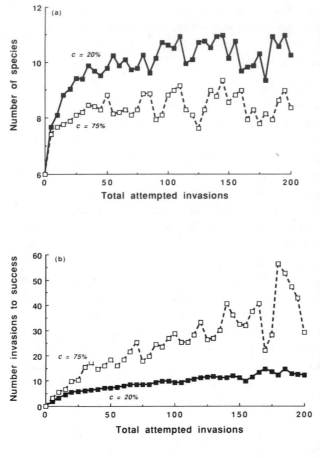

Figure 8.5. Sample results from simulations of community assembly as generated by the recipe in figure 8.4. In (a), the number of species present increases quickly and then tends to level off as the number of total attempted invasions accumulates. In (b), the number of invasions attempted before one is successful continues to rise. The two sets of results differ in how connected are the food-web models. Models that for the number of species present (dashed lines, open squares; connectance 75%), have a high proportion of their species interacting, tend to have fewer species but are harder to invade than those models that have a smaller proportion of the possible interspecies interactions (solid lines, solid squares; connectance 20%). (From Pimm 1989, based on an extensive set of simulations in Post and Pimm 1983.)

average includes species additions, simple one-to-one replacements, and species which cause several extinctions on their entry into the community.

Persistence can be measured in a simple way: the greater the number of attempted invasions that fail, the more difficult the community is to invade and the more persistent it will be. The difficulty of invasion increases continuously (figure 8.5b). Thus, even after a community has reached its maximum number of species, each successive invasion tends to make a community more persistent. Communities that are of only recent origins (that is, the process of community assembly has not been operating for a long time) might be easy to invade, even though their species richness is high. (And, of course, very recent communities may be easy to invade, because they also contain few species.)

Put simply, the "ecological age" of a community might be an important determinant of persistence.

Post and I also examined whether major losses of species, after a successful invasion, were more or less likely as the assembly process proceeded. There was no detectable difference—a successful invasion was as likely to cause a large loss of species to a newly assembled community as to one that had been assembling for a long time. Of course, successful invasions were much less common in the latter communities.

Finally, notice that there is an obvious difference between the simulations of figure 8.5. The community with the more species has an average of about ten species, and its connectance is about 20%. The corresponding figure for the less diverse community is about 75%. The more connected community is harder to invade and so is likely to be more persistent.

Species Richness and Connectance.   The effects that species richness and connectance have on persistence have been explored in more detail by Robinson and Valentine (1979). (Recall from chapter 1 that connectance is "the extent to which [species are] interconnected" in a model. Specifically, it is the number of actual interspecific interactions divided by the number of potential interspecific interactions.) Robinson and Valentine did not assemble communities; rather, they used a snapshot approach. They created model communities of different numbers of species and connectance and recorded the proportion of successful attempted introductions. Increasing species richness and connectance both made the communities harder to invade (figure 8.6).

Because Robinson and Valentine's communities were not the end products of a long process of assembly, such communities should be relatively easy to invade. Compare figures 8.5 and 8.6, and look at the models that have a connectance of 20%. Notice that many more attempts were needed to create a successfully invading species toward the end of Post and Pimm's simulations (nine of ten failed) than were needed in Robinson and Valentine's studies (about two in ten failed).

Why Are Species-Rich, Highly Connected Communities Persistent?

Part of the explanation for highly connected communities proving hard to invade involves the very familiar ideas about the role of competition, ideas that I will explore in the next chapter. The more connected the community, the more species will overlap in their use of prey species, and the more likely the species are to compete as a consequence. The more competition, the less likely a species will be able to invade a community.

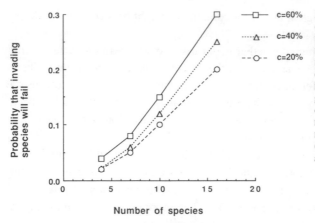

Figure 8.6. Proportion of attempted invasions that will fail versus the number of species in the model community, for communities that differ in connectance. A greater number of species and greater connectance make communities harder to invade. (From data in Robinson and Valentine 1979.)

Another part of the explanation may be more subtle. Holt (1977, 1987) has argued that the consequences of two or more species sharing *predatory* species are identical in some ways to the consequences of species sharing species of prey. When two prey species share even one species of predator, the increase in the density of one of the prey species will increase the density of the predators, and this increase will act to decrease the density of the other species of prey. Holt calls such effects *apparent competition,* for there may be reciprocal changes in the prey species densities even when the prey do not share resources. An invading prey species, if it is to be successful, cannot share too many predatory species with other prey species already present in the community.

In sum, as the connectance of a model food web increases, an invading species is likely to encounter increases in both competition for prey and apparent competition via predators. But why are species-rich communities harder to invade than species-poor communities? We might well expect the result that, the more species there are in a community, the more ways there would be to add on new species.

Consider the simple communities portrayed in figure 8.7. There are two species of plant; one community (left-hand side) has one monophagous herbivore, and the other community (right-hand side) has two. An herbivore can invade the three-species system and not encounter competition, but this is not true of the four-species system (top). We conclude that, the more species there are, the fewer ways there will be to invade the community. Unfortunately, this conclusion is true only if we restrict the number of trophic levels. If we remove the restriction on the number of trophic levels, the number of ways to invade these communities, free from competition, is the same (middle and bottom). We can add a species to either food chain (an herbivore or a predator in the three-species system, two predators in the four-species system), or we can add a species that feeds on both food chains (in the three-species system, a species, that feeds on the plant and the herbivore, and, in the four-species system, a top-predator. In general, if increased species richness makes invasions more difficult, it must be that increased richness is accompanied by severe constraints on the ways in which species might be added to a community.

Implications. These results may appear simple, but they motivate a two-pronged attack on persistence. First, persistence depends on aspects of food-web structure—the number species there are and how tightly they are connected, which species share prey, which species share predators, and how many trophic levels there are. Aspects of this are familiar; the more species there are, the more an invading species is likely to encounter a competitor that will prevent its invasion. Other aspects are less familiar, but they still involve *structural* factors that affect persistence. Second, the assembly process itself plays a role. Communities of the same connectance and number of species may differ in how difficult it is for species to invade them, depending on how long the communities have been assembling. The effects that both different structures and the assembly process itself have on persistence will need to be explored.

SUMMARY

The length of time a community retains its current composition depends on the rate at which the community is losing species (as discussed in the preceding

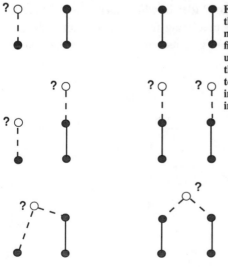

Figure 8.7. Illustration of why, so long as there is a limit on number of trophic levels, more prey species make invasion easier (left figures) than do fewer prey species (right figures). Open circles and dashed lines indicate the invading species and their potential interactions, and closed circles and solid lines indicate the already present species and their interactions.

chapter) and gaining species (this chapter and the next three). Moreover, extinctions or introductions can cause secondary extinctions and introductions (chapters 12–15).

Communities gain species at different rates. Physical features of the environment—such as the extent of its isolation—play a part in determining the rate of invasion. Species differ greatly in their abilities to reach communities. Many species may fail because the community's environment is unsuitable. When the species arrives, it may fail for one of the many reasons that very small populations become extinct. Analysis of game-bird introductions show that even species that will eventually be successfully introduced will usually fail. In short, the population-level explanations for a community not being invaded are many.

In addition, there are important community-level explanations for persistence. Models showing how diverse communities assemble from a small number of species tell us that food-web structure plays an important role in determining persistence. Some aspects of this idea are quite familiar—for example, the idea that, the more species existing in a community, the more likely the invading species will encounter a competitor that prevents it from being successful. There are other features of food-web structure whose effects on persistence are much less obvious. Finally, these models also show that the assembly process itself plays a role, for, even when the model communities are not increasing their number of species, each species replacement makes a subsequent replacement less likely.

Separating the various explanations for the reasons communities persist will be difficult. If we are to conclude that biological interactions prevent a species from invading, then we must have a good idea of which species really have the potential to invade. The difficulty of predicting failed introductions (the successful ones are obvious, of course) will be a recurrent problem in the chapters that follow.

9

Patterns in Species Composition

ALTERNATIVES

In the last chapter, I introduced two hypotheses which explained the reasons for a community not being invaded. The first hypothesis postulates that either (*a*) there are no species capable of reaching the community in sufficient numbers to avoid the dangers of small population size or (*b*) the species may arrive in force but are unable to survive in the physical and chemical environment the community affords. The second hypothesis is that invasions do not occur because biological interactions prevent them. For example, too many predators or competitors may make it impossible for the colonist's population to increase. There are two ways to separate these hypotheses. The first and direct way would be to experiment. We would attempt to introduce species to communities, then record both which species succeeded and which failed and which communities admitted species and which rejected them. Using this approach, we can ask whether the introduced species that have the appropriate adaptations to the environment routinely fail when introduced in large numbers and, if so, whether this failure can be related to the presence of competitors and predators. This is such an obvious way to separate the hypotheses, but ecologists have given more attention to a second and indirect approach—looking at the structural consequences of the two hypotheses.

A community is considered to have structure or to be patterned if the species it contains are a nonrandom subset of those that could occur in it. If the community is patterned, species may be actively excluded because the community is not allowing some species to colonize. This indirect approach of looking for patterns attempts to infer process from those patterns. Clearly stated over forty years ago, this approach has always been highly controversial (Lewin 1983a, 1983b). There are abundant data on community patterns, making it easy to speculate on their causes. The controversy stems from the difficulty in predicting the "ghosts"—the species that tried to invade a community and failed. We have to erect hypotheses identifying those ghosts, and I shall present, in a

189

historical sequence, the attempts to do this. First, however, I must introduce a little-discussed problem concerned with inferring the rate of species invasions on the basis of community patterns.

Do Communities Have a "Home-Field Advantage"?

There is a long tradition equating community pattern with persistence, and the debates I shall presently introduce will be familiar to you. (One of the classic arguments for "complex" communities being more "stable" was Elton's [1958] discussion of how species-rich communities were less readily invaded than species-poor communities; see chapter 1.) Let us suppose that communities are patterned in the various ways we will encounter later in this chapter. Further suppose that the communities are challenged frequently by potential invaders. To equate pattern with persistence requires one further assumption: that the community has a home-field advantage. That is, in a contest between the community and the invader, the community wins. The species invades only by fitting into the community in a way that maintains the community pattern. For example, ecologists have argued that morphologically similar species do not coexist. If a species too similar to a resident species tries to invade, it will fail, and the pattern of relative dissimilar morphologies will be maintained.

Of course, there is another possibility. Invading species may have the ecological habits of Ghengis Khan and effectively destroy any community they reach. Community patterns would then appear because of changes wrought by the invader and not by any feature of the community. Community patterns would persist despite high species turnover.

The key feature separating the home-field-advantage scenario from the Ghengis Khan scenario is *resistance* (chapters 11–15): community composition may or may not be resistant to a successful invader. Notice that determining resistance is not the same as determining whether the community will persist. We have to examine the effects of the successful invader, not whether it joins the community. Later in this chapter I will present some evidence that communities are resistant to invading competitors, but most of the evidence appears in chapter 15. There is not a great deal of evidence in any case, and what evidence there is deals with species usually deliberately introduced outside their normal ranges. The relevance of these studies is subject to an obvious caveat. Deliberately introduced species may well cause havoc to the structure of their new communities, while the great majority of natural introductions may not. Deliberately introduced species may be chosen from a species pool that is worldwide. Your team may have a home-field advantage when playing against other members of its league but may be beaten badly when the international champions play an exhibition match against them. In short, how large the home-field advantage is (and, therefore, how much community patterns tell us about persistence) must depend on how large is the species pool from which potential invaders are drawn.

The difficulties we encounter in studying persistence are simply ones of scale. We wish to understand the long-term, large-scale effects of multispecies systems and often cannot or have not collected enough data for long enough periods of time. Consequently we are left with sometimes tenuous arguments about the cause of the patterns we can observe. These patterns, at the very least, are important sources of hypotheses, and it is in this spirit that I shall now discuss them.

"COMPETITION AND THE STRUCTURE OF ECOLOGICAL COMMUNITIES"

The Number of Species per Genus

In March 1944, the British Ecological Society held a meeting entitled "The Ecology of Closely Allied Species" (Anonymous 1944). The meeting was motivated by Gause's (1934) ideas that two species with similar ecologies could not coexist because the competition between them would be too intense. Among those present, Lack and Elton argued for the importance of interspecific competition in shaping community pattern; others disagreed. Elton's views appeared as a paper in 1946, entitled "Competition and the Structure of Ecological Communities." His observations were simple: "If one peruses the lists of species recorded in various ecological surveys of clearly defined habitats, the thing that stands out is the high percentage of genera with only one species present. This is quite a different picture from a faunal list of a whole region or country, in which many large genera are to be found" (1946, 54).

Elton continues by introducing fifty-five surveys of animal communities. Some surveys were grouped "to give more reliable figures," and this results in forty-nine samples from land, fresh water, estuaries, and oceans and the arctic, temperate, and tropical regions and including free-living and parasitic species. Elton also assembled data on twenty-seven plant communities. These were mainly from Britain, but terrestrial arctic communities were included, as were marine, intertidal, and planktonic communities. Elton also examined the number of species per genus in the orders of British insects. He found that, although communities are characterized by 1–2 species per genus, nearly 60% of the insect orders have 3 or more species per genus. For the British flora, an early classification (Bentham 1924) divided 1,251 species into 479 genera— that is, 2.61 species per genus. So why the discrepancy?

When Elton turned to species lists for counties, he found that individual county records included the majority of the species recorded for the entire country. For example, he reported that 58% of all the British species of microlepidoptera had been recorded within about 15 km of Oxford. So the fact that a community is short on species cannot be explained by the missing species not having reached the community. Nor did Elton feel that the slight differences between species in the same genus could be related to major physiological differences and so explain the species' use of different areas. Different "genera of the same consumer level . . . are capable of living in a particular area . . . whereas it is unusual for species of the same genus to coexist" (1946, 64).

Regardless of how his results are judged, Elton proposed a clever hypothesis: communities contain fewer species than those which regularly land in the community and which could survive the physical and chemical conditions there. That so many species fail is due to interactions between species: communities are relatively persistent because they exclude species. Elton stressed competitive exclusion, but it is within the spirit of his paper to expand the hypothesis to include all kinds of species interactions which restrict community membership.

Not everyone was happy with Elton's ideas. The meeting's reporter noted that Captain Cyril Diver "made a vigorous attack . . . Mathematical and experimental approaches had been dangerously over simplified and omitted considerations of many factors" (Anonymous 1944, 176–77). Diver was especially concerned about neglecting predators and details of life history, and he pointed to the difficulties of defining "similar ecologies," noting that many similar species do coexist. Furthermore, he argued that many species are limited by factors which kept them from reaching densities at which resources become limiting. Captain Diver would be quite at home at any recent meeting of community-theory ecologists.

Lack's Analyses of Darwin's Finches

Lack's contribution to the 1944 meeting was a discussion of his work on Darwin's finches, done six years earlier. Lack's subsequent book (Lack 1947) was an explanation for the diversity of these species, all of which are thought to be derived from a single ancestor. Differences between the species must have originated when the species were isolated on different islands. Lack felt that such differences would remain small because the islands are all similar in their habitats. Marked differences only developed when the once-isolated populations resumed contact. If the differences were insufficient for coexistence, there would be either extinction of one species or a rapid evolutionary divergence in the species' feeding habitats.

This introduces a new idea: "character displacement." If communities are patterned, it may be because of species extinctions *or* because the species present may have shown evolutionary changes in their morphologies. Rummel and Roughgarden (1983) have called these "invasion-structured" and "coevolutionary-structured" communities, respectively; they are not mutually exclusive.

As evidence for his views, Lack pointed to the regularities in the morphologies of species. Bill sizes of coexisting congeners are approximately regularly spaced. If bill differences correspond to diet differences, then the species may be at least reducing competition. Lack found that two species of Darwin's finch with very similar bills did not overlap in ranges. Another species was significantly smaller on an island where a smaller, putative competitor was absent.

Darwin's finches have also been the subject of a long-term study by the Grants and their colleagues (Grant 1986; Grant and Grant 1989). In the ground finch genus *Geospiza*, there are six species. The males are black, the females are spotted brown, and the species differ principally in their size and in the shapes of their bills (figure 9 1) In bill depth, *G. difficilis* is very similar to *G. fuliginosa*, and *G. conirostris* is very similar to *G. fortis*. Species pairs involving morphologically very similar species occur on many fewer islands than do those pairs involving rather different species. Grant and Schluter (1984) concluded that morphologically similar species tended not to coexist.

REJECTING COMPETITION AS AN EXPLANATION

Williams's Statistical Analyses

Elton and Lack argued that communities are short of similar species—either species in the same genus or morphologically similar species. Other ecologists were quick to criticize them. Williams (1964) noticed an obvious problem with Elton's results. Suppose there is only one species in the community. It is trivial that the community contains only a single genus and that its species-to-genus ratio is also 1. There are nearly 500 genera of British plants and about 1,250 species. If I pick 10 of those species, the chances are excellent that I will also have picked 10 different genera. Randomly picking 10 species gives a species-to-genus ratio of 1, or at least some number close to it. If I fix the number of species, setting it equal to the number of species observed in a particular community, I can calculate how many species per genus I should expect. This ratio is going to be less than the ratio for all the country as a whole—and progressively less the smaller the number of species in the community.

Williams analyzed nine sets of data—three plant communities, five insect communities, and a study of East African birds by Moreau (1948). Williams's

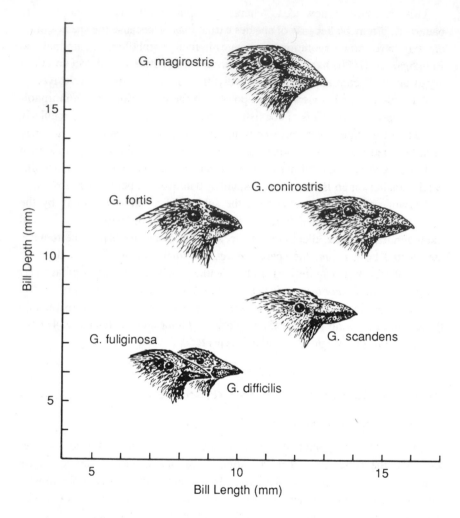

Figure 9.1. Bill morphologies of six species of *Geospiza*. The geographically restricted *G. conirostris* has a bill depth similar to that of the more widespread *G. fortis*, and the relatively restricted *G. difficilis* has a bill depth similar to that of the more widespread *G. fuliginosa*. Birds' heads are placed at the coordinates of their bill dimensions. (From data in Grant 1986.)

analysis shows that, when corrected for sample size, there are *more* species per genus in a community than one would expect. For example, Moreau recorded 172 species in 92 genera. He then placed these species into a habitat classification that recognized three major divisions (lowland, intermediate level, and highland), various habitat types within these divisions (forests, grasslands, etc.), and different places (ground, treetops, etc.) within these habitat types. Williams calculated the chance that 2, 3, or 4 species per genus should occur together; the observed numbers of such occurrences are 2.3, 8 and 15 times *more* likely than would be expected. Eliminating habitats where species might be aggregating to sporadically abundant resources did not alter the results.

Williams concluded that there are more congeneric species and fewer genera in communities than one would expect and that this excess is probably due to congeners having similar adaptations to physical conditions. Such adaptations, he argued, override any tendency for similar species to competitively exclude each other. Williams is also stating a compelling hypothesis. Communities contain as many species as we would expect, given the difficulties species encounter in making their way to a given community and being able to survive the physical and chemical environment when they get there. Communities only persist because they are not being challenged by potential colonists.

Simberloff's Criticism

In a similar way, Simberloff (1984) has argued that the distributions of Darwin's finches also fails to provide evidence for the competitive exclusion of morphological similar species. In performing their statistical tests, Grant and Schluter (1984) assumed that all of the possible pairs of the six *Geospiza* species are equally likely to be found. Only when this assumption is made is the rarity of certain pairs of similar species found to be statistically unusual. Specifically, it is the species pairs involving *G. conirostris* and *G. difficilis* that are the rare ones. Simberloff argues that this is because these two species are themselves rare. In the extreme case, if a species only occurred on one island, then it could only coexist with the other species found on that island. Conversely, if a species occurred on all thirty-two islands, then it would have to overlap with every other species. The species and the number of islands on which they occur are *G. fuliginosa*, 27; *G. fortis*, 17; *G. scandens*, 14; *G. magnirostris*, 10; *G. difficilis*, 7; and *G. conirostris*, 3. It is hardly surprising that *G. conirostris* overlaps with so few species—it occurs on very few islands. Similarly, *G. difficilis* also overlaps with few other species, and it is the next rarest species.

THE DEBATE CONTINUES

The arguments I have presented are merely the opening salvos in the debate about community patterns. Colwell and Winkler (1984) noticed that the community patterns from which we must draw inferences about competition may already reflect the influence of competition. Consider the species-to-genus ratio results. Williams argued that communities have as many species per genus as one would expect on the basis of the number of species that occur in them. But Elton argues that this expectation begs the question of why there are so few species in the first place. He argues that the reason more species do not occur in any one place is that many species are competitively excluded; it is not that many species could not survive and flourish in a given place. Perhaps it is hardly surprising that, when we attempt to detect competition by using data already containing the effects of competition (for example, reduced species numbers), we fail to detect competition.

Grant and Schluter (1984) make this point forcibly in their rejoinder to Simberloff's criticism. They argue that the reason *G. difficilis* and *G. conirostris* are present on so few islands is competition. If one tests for nonrandom patterns of distribution on the basis of the numbers of islands on which the species occur, it is not surprising that one cannot detect it: competition influences the numbers of islands on which species occur.

Grant and Schluter point to the distributions of the two species, which I present in figure 9.2. The rarest species, *G. conirostris,* is present on three islands—including one small island close to another, larger one. It is crucial that these two groups of islands are over 150 km apart, almost at opposite ends of the archipelago. The species is absent from all the islands in between. The range of *G. difficilis* is even more interesting. It, too, is present on scattered islands and is absent from islands in between. Moreover, it has been known to breed on four other islands. All told, its known breeding attempts span the entire east-west axis of the archipelago and nearly the north-south axis too. Making the assumption that these two species could only reach three and seven islands, respectively, is plainly incorrect for *G. difficilis* and very improbable for *G. conirostris*. As Grant and Schluter put it, using "the number of islands occupied by a given species as an estimate of 'colonizing ability' . . . [and to incorporate] this estimate into calculations of expected co-occurrences . . . is likely to be self-defeating." Such an analysis "is restrictive as it evades the question of why certain species are uncommon to begin with" (1984, 210).

The essential feature of this debate between Simberloff and Grant concerns what constitutes a reasonable null hypothesis. Statistical tests work in a way that always seems odd to those encountering them for the first time. Tests set up pairs of hypotheses. The null hypothesis, by definition, is a simple one (that is, X and Y have no correlation); the alternative hypothesis a complex one (that is, X and Y are correlated, with some value we do not know certainly). When we chose our alternative hypothesis, we do not make our decision on the basis that the hypothesis is highly probable in light of the available data. Rather, we make that decision by showing that the null hypothesis is very improbable.

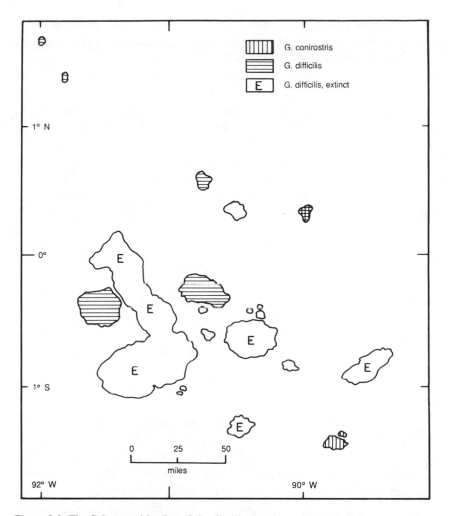

Figure 9.2. The Galapagos islands and the distribution there of two relatively rare species of Darwin's finches in the genus *Geospiza*. *G. conirostris* occurs on three islands in two groups, but the two groups of islands are widely spaced at the northeast and southeast of the archipelago. *G. difficilis* also occurs on few islands, but these islands are also widely spaced. Moreover, this species is known to have bred and become extinct on several other islands (marked with an E). (From data in Grant 1986.)

The reason for this asymmetry is that, unless we have a very simple hypothesis, we are totally unable to say anything about its chance of being true.

It follows that, in order to perform a statistical test, we must have a simple hypothesis. We can often live with this restriction. In the analyses of the morphological patterns of Darwin's finches, we have not one null hypothesis but two. That in itself is not a problem. What causes the difficulty is that, regardless of the data, both null hypotheses are surely false. Grant's null hypothesis is that, in the absence of competition, the six species have equal chances of occurring on all the islands. Simberloff finds this extremely unlikely; so do I. Simberloff's null hypothesis is that species can only occur on the number of islands on which they are actually found (that is, three for *G. conirostris*). Grant considers this to be unlikely; again I agree. I find it hard to believe that either of these two hypotheses will ever be correct, no matter which group of species we analyze or where we chose to do our analyses. What emerges from all this is that there is no plausible null hypothesis to predict on which islands the species failed. With no null hypothesis, we cannot draw any conclusions. We need alternative means of making statements regarding which invasions failed.

PATTERNS AS HYPOTHESES: A CATALOG

Whether one finds the evidence for community pattern compelling or dubious does not deny that the patterns themselves are interesting hypotheses about the factors that determine persistence. With this in mind, I review a set of case histories that, at worst, provide a list of hypotheses about community composition and persistence.

Taxonomic

Elton's hypothesis was that communities are taxonomically restricted. Communities contain, in a given genus, only a small number of the species that might reasonably be expected to find their way to that community. All species similar enough to be in the same genus might be expected to be able to survive in a particular community. That they do not indicates the effect of competitive exclusion: taxonomically similar species may be too similar ecologically to coexist. There is nothing special about using species and genera in this argument: we might make identical arguments by using any other pair of taxonomic categories.

Morphological

Lack's hypothesis was that morphologically similar species do not coexist. This argument was extended by Hutchinson (1959), who argued that competitive exclusion could lead to a minimum size ratio (The Hutchinsonian ratio) between adjacent members of size-ranked coexisting species. Thus size ratios of coexisting species should be larger than one would expect: Hutchinson suggested a ratio of 1.3:1. There have been many claims of character displacement that have not withstood the scrutiny of these strict statistical tests, or indeed, careful ecological analysis (Grant 1975; Simberloff and Boecklen 1981). Even when statistical significance is claimed, the difficulty is that only the significant test is likely to be published; the other nineteen of twenty studies may languish.

Even at the 1944 meeting it was recognized that deciding how similar was too similar would be difficult. One solution is to look for community-wide character displacements: patterns of equal spacing in the size of some morphological character. Limits on the similarity of coexisting species should prevent species from being too close, and large spaces should be readily filled by invading species. Generally, it has been assumed that adjacent species would have constant ratios—and this equal spacing on a logarithmic size. (I shall return to this assumption later.) The alternative (and statistical null hypothesis) would be for sizes to be placed randomly, resulting in some spaces being larger than others. Obviously, to compare the equality of spaces, one needs at least two of them—and so at least three species.

Dayan et al. (1989, 1990) provide three particularly compelling examples of equally spaced sizes; weasels in North America, mustelids, and small cats in the Middle East. The results and their implications are worth considering in some detail. All the species involved are sexually dimorphic, and so each sex of each species was treated as a separate "morphospecies." Dayan et al. examined two characters: skull length and the diameter of canine teeth; the former reflects overall size, and the latter has a significance I shall discuss presently. In general, skull lengths were not evenly spaced, but the patterns for canine diameters are visually impressive and far from what we would expect by chance alone. The results are particularly surprising, given both the statistical rigor and their occurrence in such different groups as felids and mustelids. Perhaps equally surprising is that Simberloff is a coauthor on these papers. As we have seen, with regard to community-wide character displacement, he is known as a skeptic—not a proponent. There is much more to these results than historical irony, however (Pimm and Gittleman 1990).

There are three species of *Mustela* in North America (figure 9.3). *M. erminea* coexists with either *M. nivalis* or *M. frenata* or both (Ralls and Harvey 1985). The species vary considerably in size over the ranges, and Dayan et al. measured them at eight locations. At all but one of these, spacings were more even than one would expect by chance. Females of the species are always smaller than the males of the same species. In general, male *M. nivalis* are smaller than *M. erminea* females, and *M. erminea* males are smaller than *M. frenata* females. Notice that in Michigan there is still even spacing, despite a reversal. In increasing order of size there are *M. nivalis* (females and then males), *M. erminea* females, *M. frenata* females, *M. erminea* males, and *M.*

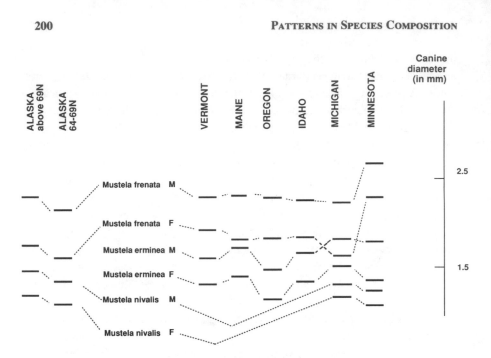

Figure 9.3. Canine diameters of male and female weasel species in various places in the United States. The data are more evenly spaced than one would expect by chance. (From Pimm and Gittleman 1990, from data in Dayan et al. 1989).

frenata males. Notice also that the actual spacing is not even; the distances between species become larger with larger sizes. So it is the ratios that are evenly spaced.

In the Middle East, there are four extant mustelids and one recently extinct species, *M. nivalis,* and there is one species of viverrid. Both with and without the extinct *Mustela,* the pattern of spacings is unusually even (figure 9.4).

Three species of *Felis* are also found in the Middle East (figure 9.4). Again, the spacing is unusually even, and, as with the Michigan mustelids, the two largest morphospecies are males of two species, and the next two largest are the females.

These patterns of equal spacing are not just for any character in any set of species. To a first approximation, the species form natural guilds, tending to share prey with other members of the guild more than with other species. Moreover, the species kill their prey in the same way, by biting. This is why the measurement of canine diameter is such a sensible one. Behavioral studies show that weasels aim their canines at the nape of their prey. Canines might well reflect the typical size of their prey and be adapted to the vertebral size of the most common prey species. The skull sizes are not evenly spaced; the length and shape of the skull is likely to be influenced by factors other than

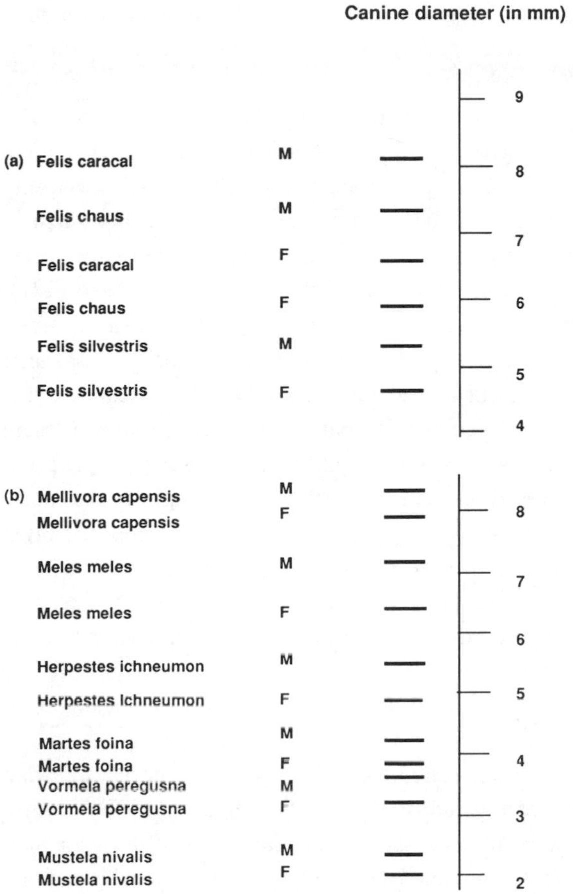

Canine diameter (in mm)

Figure 9.4. Canine diame-
ters of cats (a) and mustelids (b) in the Middle East.
When males and females
are plotted separately, data
are more evenly spaced
than one would expect by
chance. (From Pimm and
Gittleman 1990, from data
in Dayan et al. 1989, 1990).

(a) Felis caracal   M

    Felis chaus   M

    Felis caracal   F

    Felis chaus   F

    Felis silvestris   M

    Felis silvestris   F

(b) Mellivora capensis   M
    Mellivora capensis   F

    Meles meles   M

    Meles meles   F

    Herpestes ichneumon   M

    Herpestes Ichneumon   F

    Martes foina   M
    Martes foina   F
    Vormela peregusna   M
    Vormela peregusna   F

    Mustela nivalis   M
    Mustela nivalis   F

feeding. If competition for food is a prime selective pressure, then each sex of
each species is competing with its conspecific opposite sex, as well as with
both sexes of all other species.

In these studies, it is the natural history—not a large-scale search of every
possible character and species combination—that suggests the choice of species and characters. The moral is that, to demonstrate patterns convincingly,
sophisticated statistics cannot be turned loosely willy-nilly on characters that
are irrelevant to species interactions. Nor can one choose a group of species
that are trophically heterogeneous. Both the statistics and the natural history
have to be correct. All this seems neat and tidy; echoing Dayan et al.'s concerns, Gittleman and I suggested that perhaps it is *too* neat (Pimm and Gittleman 1990). On ever closer inspection, the existence of the guilds, the justifi-

cation of canines as a character largely determined by competition for prey species, and the pattern of equal spacing of the canines seem less and less expected.

Consider the idea that the species form guilds. Carnivores are unusually flexible in their behavior and ecology, and their diet is no exception. The mustelids and viverrids in Israel, while largely carnivorous, show much variability in their diets. Although the diets of these taxa have not been well studied in Israel, in other areas *Mellivora* and *Vormela* are less specialized carnivores, with up to 20%–40% of insects in their diet (Harrison 1968; Smithers 1971). Delimiting a guild often becomes harder, the more one knows about a group of species. Most puzzling of all is why the spacing is even—on a logarithmic scale. The prey species of these carnivores are not some hypothetical resource axis; rather, they are individual species with particular sizes. Moreover, there may be relatively few species of prey. One might reasonably expect the predator's canines to match the size of particular prey species. It might be that those prey have sizes that are evenly spaced on a logarithmic scale, though I think this possibility is unlikely. If so, then this begs the question of why the prey are evenly spaced! For the canines to be evenly spaced, there has to be some process that generates this pattern that is insensitive to the great variability in the diets of the carnivores but not to the presence of other carnivores. To my knowledge, such a result is not predicted by current theoretical studies of competition.

Dayan et al.'s results are fascinating. Superficially, the patterns are what ecologists have often expected but rarely shown. Now that the statistics *have* uncovered the pattern, we are still uncertain about the process of how the species compete. Further thought about the patterns show that our understanding of both community-wide character displacement and its implication for persistence is inadequate.

Trophic

So far, I have used taxonomy and morphology as surrogates for ecological relationships. Why not use ecological similarity directly? To do this would require an understanding of the natural history of the organisms that is detailed enough to be able to construct categories that reflect ecological relationships. One such set of data comes from a Lawton's (1976, 1984) fifteen-year study of the herbivorous insect communities on the bracken fern.

So far, I have argued that the existence of pattern indicates that species invasions are unlikely or, at least, less likely. Lawton has argued that there is a lack of pattern in his communities and that the gaps in the community are indications of where species might invade. Lawton defines four major categories that exploit bracken: (1) chewers, which live externally and bite large pieces out of the plant; (2) suckers, which puncture individual cells of the vascular system; (3) miners, which live inside the tissues; and (4) gall formers, which also live inside the tissues but induce the tissues to form galls. Insects do these four things to different parts of the plant, attacking the pinnae ("leaves"), rachis (main stem), costae (main stalks of the pinnae, arising from the rachis), or costules (main "leaf veins" arising from the costae). The combinations of modes of exploitation and parts attacked are the *trophic niches*.

Between communities there are major differences in the number of pattern of occupied niches (figure 9.5). There are no rachis miners in Lawton's main site in England, Skipwith Common, but there are five such species in New Guinea. A rachis miner does exploit bracken in England, but not at Skipwith. New Guinea is relatively short of external foliage chewers, though it does support two rachis chewers, a group entirely absent from Skipwith. New Mexico bracken is full of unexploited niches, and Hawai'i (not shown in the figure) has no bracken-exploiting species at all.

Do species in one niche competitively exclude species from exploiting other niches? That is, does the plant itself constitute the niche, such that the communities in England and New Guinea are equally saturated with species? Lawton thinks not. There is no evidence that the species compete. Among other

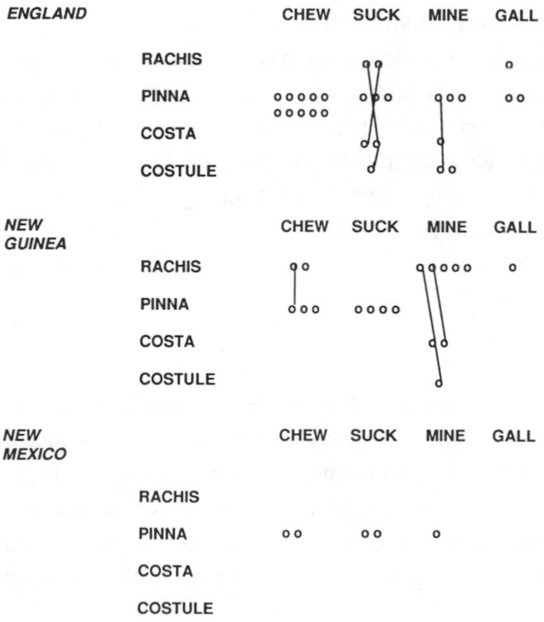

ENGLAND

		CHEW	SUCK	MINE	GALL
	RACHIS		o o		o
	PINNA	o o o o o o o o o o	o o o	o o o	o o
	COSTA				
	COSTULE		o o		

NEW GUINEA

		CHEW	SUCK	MINE	GALL
	RACHIS	o o		o o o o o	o
	PINNA	o o o	o o o o		
	COSTA				
	COSTULE				

NEW MEXICO

		CHEW	SUCK	MINE	GALL
	RACHIS				
	PINNA	o o	o o	o	
	COSTA				
	COSTULE				

Figure 9.5. Possible niches for herbivorous insects on bracken fern. Niches are defined by where on the plant the insects feed and how they choose to feed there. Circles indicate species, placed in their observed niches in three areas on three different continents. In New Mexico, for example, there are no gall-forming insects on bracken, while in New Guinea there are five species of insects that mine the rachis. Lines join the same species when it occurs in more than one place on the plant. (Redrawn from Lawton 1984.)

things, the individual species are not more abundant in New Mexico, where there are fewer species, than in England, where there are more. In each area, similar species do not show either compensatory changes in abundance or mutually exclusive distributions. Bracken, one might conclude, has many vacant niches; its insect communities might be readily invaded, particularly by species chosen to exploit those niches currently vacant. Lawton argues that insects from outside England could eat bracken in new ways, and he is planning introductions of such species to the apparently vacant niches, in an attempt to control bracken (which is a major weed)—and, in doing so, to test the "vacant niche" idea.

Biogeographical

In archipelagos, which one of a set of similar species occurs on a given island may depend on which of the species got there first. We might see islands with species A or species B, but never with both and only rarely with neither. These are called *checkerboard distributions,* and Diamond (1975) provides several examples for birds in the Bismarck archipelago off New Guinea. For example, two kinds of fruit pigeon (figure 9.6) occur on thirty-two islands: they are absent from only three islands; one species is present on eleven islands, the other is present on nineteen islands; and both are present together on only two islands. If their ranges are totally independent, the species should only co-occur on six islands, and eight islands should have neither species. So, the species tend to co-occur less often than we would expect.

The debate about biogeographical evidence for competition has been every bit as heated as those about morphology and taxonomy. Many of the conceptual issues are identical, and so there it no point in repeating them. The distributions of the fruit pigeons reject the statistical null hypothesis, but Diamond might have selected this example from many other distributions; just how many is very controversial. Certainly I have selected it from his work as being the most obvious checkerboard distribution. At the very least, Diamond is proposing an interesting and important hypothesis.

Temporal

Species' activities may be distributed in time so that temporal overlap in activities is minimized. As an example, consider bird- or insect-pollinated flowers. These plant species might have to compete for pollinators, for, if the plants flowered at the same time, they would be sharing those pollinators. Competition might be expected to result in staggered flowering times—and thus in a uniform spacing of flowering peaks. Plant species with staggered

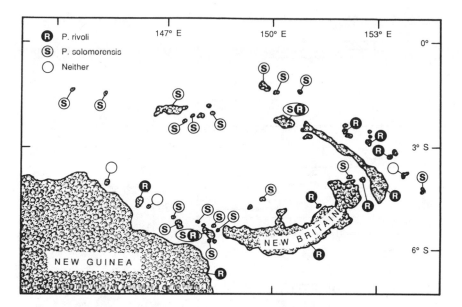

Figure 9.6. Distribution of two species of fruit pigeons (*Ptilonopus rivoli* and *P. solomonensis*) on islands off the coast of New Guinea. With few exceptions, each island has one and only one species. (Redrawn from Diamond [1975], who calls this a "checkerboard" distribution.)

flowering times would be minimizing competition for species that are their mutualists, rather than their prey.

Such staggered flowering times might be less obvious at high latitudes, where pronounced and regular cycles of photoperiod and temperature impose restrictions on flowering times. But what about the tropics? Stiles (1977) collected data in the less seasonal tropics, in a rain forest in Costa Rica (approximately 10° N), over a four-year period. His data can be grouped according to whether the flowers were in "good" or "full" bloom (figure 9.7). At no time were there more than three species in full bloom or more than five species in good bloom.

Are the flowering times staggered? There are obviously fewer species flowering during the end of the dry season, so we should separate dry and wet seasons. Then, we should ask: How many days of overlap do we expect if the flowering periods are arranged randomly? Cole (1981) uses this recipe and finds that in each season, in each of the four years, the observed overlaps are less than the expected overlaps. In short, the flowering periods tend to overlap less than one would expect.

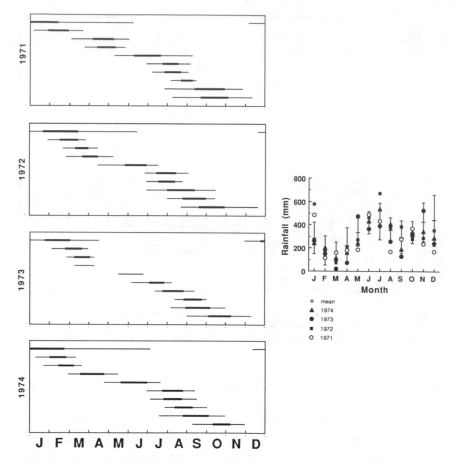

Figure 9.7. *Right,* Monthly rainfall at La Selva, Costa Rica, 1971–74. *Left,* Main periods
(thick lines) and good periods (thin lines) for the flowering of ten species of plants on which
hermit hummingbirds depend. These flowering periods tend to be staggered throughout the
year, with only one or two species' main flowering periods overlapping, suggesting that the
plants are competing with each other for the services of their pollinators. (Redrawn from
Stiles 1977.)

Summary

There are perhaps three lessons that we should learn from these debates
about community patterns. First, examining existing community patterns is a
powerful way of generating interesting hypotheses about community processes. One cannot deny either the appeal of the ideas presented or their potential implications for community persistence. Second, notice that the debate
concerns the level of explanation. Do we interpret the data as consequences of
the species' natural history or as consequences of community structure? Ob-

viously, the debate about the influence that community structure has on the various meanings of "stability" is as old as modern ecology itself.

The third lesson is that science advances when we choose one hypothesis from two or more alternatives. Using statistics means that this choice requires the rejection of null hypotheses that, by their very nature, are simple ones. If a null hypothesis is a priori unbelievable, it is axiomatic that we are not advancing. We have not made a choice in rejecting something already considered implausible. In the context of the debate over community structure, the null hypotheses involve predicting the "ghosts": the species that colonized but that subsequently become extinct. Few hypotheses make these predictions in ways that are completely convincing. The conclusion is simple. We must avoid the problems inherent in these studies, by looking for data where we know the ghosts. We need studies where we do not have to erect a priori improbable null hypotheses but where we know directly which species failed to invade.

EXPERIMENTS IN PERSISTENCE: INTRODUCED SPECIES

A Direct Approach: Introductions as Experiments

There is abundant evidence, from small-scale removal or addition experiments, that competitors reduce the growth rates or densities of similar species (see chapter 12). Unfortunately, competition between small sets of species says little about its extent in an entire community. Moreover, the critical problem is that, no matter how many of these small-scale experiments there are, they say little about extinction or the lack of invasion. To show that competition patterns communities, one must show that competition is simultaneously *intensive* enough to cause extinctions and *extensive* enough that the effects of these extinctions predominate at the community level.

What sort of experiment should I perform if I want to test the role of competition in determining whether a species could successfully invade? The experiment seems like a fantasy. First, take a number of islands. Islands are necessary because they are isolated and because they either can be given different experimental treatments or can serve as replicates. The islands must be remote enough that the rate of natural colonization will be extremely slow. Next eliminate the native species in the group to be studied. Now that no native species or natural invasions can cloud the results, introduce different numbers of species to the different islands. The numbers of individuals of each species introduced ought to be large enough for the problems of random extinction to be minimized. Finally, record which introductions succeed and, very important, which fail. In order to see which species fail, we probably ought to introduce some large, conspicuous group, such as birds. How long should we look at these introductions to determine which species have failed? My guess is, for at least several decades, perhaps even a century, because it may take many years for some of the introductions to fail. Once we have done all this, we can ask such simple questions as: Do introductions fail more often when there are more species? Do species fail more often if species in the same genus already are present? And do species fail more often if morphologically similar species already are present? I do not think that the United States National Science Foundation would be receptive to my proposal to undertake this experiment and fund me for the next century or so. In fact, this is not a problem; the experiment has already been done for us in Hawai'i.

The Introduced Hawaiian Avifauna

It was no accident that Michael Moulton chose this avifauna as a subject of study (Moulton and Pimm 1983, 1985, 1986, 1987; Moulton 1985). The introduced Hawaiian avifauna is an experiment in community persistence that almost exactly fits the conditions of my fantasy. The remnants of native Hawaiian forests harbor the remains of a large, bizarre endemic fauna and very few exotic species. Below 600 m, however, only the most dedicated observer will find native species in the almost totally exotic forests. Bird introductions began in the mid-1800s and, through accidental escapes, continue to this day. Different numbers of species have been introduced to each of the six largest islands: Kaua'i, O'ahu, Moloka'i, Lana'i, Maui, Hawai'i. Most species have been introduced to O'ahu, where most of the people live. Professional ornithologists as well as many amateur bird-watchers have kept records over the past century, most actively since the 1940s. More species of birds have been introduced to the Hawaiian Islands than anywhere else (Long 1981). Fifty species of passerine birds were introduced to the six main islands a total of 125 times between 1860 and 1983. Of these introductions, about two-thirds were direct, and the remainder were subsequent, natural colonizations of other islands. Many of the direct introductions were made intentionally and with substantial numbers of individuals, rather than being birds' accidental escapes from captivity. Indeed, for many years there was an organization, the Hui Manu, which existed solely for the purpose of introducing exotic birds into Hawai'i (Berger 1981).

The rate of failed invasions.   Even in the absence of competition, one might expect the rate of failed invasions to increase with the number of species, because, the more species there are, the more there are to go extinct. The extinction rate increased faster than this: the *per-species* extinction rate increased with increasing numbers of species (figure 9.8). This means that the

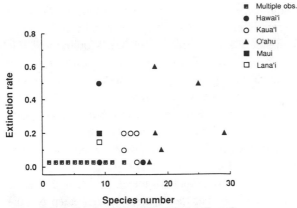

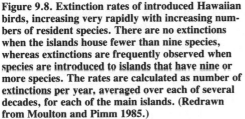

Figure 9.8. Extinction rates of introduced Hawaiian birds, increasing very rapidly with increasing numbers of resident species. There are no extinctions when the islands house fewer than nine species, whereas extinctions are frequently observed when species are introduced to islands that have nine or more species. The rates are calculated as number of extinctions per year, averaged over each of several decades, for each of the main islands. (Redrawn from Moulton and Pimm 1985.)

community was harder to invade, the more species there were. One might suppose that this result might be a consequence of the early introductions—when there were few species—being made into relatively undisturbed habitats, while the later introductions were made during modern periods of intense habitat changes. We were able to show that the reverse was true (Moulton and Pimm 1983). Indeed, habitat upheavals, such as converting lowland Hawai'i into sugar cane fields, required the importation of cheap labor, and it was often the laborers who wanted the familiar birds of their homelands after they had decided to make Hawai'i their home.

The Influence of Species in the Same Genus.   Elton's hypothesis would predict that introductions would fail more often if congeners already were present in the community. In fact, the chance of failed introductions for species introduced to islands were species of the same genus already were present (sixteen failures and sixteen successes) was almost identical to that when such species were absent (twenty failures and twenty-three successes).

The Influence of Species with Similar Morphologies.   Lack's hypothesis predicts that introductions are more likely to fail when morphologically similar species already are present. Moulton used two approaches to test this idea. The first test looked at how similar the species were in their beak lengths. Within a genus, similarity in beak length often implies similarity in diet. Indeed, for one group of birds introduced on Hawai'i, the estrildine finches, Moulton was able to demonstrate a direct relationship between beak similarity and dietary overlap (M. P. Moulton, in preparation). Overall, in cases in which one species in a genus failed but the other succeeded, the beaks differed by 1%–29%, with a mean of 9%. Where both species in a genus coexisted, the differences were 2%–38%, with a mean of 22%. The means of 9% and 22% were significantly different (Moulton 1985). The presence of morphologically similar species makes a successful introduction less likely.

The data provide a test of the home-field-advantage problem discussed above. Is the community resistant, so that the resident species keep out the ecologically similar invaders? Or is the community not resistant, so that the similar invader beats out the residents? Moulton (1985) found that the home-field advantage was substantial: in all but one case it was the most recent invader that failed.

The second way to test for the effects of morphological similarity is to compare species that are similar yet not necessarily in the same genus. The difficulty here is that a beak length of, say, 1.0 cm may belong to either a seed-eating bird or an insect-eating bird. The beak depth on the former may be 1.0 cm, also, whereas that of the latter is only a third as much. The seed eater may have short wings and hop along the ground, whereas the insectivore may have much longer wings to catch insects on the wing. To test the effects of morphological similarity, we need a multidimensional test of similarity that simultaneously includes information on beak length and depth, wing length, tail length, and other variables. When we restrict our analyses to species in the same genus, those species are sufficiently closely related that just one measurement will capture much of the essential ecological difference between them.

We measured several different features of the birds and then used a principal components analysis to find a simple two-dimensional representation of the birds' morphologies (Moulton and Pimm 1986, 1987). An example of this analysis, for the introduced species found in forests on the island of Kaua'i, is shown in figure 9.9 (top). The horizontal axis (the first principal component) correlates closely with overall size. The vertical axis (the second principal component) assesses beak width and depth. For example, high values indicate beaks that are wide and deep for their length; such species typically eat seeds rather than insects.

In the figure, I have identified which species were successfully introduced and which failed. Notice that the four failed species are always close to and often are surrounded by morphologically similar species. The resulting pattern

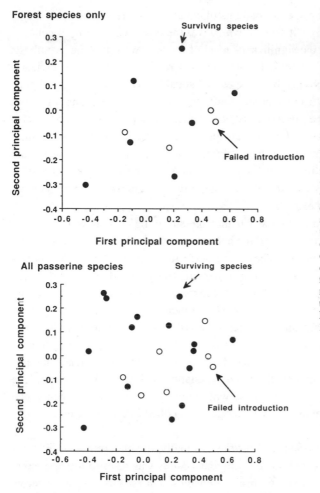

Figure 9.9. Morphologies of Kaua'i forest birds and all passerine birds on Kaua'i, which can be summarized by coordinates in a two-dimensional space by using a principal component analysis. (To a rough approximation, the first component is a measurement of overall size, and the second reflects shape.) The extinctions of species in the forest are nonrandom, and the survivors are more evenly distributed in the two-dimensional space than one would expect by chance. This is not true for the island community as a whole (see text). (From data in Moulton and Pimm 1987.)

of successful introductions looks overdispersed: the survivors appear more spread out than one would expect if one picked four failures at random. Moreover, the survivors appear to be more evenly spaced: the distances between them are more uniform than one would expect if one picked four failures at random. Both results are exactly what one would predict if morphological similarity made failure more likely. Species close to other species will fail, and, given enough time, the species will fill the available set of morphologies in ways that lead to constant differences between them.

The statistical tests associated with these patterns are quite simple. Using a computer, Moulton and I picked all the possible combinations of four failures and seven successes (330 in all). We then looked at which of these 330 combinations had surviving-species patterns of equal to or more widely spaced than the one observed. Only five (one of which was the observed pattern, of course), or 1.5%, did. This is the significance level of the test; we reject the hypothesis that the patterns are random, because such a random pattern is sufficiently improbable. The surviving birds were statistically more widely spaced than one would predict by chance alone. Only one of 330 combinations was more evenly spaced than the observed species. Thus, the surviving birds were also statistically more evenly spaced than one would predict by chance. Similar statistically significant results were obtained for the islands of Hawai'i and O'ahu. We failed to find them for Lana'i and Maui, because the numbers of species introduced there were so small that only four and seven combinations, respectively, could be explored.

Moulton and I also looked at whether the surviving species were morphologically overdispersed across *all* the habitats on an island—and not just within forest habitats on an island. They were not; the morphologies of the species that survived on each island were apparently random (Moulton and Pimm 1986). For Kaua'i (figure 9.9, bottom), the result changes from being overdispersed in forest species to being apparently random in all the island's species, because morphologically similar species coexist by choosing different habitats. I do not think that this result will surprise anyone: few would predict competition between birds of grasslands and birds of wet forests. Grasslands are also quite restricted on Kaua'i, and several grassland species failed to invade. The result is an obvious caution to those who would perform analyses concerned with which species persist using heterogeneous collections of species.

Species Richness and Species Densities.   Finally, the occurrence of different numbers of introduced species on different islands was used to test the effect of competition in determining species densities. On islands where a species encounters more competitors it should be rarer than where it encounters only few. Moulton and I counted, on O'ahu (where there are many additional introduced species) and on Lana'i and Maui (where there are few additionally introduced species), the number of individuals in a set of nine species common to all the islands. In nearly all cases, the nine species were rarer on O'ahu (Moulton and Pimm 1985).

Summary and Extensions.   In sum, these studies provide compelling evidence that both increased species numbers and the presence of morphologically similar species reduce the chances of a successful invasion. There is no evidence that the mere presence of congeners has any effect.

Humans have moved large numbers of species around the world, sometimes accidentally but often deliberately. These introductions are all experiments in community persistence. The extent to which they can be used to test ideas about persistence depends, to a large extent, on how well the introductions have been documented. In particular, the theories are about which species failed. It is easy to observe which introduced species succeeded—house sparrows in much of the world are an example. The failures are not always documented. What is special about the introduced Hawaiian avifauna is that the failures have been carefully recorded. There are only a small number of other studies that allow us to ask whether the patterns found for Hawaiian birds are general.

Introduced Australian Birds

The study most comparable to that of the Hawaiian birds is by Newsome and Noble (1986) on birds introduced to Australia. They found similar results: thirty-eight introductions to the species-rich mainland of Australia failed, while thirty-four succeeded. In contrast, only fourteen introductions to islands off Australia failed, while thirty-eight succeeded. A large island (4,350 km²) only 13 km from the mainland was similar to the mainland in having a high failure rate: ten species failed, and only three succeeded.

Lizards on Tropical Islands

It is unusual to know which introduced species failed; the successes, of course, are obvious because they are still there. Case and Bolger (1991) suggest an ingenious way to look at the effect of native on introduced species, a way that makes a plausible guess at how many introduced species there ought to be on a series of tropical and subtropical islands. In general, one might expect that large islands and islands with a diversity of habitats should have more species on them. For lizards, this is the case; the larger the island is and the greater its elevation, the more species on the island. Of course, we might also expect that larger and higher islands might have more species of introduced lizards; this is also true. Case and Bolger then look at how the observed numbers of native and introduced species of lizards differ from what one would expect on the basis of the islands' area and elevation. If competition between native and introduced species is important, then one would predict that islands that (for their size and elevation) have unusually rich native faunas should have fewer introduced species than do islands that (for their size and elevation) have unusually poor native faunas. The data (figure 9.10) show that this prediction is supported. Case and Bolger also find the same pattern for a group of forty islands within the Seychelles; indeed, the result is even more significant statistically. There is always the possibility, however, that some other and unmea-

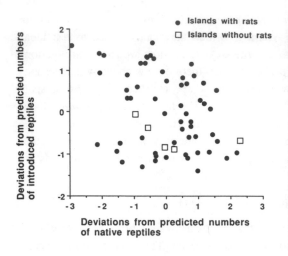

Figure 9.10. Lizards in tropical islands. Tropical islands that have more native species of lizard than one would expect on the basis of their size and elevation also have fewer introduced species of lizard than one would expect on the basis of their size and elevation. (Obviously, we expect larger, higher islands to have more species of both native and introduced species than do smaller, lower islands, so these data are corrected for these factors. Notice also that, on average, islands without rats have *less* species of introduced lizards. Possibly the predatory rats may facilitate the invasion of introduced species by reducing the densities of native species. (From data supplied by Ted Case.)

sured variable correlates with both introduced and native species. An obvious choice is the degree of human activities, which might increase the number of introduced species and decrease the number of native species. In the worldwide survey there are five islands without rats, whereas in the Seychelle islands there are eight islands without rats. Such islands must have suffered very little human impact. For both subsets of islands, the overall pattern holds.

Insect Introductions

Biological Control.   In at least one set of insect studies, there are records of which species failed. Hall and Ehler (1979) and Ehler and Hall (1982) analyzed the worldwide compilations of insects introduced to control homopteran, coleopteran, and lepidopteran pests. Their analysis looked at the proportion of successful species against how many species were introduced. The results look compelling (figure 9.11), with success rate dropping off significantly as the number of species (and the putative amount of competition) increases.

   Simberloff (1986) discussed a particularly detailed example of how failure increases with the number of species because those species compete. The example involves the control of the California red scale insect by three aphelinid wasps. *Aphytis chrysomphali* was first introduced around 1900 and was quite successful in mild coastal areas. In 1947, *A. lingnanensis* was released, and within three years it had colonized all the citrus areas of California. It gradually replaced *A. chrysomphali* in interior and intermediate climate areas, which by

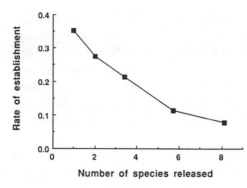

Figure 9.11. Rate of establishment—that is, the proportion of successfully introduced insects for biological control of insect pests. The rate drops as the number of introduced species decreases. Quite what this means is not clear, and various explanations are discussed in the text. (Redrawn from Hall and Ehler 1979.)

1958 had disappeared from all of southern California expect for a few small regions near the coast. *A. lingnanensis* was considered not to be effective enough in interior regions of California, so a third species, *A. melinus*, was released in 1957. It quickly displaced *A. lingnanensis* from interior areas and reduced its density in intermediate areas.

This example notwithstanding, the work by Hall and Ehler aroused considerable controversy. As Keller (1984) and Simberloff (1986) point out, there are at least three features of biological control procedures that would also explain why few, rather than many, introduced species will appear to succeed. First, there are two extreme strategies for releases: either introduce only one species after very careful evaluation of which species will work best or use a shotgun approach of introducing many species in the hope that one will succeed. Second, if a species is successful, there likely will be no more introduction attempts; if it is not successful, other species will be introduced. Third, species successful in other areas will likely be among the more probable choices for introductions. In all three cases, introductions of single species will be more successful than introductions of several species.

Biogeographical Patterns of Introduced Species.   Although there are no records of how many species introductions have failed, Sailer (1978) found patterns that are reminiscent of those discussed so far. The 1,554 successful insect introductions to North America represent about 1.7% of the continent's insect fauna. In contrast, the 1,476 species introduced to Hawai'i are 29% of the islands' total, and the remote south Atlantic island of Tristan de Cuhna has 84 native insects but 32 introduced ones (Holdgate 1960).

Simberloff (1986), who has reviewed these insect introductions with his usual intense scrutiny, doubts that these differences support the prediction that species-poor communities are easier to invade than species-rich ones. Hawai'i's insect fauna is unusual: it is not surprising that some insect orders can reach such a remote archipelago more readily than others. There are no native ants, for example. So one might predict that insect orders naturally underrepresented in Hawai'i would have more introduced species, relative to orders that are naturally overrepresented. Indeed, in Hawai'i there are proportionately fewer native Diptera than there are in North America, and the fraction of introduced Diptera is greater for Hawai'i than for North America. On the whole, however, one sees the same proportions of introduced species in Hawai'i as in North America.

Simberloff also points out that the *number* of species introduced to Hawai'i is very similar to that of species introduced to North America. The greater proportion of introduced insects reflects the smaller number of native species in Hawai'i. So perhaps there is a pool of jet-setting insects that can reach nearly anywhere and be successful when they get there. Then all but the most remote places would have this same set, and places with fewer species to begin with would have proportionately more. When I first heard this idea, I remarked that Honolulu International Airport is a smaller target that the entire continental United States. It would be more interesting to compare how many species had entered Hawai'i through Honolulu with how many species had entered the continental United States through, say, Los Angeles. If many more species had entered through Honolulu, then one might conclude that they were afforded a warmer welcome there. Perhaps the best conclusion to draw is that these data provide only equivocal tests of the theories, because we have to guess which species failed to become introduced.

The insect data are consistent with the hypothesis that species-rich communities are harder to invade than species-poor communities. Unfortunately it is impossible to eliminate other hypotheses that do not invoke community composition as an explanation.

Plant Introductions

Few taxa afford such good opportunities to test ideas about which species have succeeded and which have failed to invade as plants. The species that have failed are those that just grow in our gardens; the successes have "jumped the fence." Crawley (1987) notices that *The European Garden Flora* (Walters et al. 1984) lists over twelve thousand species—about eight times Britain's approximately fifteen hundred native species of flora. So why have some species jumped the fence, while most are still only found in gardens? Crawley compiled data on how plant communities differ in the degree to which they are penetrated by introduced species. While he appreciated both the potential to contrast which species succeeded versus which failed to jump the fence and how the species richness of the community affects the outcome, he did not pursue this issue—and, to my knowledge, no one else has either. Unfortunately, we are left with comparisons of what proportions of the existing floras are composed of introduced species. Fox and Fox (1986) found for Australian floras much the same relationship as I discussed earlier for insects. Communities with few native species had a greater *proportion* of introduced species; the *number* of introduced species was similar across communities differing in the number of native species. Again these data are consistent with the hypothesis that increased species richness decreases the chance of invasion, and, as before, they are also consistent with other, noncompetitive explanations.

Summary

There is an extensive literature on introduced species. Although there are some studies that use these introductions as experimental tests of hypotheses, many more such analyses are possible. The critical information concerns which species failed. Studies on birds do have data on which species failed, and they provide evidence that increased species richness decreases the chance of success. Although comparable studies of insects find a similar result, the different protocols of insect introductions admit other, noncompetitive explanations. There is abundant evidence that species-poor communities have a greater proportion of introduced species than do species-rich ones. Often this means no more than that there are approximately the same number of introduced species throughout the different communities.

PERSISTENCE AND PREDATION

So far, I have discussed species that are excluded from a community because of competition. Equally, a species might fail because the community lacks the mutualists necessary for the species' survival. What should the role of predation be? The obvious argument is that predatory species may make it harder for a species to invade. Nature may be much more complex than this, for there are several arguments for why the presence of predators may make it easier for a species to invade (and, indeed, for why increasing the diversity of predators may make it easier to invade; chapter 10). Obviously, predators could also make it easier for a species to invade if the predator feeds heavily on resident species and so reduces their competitive effects. The arguments get more complicated than this, however, and to introduce these complications I must talk about the phenomenon of apparent competition.

As we have seen, if two species are similar in some aspects of their ecology, then they are likely to compete. Consequently, the species that coexist in a community will be less similar than we might otherwise expect. The more species that are in a community, the more likely an invading species will encounter one already present that is sufficiently similar that it prevents the invading species' colonization. If similar species do not coexist, there is an ingenious alternative explanation for this that does not require competition for resources—or, indeed, competition in the traditional sense that ecologists use that word. Williamson (1957) and Holt (1977) have shown that shared predators may have dynamical consequences similar to those of either shared prey or shared resources. (The argument applies quite generally to a range of different kinds of enemies, such as parasites or diseases; Holt and Pickering 1985.) Suppose two species, A and B, share a common enemy, P. If A increases in numbers, so too (eventually) will P. If P increases, one would expect B to decrease. Increases in B will, similarly, cause decreases in A. Because the two species A and B mutually harm each other, Holt calls this "apparent competition."

Expanding on this, Holt has shown theoretically that polyphagous predators can play an important role in preventing the colonization of species that too closely resemble the predators' normal prey. An invading species that shares too many predators with a species already present may suffer too much apparent competition to succeed. So the result—invasion difficulty increasing with increased species richness—may be due to invader and resident sharing predators, not just sharing resources. Because species that share resources may be more likely to share many other features of their niches, including predators, separating the effects of shared resources from those of shared predators may be very difficult or impossible.

Indeed, all the various community patterns I have discussed earlier in this chapter can be the result of the effects of shared predators, as well as the result of effects of the more familiar competition for resources. Lawton (1984) has argued that such predator-mediated effects are the mechanism behind the niche differences in the bracken insect communities (figure 9.5), as well as an explanation of community patterns for herbivorous insects in general (Lawton and Strong 1981). Shared predators as well as shared resources may involved in the phenomenon of community persistence.

In short, a community with predatory species could be either easier or harder to invade than one without them. Alternatively, the presence of predatory species may make little difference to a species' chance of invasion, even when we know that the predators have a marked effect on the numbers of their prey. Here are two studies that explore these ideas.

Rainbow Smelt in the Great Lakes Region

Evans and Loftus (1987) have analyzed the distribution of rainbow smelt. This fish is indigenous to freshwater lakes of eastern North America, but it has been widely introduced into other lakes. The Ontario Ministry of Natural Resources conducted a detailed survey of 8,842 lakes in the province; the smelt is found in 195 of them; it is not known into how many lakes it was introduced unsuccessfully. The smelt have invaded nearly all the lake types found in Ontario: shallow to deep, small to large, productive to unproductive, and fish poor to fish rich. Most relevant to my story is that these lakes contain between zero and six species of fish that eat the smelt. The smelt are found more often in the lakes with the *greatest* number of predatory species.

Now, of course, this result might occur because predatory fish might like special kinds of lakes (deep, large, acid lakes, for example) also preferred by the smelt. If this were true, then within each kind of lake we would not encounter smelt more often in lakes that also had more species of predator. To examine this possibility, David Evans, Jeffrey Hyman, and I, in an unpublished study, performed multiway contingency analyses to see whether the increased chance of finding smelt in predator-rich lakes remained the same even within different kinds of lakes. It did; even particular kinds of lakes were more likely to have smelt in them if they also had more species of predatory fish. We could not rule out the explanation that fisherman introduce the smelt into different kinds of lakes only if these lakes also have unusual numbers of predator species. We do not know in which lakes the smelt was unsuccessfully introduced. Nonetheless, the results are at least consistent with the theoretical prediction that an increased diversity of predators may facilitate invasion.

The Effects of Rats on Lizards

Figure 9.10 shows that islands with unusually rich native lizard fauna tended to have fewer introduced lizards than one would expect. Most of the islands studied by Case and Bolger (1991) had introduced rats that prey on these reptiles; but five islands did not. These five islands *without rats* always had (for their size and elevation) *fewer* species of introduced lizards (figure 9.10). (A comparable result was obtained when Seychelles islands that have rats were compared with Seychelles islands that do not have rats.) One can suggest an explanation for these results that is similar to that used in the case of the introduced smelt: inaccessible islands may have relatively few introduced lizards and no rats just because they are inaccessible. Nonetheless, the data are consistent with the prediction that predators facilitate invasions. Certainly, the data exclude the hypothesis that predators have a strong effect on preventing invasions. In the case of island lizards, the mechanism seems to be that predators greatly reduce the abundance of native lizards, while the introduced lizards are from places with a long history of rats being present (Case and Bolger 1991). The rats make it easier, not harder, for the introduced lizards.

SUMMARY

Elton (1946) suggest that communities might be structured; that is, for each genus in any community, there might be fewer than the expected number of representatives because taxonomically similar species tend not to coexist. Communities exclude many species through ecological interactions. Elton's ideas were extended to morphological similarity by Lack (1947). In contrast, Williams (1964) argued that, given the observed number of species in a community, there are actually more species per genus than one would expect. This is because species in the same genus occur together because they share similar adaptations to the environment's physical and chemical variables. Taxonomically similar species tend to share adaptations and thus habitats. If communities persist, Williams argues, it is because there is a dearth of species able either to reach the community or to survive when they get there. The debate about these hypotheses continues. For example, Grant (1986) observed that morphologically similar species of Darwin's finches rarely occur on the same islands in the Galapagos archipelago. However, Simberloff (1984) argues that this is because two of those species are rare and, because of this fact alone, cannot be found on many islands. Grant replies that such arguments are self-defeating. The two rare species are probably rare because competition from morphologically similar species restricts their distribution. Grant's evidence for this assertion includes the very scattered distributions of the two rare species and known failed breeding attempts on islands where morphologically similar species occur. The two species can clearly disperse well beyond the limited numbers of islands on which they occur.

There are two conclusions that emerge from these studies. First, regardless of whether the patterns of community composition are evidence for persistence, they are interesting hypotheses. The patterns suggest the hypotheses that community composition may be limited either taxonomically or morphologically. Similarly, communities may show geographical patterns, such as checkerboard distributions. In an archipelago, for example, some islands may have one species, and other islands may have another species; the two species may only very rarely occur on the same island because they cannot coexist. Communities may show temporal pattern with species activities being staggered to reduce overlap. The alternative to communities being patterned is that community composition might have gaps which any qualified species might fill, provided it can reach the community. Lawton (1984) suggests that there are gaps in the community of herbivorous insects on bracken fern.

Second, the most compelling evidence that competition restricts community membership comes from the known failed breeding attempts of the species. It is this actual demonstration of failure that is the critical evidence. Where there is contention, there is doubt about which species failed to invade. The obvious extension of this is for us to study persistence by looking to the studies that document which invasion attempts failed—rather than trying to infer them.

Introduced species provide alternative ways and sometimes direct experiments to test ideas on community persistence. Studies of plants, insects, and birds show that the proportions of introduced species are greatest in communities with fewest species. Such results are consistent with the hypothesis that increases species richness makes invasion more difficult. But such studies typically find that the numbers of introduced species are constant across different communities. As Simberloff points out, this observation could be a consequence of communities with different numbers of native species being equally easy to invade and the existence of a fixed number of species available to invade.

The most convincing evidence for invasions being more difficult in species-rich communities comes from studies of introduced birds on islands that report which species failed. The proportion of failures is greatest on islands that have more species and is lower on islands than on adjacent mainlands (that also have more species). For the introduced Hawaiian birds, species fail more often when the island avifauna already contains morphologically similar species—but not when it merely contains species of the same genus.

Predation has a complex role in persistence. Certainly a predator may make it impossible for a species to invade. However, a predator may also make it easier, by reducing the density of the native species and so making that native species a less effective competitor. In one study, an introduced species of fish is more common in lakes with unusually large numbers of predators. In a second study, the presence of rats on islands was associated with relatively high numbers of introduced reptiles.

10

Food-Web Structure and Community Persistence

DOES PATTERN INDICATE PERSISTENCE?

I have been discussing the length of time the species composition of a community persists before a species invades. The obvious way of exploring this topic would be to measure the rates of invasion for communities with different features. We cannot do this often, of course, because the time scale of our observations is so short. As a result, ecologists use an indirect approach proposed by Elton (1946), who argued that community patterns might imply that species are being excluded, and, if species are being excluded, then the communities must be relatively persistent. I continue with this "pattern equals persistence" argument by considering food-web patterns. Food webs are diagrams depicting which species in a community interact, and recent studies have developed a long catalog of food-web patterns. Only certain statistically rare food-web patterns are found in nature.

What the existence of community patterns tell us about persistence is potentially important but ambiguous. Each invading species will have a particular pattern of species interactions with the other species in the community; it will be the predator of some species, the competitor of others, the prey of yet others. Suppose the community has the "home-field advantage." Then species can only invade communities where their presence will not violate the known food-web patterns. The existence of those patterns restricts which species can invade. Equally in accord with the existence of food-web patterns is the possibility that a species may invade and then exterminate other species whose presence would lead to violations of the patterns. The presence of food-web patterns would then tell us how communities may disintegrate—that is, how they lack resistance to the invader.

In the last chapter, I present evidence that the community has the home-field advantage vis-à-vis potential competitors, and I anticipated material (to be presented in chapter 15) on resistance to invading competitors that supported the same conclusion. For example, the pattern of overdispersed morphologies in introduced Hawaiian birds arises because the species resident in the community repel morphologically similar invading species; the patterns do not arise because the invaders eliminate morphologically similar resident species, and consequently the patterns indicate the community's persistence.

The evidence that communities repel species which would violate the food-web patterns is very much weaker than that for studies of competitors. In chapter 15, I shall argue that, although the evidence that invading species eliminating other species competitively is rather tenuous, successfully invading species *often* cause extinctions at different trophic levels. Consider, for example, Zaret and Paine's (1973) study of the effect of a cichlid fish introduced into Lake Gatun in Panama. The cichlid radically changed the lake's species composition: yet the food webs before *and* after satisfy the catalog of food-web patterns (Pimm 1982). Of course, one can argue that the species pool of deliberately introduced species is so large that we should not be surprised when we find some species that destroy all before them (chapter 9). The existence of these ecological Ghengis Khans does not necessarily refute the idea that, in general, food-web patterns give important clues as to persistence.

In short, it is very difficult to decide whether food-web patterns tell us about persistence or about the ways in which communities disintegrate. Nonetheless, the existence of food-web patterns provides fascinating hypotheses about persistence, and it is in this spirit that I present them. Of the many food-web patterns (see Lawton 1989), five groups have obvious implications for persistence: The number of trophic levels, the patterns of feeding on different trophic levels, compartments, predator-prey ratios, and various subtle features that describe the feeding relationships between sets of predators and their prey.

FOOD-CHAIN LENGTHS

Most food chains are three or four trophic levels long; the percentage of food chains with five or six trophic levels is small; and food chains with only two trophic levels often represent incomplete studies (Pimm 1982, 1988a; Schoenly et al. 1990). While some food chains do appear to be unusually long, we are faced with surprisingly little variation across marine, freshwater, and terrestrial systems (Cohen et al. 1990). Schoenly et al.'s (1991) compilation of a large number of arthropod-based food webs shows that their food chain lengths are remarkably similar to those in which vertebrates dominate at the higher trophic levels. What little variation in food-chain lengths exists is interesting because it suggests ways to resolve the various conflicting hypotheses.

Implications for Persistence

I have discussed in other publications the various conflicting hypotheses for food-chain lengths (Pimm 1982, 1988a). The material developed in this book allows me to bring various arguments together in ways that space has prevented before. Ultimately, whether a given trophic level can persist in a community depends on the survival or extinction of a top-predator. Species become extinct or fail to invade when their numbers are rare, when the population density is likely to recover very slowly from disasters, and when the population density fluctuates considerably from year to year (chapter 7). Populations differ greatly both in how much they fluctuate (chapters 3–6) and in how resilient they are; population resilience depends on, among other things, the length of the food chain in which the species is embedded (chapter 2). As we increase food-chain lengths, resilience decreases, and, for vertebrates at least, an increase in resilience means an increase in population variability. If one wishes to know why food webs are so long or so short, one must go further and ask which factors determine abundance, variability, and resilience.

Species' numbers will be limited by the space available. Populations will be smaller on small islands, and we should expect food chains to be short on small islands; they are (Schoener 1989).

Species at progressively higher trophic levels get less energy and so will be rarer. This explains the well-known pyramid of numbers. By extension, we obtain an *energy-flow* hypothesis about food-chain lengths that makes two related predictions. First, if we increase the energy entering the food chain (that is, the primary productivity), then we should see an increase in food-chain length. This prediction is not supported. Across ecosystems, there is considerable variation in primary productivity, yet, despite this variation and the variation in the proportion of energy transferred between trophic levels, the number of trophic levels is relatively constant. Moreover, attempts to correlate the number of trophic levels with energy flow either fail (Pimm 1982, 1988a) or, as in Ryther's (1969) study of marine food chains, find that food chains are shorter in the most productive habitats. The second prediction is that, if we increase the proportion of energy transferred between trophic levels, we should also increase food-chain lengths. This prediction also fails. Food chains composed of insect species are *not* longer than those composed of vertebrate species, despite the fact that insects transfer energy to the next trophic level much more efficiently than do vertebrates (Schoenly et al. 1991).

Alternatives to the energy-flow hypothesis stem from the interlinked consequences of resilience and variability. Food chains are short, goes the argument, because the variable nature of the environment causes top-predators to spend much of the time recovering slowly from population crashes. The longer the food chain is, the longer the time spent recovering; eventually, recovery time exceeds the interval of time between major population crashes. Food chains should be shorter in unpredictable environments, and there is evidence to support this. Detailed studies of communities specifically designed to detect variation in food-chain lengths often find a small amount of variation (from two to four trophic levels). This variation, however, cannot be attributed to patterns of energy flow (Beaver 1985; Kitching and Pimm 1986; Pimm and Kitching 1987). What little variation is in food-chain length in mainland communities appears to correlate best with the frequency and severity of environmental disturbances.

Food-chain lengths should be longer in communities composed of species with resilient populations and should be shorter in communities composed of species with variable populations. As we have seen in chapter 2 and 3, resilience and variability are closely related; over all animal species, the most resilient species are the most variable species. Food webs composed mainly of insect species do not have more trophic levels than do those composed of vertebrate species (Schoenly et al. 1990), I suspect because, although insects have resilient populations, they are extremely variable in density.

Why is energy flow such a poor predictor of food-chain lengths? It is not that the argument is conceptually flawed. Recall the large ranges of population variability (figure 3.1) and the many factors determining variability. Consider also the large range of island sizes harboring terrestrial organisms. It should come as no surprise that food-chain lengths can be easily observed to vary with these variables. In contrast, the range of energy flows through ecosystems is much smaller (Pimm 1988b), and its effect should be much harder to detect. In sum, the evidence suggests that the factors which render species rare, variable in their densities, and slow to recover from disasters either prevent the establishment of species at higher trophic levels or increase the chance of extinction of the existing species at these higher levels.

FEEDING ON MORE THAN ONE TROPHIC LEVEL

For the purposes of food-web studies, John Lawton and I defined *omnivory* to mean feeding on more than one trophic level (Pimm and Lawton 1978). In a food chain with four trophic levels, there could be three omnivorous interactions—trophic level one to trophic level three, one to four, and two to four. Yet, in food webs dominated by vertebrates, there is an average of about one omnivorous interaction per food chain (Pimm 1982), and this is statistically rarer than one would expect on the basis of a computer-generated random set of food webs. I have always argued that there are many exceptions to this rarity of omnivory. In the most detailed empirical study of omnivory, Sprules and Bowerman (1988) found omnivory commonly in the zooplankton communities in glacial lakes. Omnivory is also common in food webs of insects and their parasitoids (Pimm 1980a). Fish that in their larval stage feed low in the food chain but end their lives as top-predators are also well known (Hutchinson 1959; Pimm and Rice 1987). (I must confess my bias toward terrestrial vertebrates; fish, insects, and zooplankton each outnumber terrestrial vertebrates in diversity and biomass. Clearly, it is the terrestrial vertebrates that should be considered the "exceptions.")

The relative rarity of omnivory in some communities must be affected by the difficulties of feeding on different trophic levels; exploiting both seeds and insects efficiently may be difficult, for example. Even so, there are advantages to feeding on more than one trophic level. We humans like to have fish with our chips, and many animals follow our example. The difficulties of feeding on more than one level do not seem to adequately explain the patterns of omnivory. One might predict that feeding on plants and animals would be more difficult than feeding on animals at different trophic levels. In fact, omnivores feeding on plants and animals are at least as common as omnivores that feed on animals at different trophic levels (Pimm and Lawton 1978).

Implications for Persistence

Omnivory can arise in two ways. In the first way, an omnivorous species enters at the top of a food chain but also feeds lower in the chain. Food chains are short, in part, because of the difficulties of adding to the top of an existing food chain (see previous section), and those difficulties apply to this strategy too. The second way in which omnivory can arise is when a species tries to join by inserting itself between tropic levels, causing a species already in the community to become an omnivore. This strategy makes the invading species both prey and competitor of an existing species; it ought to be a difficult strategy to adopt and likely to occur only in special circumstances. Consider the simplest pattern of omnivory: species 3 eats 2, 2 eats 1, and 3 also eats 1 (figure 10.1a). Clearly, species 2 is going to have a hard time of it—species 2 competes with species 3 for species 1, and species 2 also suffers predation from species 3. If species 2 is to survive, then species 3 cannot be simultaneously a good predator and a good competitor. We might expect that only special conditions will permit species 2 to invade a community containing species 1 and 3.

When omnivory is not rare, the situations, as we would predict, involve situations in which the species' attributes *are* special in various ways. Many

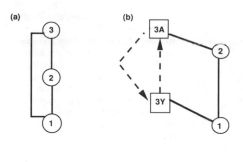

Figure 10.1. (a), Simple food web to demonstrate omnivory. Circles indicate species, and lines indicate feeding interactions between species with prey lower in the figure than their predators. Thus species 3 feeds on both species 2 and 1, and species 2 also feeds on species 1. Species 2, therefore, suffers simultaneous competition and predation from species 3. (b), Life-history omnivory. Young of species 3 (3Y) feed on species 1, and adults of species 3 (3A) feed on species 2. The density of 3Y affects the density of the adults through maturation, and, of course, the density of the adults affects the density of the young through recruitment.

aquatic communities are dominated by fish species which as they mature feed on several different trophic levels. Rice and I call such species *life-history omnivores* (Pimm and Rice 1987). There is an obvious reason why life-history omnivory is relatively common. Figure 10.1a shows a simple model of omnivory which would often lead to species 2's demise. In figure 10.1b, however, we recognize that species 3 has two stages—adults feed on species 2, and young feed on species 1. Species 2 can no longer be eliminated so readily from this system, because it is now the sole food supply of the adults of species 3. Species 3 cannot prevent species 2 from invading; indeed, species 3 *requires* species 2 before it can invade.

Food webs with parasitoids and their hosts contain many omnivores (Pimm 1980a). Unlike many true parasites, parasitoids inevitably kill their hosts, but like parasites they are smaller than the hosts on or in which they live. The per-capita effect of a parasitoid on its hosts, then, is likely to be much smaller than the effect of, say, a large vertebrate predator on its small invertebrate prey. These smaller per-capita effects of parasitoids on their hosts—and the comparably small effects of secondary parasitoids on the parasitoids they feed on—led Lawton and me to predict that food chains composed of parasitoid species would have greater amounts of omnivory (Pimm and Lawton 1978). It should be easier for a species to invade a community, as both the competitor and prey of another species, when the effects of the predation are relatively small.

COMPARTMENTS

Food webs are compartmented when interactions between species are grouped in such a way that they are either numerous or unusually strong within the

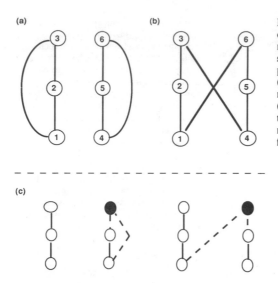

Figure 10.2. Food webs and food chains. One food web is compartmented (a), and the other has the same connectance but is not compartmented (b). The food chains (c) show that a species can more readily add on between chains (right) than within chains (left); in the latter case, the species gains resources from two independent food chains.

compartment yet either few or unusually weak between compartments (Pimm and Lawton 1980). Figure 10.2a shows a compartmented food web; the food web in figure 10.2b has the same number of species and the same number of interspecific interactions, but it is not compartmented.

Each habitat requires a set of adaptations from its component species. These specialized adaptations may preclude extensive interactions between species in different habitats and encourage more interactions within each habitat. This process of adaptation will create compartments, and, when created, their boundaries will correspond to the boundaries of the habitat. We might expect that interactions between animals and plants will be fewer across the boundary dividing a pond and a field than they are within the pond and field. For similar reasons, feeding on plant detritus may preclude feeding on live plant material, and vice versa—hence Odum's (1963) suggestion that species in the grazing-based food chain should be a compartment different than that of species in the detritus-based food chain.

All this discussion suggests that communities may have compartmented food web across habitat boundaries. Should there be compartments within habitats? At least one theoretician May (1972) has argued that there might be; I have argued the opposite (Pimm 1979b). By developing a statistic to measure the degree to which food webs are compartmented, Lawton and I analyzed a dozen natural food webs, covering terrestrial, marine, and freshwater communities (Pimm and Lawton 1980). Within these communities we did not find evidence for compartments; the food webs did not differ systematically from a computer-generated set of model food webs. We *did* find evidence for compartments across major habitat boundaries such as those between prairies and forests and between marine and terrestrial ecosystems. We also found several studies which supported Odum's suggestion of a separation between grazing- and detritus-based food chains.

Implications for Persistence

There is probably a simple reason why compartments exist only where they are imposed by habitat boundaries. Unless such behavior is made difficult for the reasons already discussed, there is simply no reason not to feed across compartments. Moreover, there are good reasons to believe that the strength of interspecific interactions can be greater across than within compartments. Suppose that an invading species A needs to feed on two trophic levels in order to succeed: plant and animal food are required. Figure 10.2c shows two options available to species A entering a system that is compartmented. Only if A's interactions are in the same food chain will the food web remain compartmented, so should those interactions be in the same chain? In the option where A feeds in the same food chain, it will be competing with one of its prey species B for its other source of food C. In competing with species B species A reduces species B's density and so the amount of food it can get from B. Put simply, there is a limit to how much energy can be extracted from just one food chain. By and large, feeding across different food chains in the same habitat ought to be an easier option available to an invading species, because energy can be extracted from both compartments.

PREDATOR-PREY RATIOS

How is the number of species at one trophic level determined? This is a familiar question for community ecologists. When we look beyond a single trophic level, an obvious question is how the species richness at one trophic level compares with the species richness at the adjacent tropic level—that is, their prey or their predators. The *predator-prey ratio* is the ratio of the number of species of predators to the number of species of prey. It appears to be relatively constant and independent of the total number of species in the community. Cohen (1977, 1978) found that the predator-prey ratio was 4:3. I criticized this value on the grounds that the data available to Cohen were severely prejudiced against plants and invertebrates (Pimm 1982). The top-predators of many food webs are described to species level by the ecologists who reported them, but the lowest trophic level is often described as "plants," "detritus," or "phytoplankton." Jeffries and Lawton (1985) addressed the criticisms with a new data

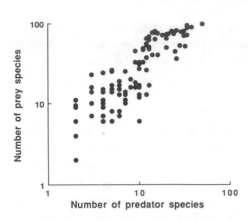

Figure 10.3. Number of predatory
and prey species are closely correlated in samples taken from communities of freshwater invertebrates. (Redrawn from Jeffries
and Lawton 1985.)

set of almost one hundred studies of freshwater habitats assembled with special
care taken to address the taxonomic difficulties (figure 10.3). They found a
predator-prey ratio that varied from 1:2 in communities with few predators and
prey species to 1:3.5 in communities with many predators and prey species.
Between communities in stream, river, and lake, they found no differences in
the ratio, though the ratio is likely to vary across a wide range of communities.
It is not the value of the ratio itself that is perhaps so interesting—but its
constancy within any broadly defined habitat type.

Implications for Persistence

How prey diversity controls predator diversity. In one way, the constant of
predator-prey ratio is not surprising, and it indicates an obvious result about
persistence. The more species of prey there are, the more predatory species
there can be, if those predators are not going to overlap extensively in their use
of prey species. The more similar the predators are in their diets, the more
likely they are to compete, and the more difficult it will be for other predators
to invade (preceding chapter). However, the constant predator-prey ratio probably means more than that the number of prey species controls the number of
predatory species.

How predator diversity controls prey diversity. This is a complicated topic.
First, there is a simple population-level explanation for why increased predator
diversity may make it *harder* for prey species to invade. Clearly, the more
predatory species there are, the more likely one of them will be able to prevent
the successful invasion of a particular prey species. Next, consider the complications imposed by food-web structure. It becomes less obvious what the effect
of a large number of predators will be, because there are several mechanisms
that suggest that communities rich in predatory species might be relatively
easier to invade.

Looking at the food webs in figure 10.4, consider the effect of a predator that is not shared. The food web with just the herbivore and the plant species (left) might not readily be invaded by another herbivore. The food web with an additional predator (right) will be easier to invade: the predator reduces the competition for the plant species by keeping down the numbers of the herbivore already present. In short, adding a predator makes it more likely that an herbivore can enter this community. It is possible that, the more predators there are, the more they are likely to suppress their prey that are the potential competitors.

Next, consider what happens when the prey species do not compete for resources but *do* share predators. How does apparent competition (chapter 9) affect the relationship between predator diversity, and how easy is it for a species to invade? Certainly, the presence of a shared predator can make it hard for a species to invade, and, the more polyphagous the predator, the more any increases in its diverse prey may cause difficulties for the invading species. It is quite possible, however, that, the more species of predators there are, the more likely they are to be trophically specialized and so perhaps prove less of a threat to invading species. Another complication is that, the more the species of predators overlap in their diets, the more they will likely compete, the less abundant they will be, and the easier it may be for an invading species to succeed. The likelihood, then, is that a community with few predators will be difficult to invade. In contrast, a community with many monophagous predators may be easy for the prey to invade.

Mithen and Lawton (1986) show that these simple results can be extended to much more complex models. They use food-web-assembly models (discussed at length in chapter 16) and constrained the food webs to contain only two trophic levels. Both predators and prey were allowed to invade or become extinct from the communities, and the assembly process stopped when fifty attempted species invasions had failed to add a species to the community. The original communities converge to communities that have an approximately constant predator-prey ratio (figure 10.5). Certainly, the more prey species that

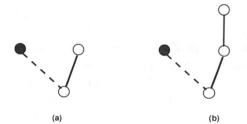

(a)   (b)

Figure 10.4. Herbivore invasion. An herbivore may more readily invade a community when a predator (on the existing herbivore) is present (b) than when it is absent (a). The predator reduces the competition from the herbivore.

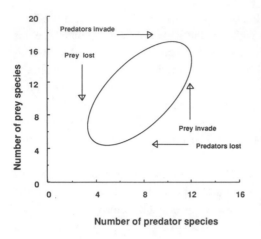

Figure 10.5. Synopsis of results of simulations by Mithen and Lawton (1986) on the numbers of predatory and prey species. The oval suggests the final values of numbers of species of prey and predators after the process of community assembly. Lines indicate how these points were obtained from the initial values of the numbers of predatory and prey species selected by the program. When the number of predatory species is relatively high, some communities gain species of prey, and others lose species of predators. Other communities gain species of predators when there are few to begin with.

there are, the more predatory species might be able to invade, and, if there are too many predators for the number of prey species, then predators will be lost. From our discussions of competition, we would expect these effects. These effects alone ought to produce a roughly triangular scatter of points, with the number of predatory species varying up to some constant times the number of prey species. Both the model data (figure 10.5) and the actual data on predator-prey ratios (figure 10.3) are more tightly correlated than this. When Mithen and Lawton compared the communities at the start of the assembly with the final communities, they found that communities with a relatively larger number of predatory species were more likely to gain prey species and less likely to lose the ones they already had. Thus, high numbers of predatory species make high numbers of prey species (and high numbers of both species) more likely.

In sum, a community that has both few prey species and many predatory species might lose predatory species, but it is equally possible that prey species may be more likely to invade. Only communities with the "right" predator-prey ratios are likely to persist.

NICHE GEOMETRY

Cohen (1978) and Sugihara (1984) have developed some simple concepts which lead to powerful ways of summarizing the interactions between predators and their prey. What the patterns of these interactions mean for persistence I shall explain after I have introduced them. The techniques for their work comes from the field of topology, and the ideas and terminology are exotic. While some may find the ideas immediately transparent, they only became clear to me when I built with sticks and modeling clay the structures I am about to describe.

Rigid Circuits

Technically, a food web is a graph—that is, something describing relationships between objects. In a food web, the objects are species, and, in my definition (Pimm 1982), the relationship is the interaction (feeding, direct competition, or direct mutalism) between the particular pair of species. A food web is not the only graph that describes communities; other graphs are produced by using different relationships. One of these is the *predator overlap graph:* the objects are the predators, and the relationship involves the question of whether they share prey species—that is, overlap in their diet. To produce these graphs, we draw lines connecting each pair of predators that share one or more species of prey. Because these graphs show dietary overlap they indicate the potential for indirect (that is, exploitive) competition.

In nature, predator-overlap graphs show surprising regularities (figure 10.6a). Consider an approximate physical analogy. Make a physical model of the predator-overlap graph with spheres to represent the predators and with rods to connect those predators that share prey. The physical model of figure 10.6a would be rigid, not flexible, because when there are connections around four or more predators these connections are triangulated. Note that this is not the case for the predator-overlap graph in figure 10.6c. Also notice that species 8 in figure 10.6a does not violate this condition, because it is not part of a circuit around four or more species. The two unconnected parts of the graph (species 1–3 and 4–8) do not violate the condition, for the same reason. If the connections from species 4 to 7 and 5 to 6 were missing, then the graph would not be triangulated. Technically, this property of being triangulated is called a *rigid circuit.* Compared with sets of computer-generated model food webs (Sugihara 1984), the predator-overlap graphs of real food webs contain an overwhelming preponderance of rigid circuits.

Intervality

There is another pattern in communities that is found by considering the overlap in the predators' use of their prey species. A food web is defined to be *interval* if the overlaps in the predators' use of prey species can be expressed as possibly overlapping segments of a line (Cohen 1978). Figure 10.7a shows a simple interval food web and the line segments that describe the patterns of

(a)

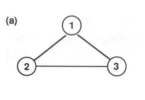
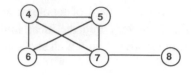

(b)

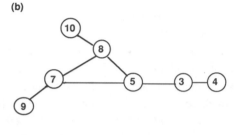

(c)

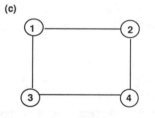

Figure 10.6. Predator-overlap graphs connecting predators that share one or more prey species. Two circuits (a and b) are rigid, and one (c) is not. Only one predator-overlap graph (a) has an interval representation, whereas one (b) is asteroidal (see text for definitions). In the graph from Bird's (1930) study of a Canadian willow forest (a), the numbers are: 1, a fungus; 2, insects; 3, another group of insects; 4, three species of birds; 5, another three species of bird; 6, spiders; 7, a frog; 8, garter snake. In the graph from Kohn's (1959) study of predatory gastropods of the genus *Conus* on the subtidal reefs of Hawaii (b), all are species of the genus *Conus* and the numbers are: 3, *ebraeus;* 4, *chaldeus;* 5, *miles;* 6, *rattus;* 7, *distans;* 8, *vexillum;* 9, *vitulinus;* 10, *imperialis.* One of the patterns (c) is so rare in nature that I have no example; this graph is hypothetical.

overlap. Figure 10.7b shows an only slightly more complex food web. The overlaps cannot be expressed along a line, though they can be expressed in two dimensions. The second food web is not interval.

Cohen (1978) found that food webs are interval more often than chance alone dictates. Whether a model food web should be interval depends a good deal on the number of predators in it. Nonintervality is only possible with four or more predators—if we take any predator away from the noninterval food web of figure 10.7b, it becomes interval. Even with four predators, intervality is hard to achieve. The noninterval pattern of figure 10.7b becomes interval if more overlaps between the predators are added or if they are placed differently.

Using Cohen's recipe for generating model food webs, I plotted, versus the number of predators in the web, the probability that a model food web would be interval (Pimm 1982). The probability shows a sharp transition. With fewer then twelve predators, the real food webs were all interval, and the model food webs were almost always interval. With more than twelve predators, most of the model food webs and most of the real food webs were noninterval. Among food webs with more than twelve predators, there were a few interval food webs, and it was these that gave Cohen his result of the predominance of interval food webs.

For many arrangements of predator-overlap graphs, the rigid-circuit property ensures that the overlaps will have an interval representation. Yet, the rigid-circuit property does not *guarantee* that food webs are interval. It is possible to draw a noninterval, rigid-circuit predator-overlap graph; figure 10.6b is an example. The graph is a rigid-circuit because there are no circuits around four or more points, yet it is still not possible to express the overlaps as overlapping segments of a line. For obvious reasons, this pattern is called *aster-*

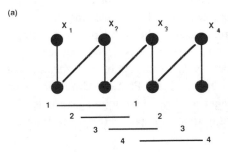

Figure 10.7. Interval food web (a) and noninterval food web (b). The lines below the food webs show the patterns of overlap among the four predatory species. (From Pimm 1982.)

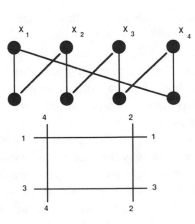

oidal. The predator overlap graph is not asteroidal and is rigid-circuit, it will be interval (Sugihara 1984). Indeed, when real food webs are noninterval, it is because they are asteroidal.

Holes

Next, let us create an entirely new kind of graph, one that combines information on both predator(s) and prey. The graph is formed by connecting prey species sharing a particular predator. Two prey sharing a particular predator would form a line, three a triangular plane, four a tetrahedron, and so on. The graphs formed in this way are the *prey-overlap graphs*.

Some prey species will be used by more than one predator, and this allows us to connect the individual graphs. Figure 10.8a shows a combination of the graphs for three predators and how they exploit six species of prey. Figure 10.8b is the food web from which I produce the prey-overlap graph. In general, I have used three prey species in the diet of each predator because planes are easier to draw than a multisided solid and because the examples therefore are easier to visualize. Again, consider a physical analogy of the prey-overlap graph. The planes or the multisided objects formed by connecting the various prey species that share a particular predator are solid in the physical analogy. Continuing the physical analogy, observe that there is a *hole* between prey species 2, 3, and 5: no predator species feeds on this set of prey species. For another example, consider the web in figure 10.8c. Predator C now feeds on prey species 3, and so there is no longer a hole. These topological holes are rare in real food webs containing small numbers of species (Sugihara 1984).

Implications for Persistence: Sugihara's Explanation

Sugihara (1984) has derived a simple assembly rule that explains both the ubiquity of rigid-circuit predator-overlap graphs and the absence of holes in the assembled prey-overlap graphs. The assembly rule, by its very nature, restricts the ways in which species can join to a community and so reduces the chance of the community being invaded. Sugihara's rule is simple: predators cannot join a community by overlapping, in their diets, with two or more predators that do not overlap in *their* diets.

Figure 10.9 provides examples of ways in which a predator may and may not be added to a community. Before I ask why this rule works, consider the implications of the bottom left and bottom right predator-overlap graphs. One sequence of species invasions is possible (1 and 3, then 4, and, finally 2); the other sequence is prohibited (1 and 2, then 3; but we cannot then add 4). Community structure obviously depends on the sequence in which the species

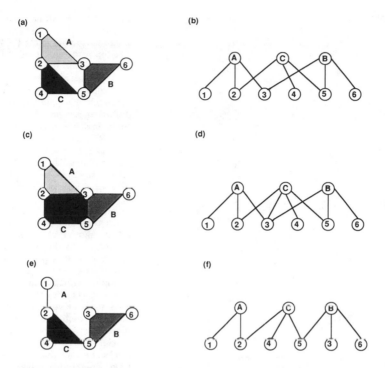

Figure 10.8. Assembled prey-overlap graphs (a, c, and e), formed by connecting prey species that share a predator, and food webs (b, d, and f) from which these graphs are derived. In one graph (a), prey species 1–3 are connected because predator A feeds on all three species. Species 1 and 4 are not connected because they have no predator in common. Prey connected in this way form "solids." (These are planes in all cases except the tetrahedron connecting 2–5 in one graph (c) and the line connecting 1 and 2 in another (e). This recipe of connecting species forms a hole between 2, 3, and 5 in graph (a). The significance of the different figures is discussed in the text. (From Pimm 1988b.)

are added to the community: 1, 2, 3, 4 or 1, 2, and 3. I shall develop this point at length in the next chapter.

Why might Sugihara's rule work? He provides one explanation, saying that it is the "conventional wisdom" for species to enter communities in order of decreasing polyphagy. What this means is that the first species to enter a community feeds on more kinds of prey than do species entering later. If an incoming species of predator overlaps with two nonoverlapping groups of predators, it will likely feed on more species of prey than does any one of the predators with which it overlaps. Sugihara's "conventional wisdom" is that this does not happen. What he suggests is happening is the alternative. If the incoming species is relatively less polyphagous, it is likely to overlap only with predators that do not overlap in their diets. This alternative of decreasing polyphagy creates the patterns in the graphs we have been discussing.

Permitted   Forbidden

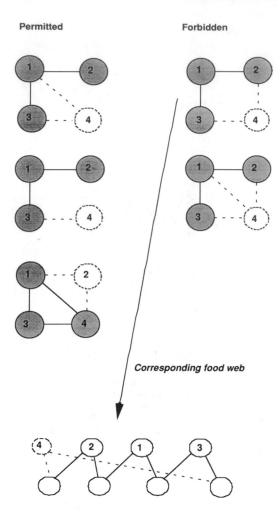

Corresponding food web

Figure 10.9. Rules for adding species to communities (after arguments in Sugihara 1984). The top figures are predator-overlap graphs (see figure 10.6); circles are predators, lines connect predators that overlap in diet, and circles and dashed lines represent, respectively, invading predators and their interactions. Predator-overlap graphs will be rigid circuit if invading predators attach only to adjacent predators—that is, if predators connected directly. The possibilities at the right are forbidden under this rule; those at the left are allowed. As a consequence, the food web at the bottom cannot be produced, and it has a noninterval predator-overlap graph (top right). Also notice that the predator-overlap graphs at the bottom of each column are the same: one cannot be formed (at right), the other (at left) can be, by using Sugihara's rules. This suggests that the sequence in which the species invade may determine the end point. Species 2 can follow 1, 3 and 4, but species 4 cannot follow 2 and 3. The bottom figure is a food web and corresponds to the predator-overlap graph at the top right. Why should the invasion suggested be forbidden, when the invading species takes those prey species on which there is least predation?

It is not at all clear to me that real communities work this way or that Sugihara's rule does represent "conventional wisdom." If this progressive trophic specialization is the mechanism, then *why* do species enter in such a strict sequence? Certainly, if this is the mechanism, it proscribes a highly restricted sequence for species invasions, with obvious consequences for persistence. Right or wrong, Sugihara's rule is an interesting conjecture.

Another way of viewing the assembly is to consider an example that, at face value, would favor a species breaking Sugihara's assembly rule. In the food web at the bottom of figure 10.9, three predators feed on a series of prey species that are strung along some resource axis. (Perhaps this axis is the size of the prey species or is some measurement of where they feed). To which prey species should the next species of predator attach? It should attach itself to the one offering the least competition, of course. *If the prey abundances are approximately equal,* then the best place for the new predator to feed is at both ends of the resource axis; this is where the prey are being exploited by only one predator rather than two. When this happens, the resulting food web is noninterval, the predator-overlap graph is not rigid circuit, and the assembly prey-overlap graph has a hole. We know this pattern of addition appears rarely in the real world. What, then, is wrong with our argument?

Perhaps it is hard to feed at both ends of the resource axis. Over long resource axes, it is clear why species do not choose the ends. What could feed only in the ocean depths and on the high Himalayas or only on tiny herbivorous zooplankton and on cows? Unfortunately for this explanation, the data sets in which the patterns of niche geometry have been found do not involve predatory species spanning such long resource axes. Furthermore, many species are polyphagous and might be able to exploit all the prey resources, yet they still choose only the resources at the ends of the resource axes, because that is where they encounter the least competition. If all this is true, then why is the pattern of invasion suggested by figure 10.9 difficult? Why is it hard to feed across the ends, when the ends are not, apparently, special? The ends *are* special, as I shall now explain.

Dietary Opportunism as an Alternative, And Its Implications for Persistence

In the literature, there are abundant tabulations of the use of prey species, habitats, or other resources by a set of predators; Schoener (1974) provides a compilation of such studies. Ecologists are practiced at interpreting them—we all know that species tend to take different resources in different proportions. Nonetheless, these differences should not obscure the obvious feature of these tabulations—namely, that the different species in these studies share not only habitats and taxonomic affinity but many prey as well. Typically, there is one or more prey species common to the diets of all the predators in these tabulations. If all of a group of predators take one or more of the same species of prey, the niche geometry might appear, superficially, to be rather dull. Any prey common to the diets of a set of predators means that each predator overlaps with every other predator. When we obtain the predator-overlap graph from this arrangement, it is rigid-circuit. The food web would also be trivially interval because the niches are represented by a point rather than by an axis. Similarly, if all the predators exploit the same set of prey, the prey-overlap graph of this arrangement would not contain any holes.

The most familiar view of niche geometry is represented in caricature in figure 10.10a. It suggests that species are strung along a resource axis that might represent some environmental gradient, prey size, or more abstract ranking of the prey's characteristics. There might be more than one dimension, but the overall idea is the same. Such a view is the basis of many of the models of species packing, resource partitioning, limiting similarity, and many other ideas. I suggest that a better caricature of niche geometry—the *flower-petal model*—is provided by figure 10.10b. Here the niches are drawn, in plan, to show both the extensive predator overlap formed by species exploiting some

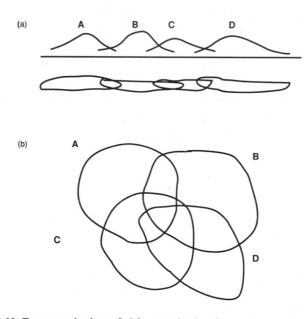

Figure 10.10. Two synoptic views of niche organization. In one view (a), the niches are strung out along a dimension, which might represent an ordering (by size, for example) of the prey species. The height of the graph represents the percentage utilization of the resources by each species. Below is shown the corresponding Venn diagram to show how the niches overlap. In an alternative view (b) of niche organization, the niches are drawn in plan—that is, as a Venn diagram—to show how predators overlap in their use of resources. The species share a core of resources, all taking abundant prey, at the center of the cluster. But each species selects some prey uniquely. The prey common to each combination of two predators are distinct sets in one view (a), but, in the other view (b), there are some (probably many) prey that are utilized by every predator.

common prey species and the resource partitioning which tends to give each species a slightly different, though idiosyncratic, set of resources (Pimm 1988b).

This flower-petal view of niche geometry is identical to that suggested by Rosenzweig and Abramsky (1986) as the way habitat choices are organized. Working on desert rodents in Israel, they suggested that community organization might be best described as "centrifugal," and they argued that the primary preference of all the rodent species is shared but that secondary preferences are distinct.

Why should the niches of the predators be arranged like flower petals? In the diet of predators, some prey species are much more abundant than others. (These prey may be actually more abundant, more available, relatively more nutritious, etc. I shall talk about "abundance" for simplicity.) These abundant prey species will be the ones selected first by the first predators to invade the community. Unless the effect of the predator already in the community is so very great that it depresses the common prey species below the abundance of the rare prey species in a community, predators will not create "doughnut"-shaped holes in prey species abundances. Only if they did so would they force the later predators to enter the community by taking a selection of rare prey species. Predators may enter a community by taking rare species of prey, but they also take the common ones. Predators thus give the impression of entering the community and overlapping with other predators where there are *more* species competing for the resources, rather than where there are *few* species sharing those resources. In short, it is the range of prey abundances that imposes a ranking on the prey species so that the predators appear to avoid the ends of resources axes. This gives the topological regularities.

This process I have just outlined does not proscribe the strict, ordered sequence suggested by Sugihara. A given species could enter a community early or late in the assembly, so long as it takes the common resources. The process certainly suggests that, the more predator species there are, the harder it will be for another one to invade; but what of the details? What the process suggests are questions about whether a species can invade, questions which differ from those usually asked. Is it the availability of rare prey species that enables a predator to invade and so also exploit some of the common prey species? Or is the opposite true—does the availability of some left-over common prey enable a species to enter the community and, incidentally, take some unexploited rare prey species? I do not know the answer to these questions, because, to my knowledge, the problem of packing species into a community has not been framed in quite this way before.

COHEN'S CASCADE MODEL

Joel Cohen and various colleagues have produced a large body of work on food webs that is most conveniently summarized by a recent book (Cohen et al. 1990). I suspect that one motivation behind this work was the large number of food-web patterns discovered between 1975 and 1985. Just how many patterns are there, and which patterns are simple consequences of the existence of other patterns? Cohen and his colleagues created several simple recipes to produce food webs and then examined the ingredients needed to produce model food webs matching the now large collection of empirical food-web studies. Several extremely simple models failed, but one—the *cascade model*—makes some surprisingly good predictions.

The cascade model requires two parameters: the observed number of species in the web (S) and the expected number of interspecific interactions in the web. It requires one assumption: that we can rank the species 1, 2, 3, . . . *i, j*, . . . S, so that a species *j* with a higher ranking than a species *i* can be the predator of species *i*, but never vice versa. If you will, there is a *cascade* of predatory effects, with this simple rule precluding loops where, for example, *j* eats *i* and *i* eats *j*, or *i* eats *j*, *j* eats *k*, and *k* eats *i*. Applied to the data base of a hundred or more published food webs, the cascade model predicts the number of species at the base of the food web, the number of species that are prey to some species and predators to others, the number of species that are only predators, and hence the predator-prey ratio in the food web, the average food-chain lengths, and the various patterns of rigid-circuit graphs and intervality. This is an impressive list of predictions for such a simple model, but there are two points I must make about what it does *not* do.

First, while the model predicts the average lengths of food chains (and, indeed, the variance of food-chain length within a food web), it consistently predicts both longer and shorter food chains than are actually observed. Unusually short food chains occur when species high in the food chain also feed low in the food chain—that is, when there is extensive omnivory. Consequently, it is a matter of debate whether the cascade model adequately explains the patterns of food-chain length and omnivory or whether additional explanations are necessary.

The second point is more important and is made explicitly by Cohen et al. Even if the cascade model were to predict all the known food-web patterns, it still begs the question of what determines the number of species in the food web, the number of interactions in the food web, and why there should be a cascade rule. As we have seen, there are many limitations on both how species may invade a community and, if they do, how their interspecific interactions can be organized. I am often asked by critics whether the cascade model replaces arguments about the dynamical processes of invasion and extinction in food webs. It does not. Rather, it makes an understanding of those processes a more critical research need (Pimm et al. 1991).

SUMMARY

In nature, food webs show a number of patterns that are statistically unusual when compared with model food webs. Real communities have relatively short food chains, and some communities have a scarcity of species that feed on more than one trophic level (omnivores). Food webs are only compartmented where the compartment boundaries correspond to major habitat divisions. Across communities with many different numbers of species, there are relatively constant predator-prey ratios. The overlaps in the prey used by predators are generally interval, meaning that the predators' diets can be arranged as possibly overlapping segments along a line. The predator-overlap graphs derived from communities are usually rigid circuit, and the prey-overlap graphs derived from communities usually lack topological holes.

All these patterns may be related to persistence, for each one implies that the ways in which a species may enter a community are restricted. The limitation on food-chain lengths means that species cannot simply join a community by feeding on all the species below them in the food chain and so avoiding predation and competition. Nor can species easily insert between trophic levels, because such species would suffer simultaneous competition and predation. Unless habitat boundaries prevent such behavior, it behooves a species to feed across rather than within food chains. Within a habitat, compartmented food webs ought to be easily invaded. Predatory species can more readily invade communities that have more prey species, and, under some circumstances, prey species can more readily invade communities that have more predator species. The patterns in the predator and prey overlap graphs may be a consequence of a strict assembly rule that requires species to enter a community in order of increasing dietary specialization. Alternatively, the patterns may be a consequence of the tendency for species to share similar preferences for common prey species while retaining an ability to use different rarer prey species. This alternative forces the question of what circumstance allows a species to invade a community—is it the ability to exploit the common and heavily utilized prey species or the ability to exploit a special set of rare species?

While these patterns may indicate how difficult it is for a community to be invaded, what is to stop a potential invader from violating the food-web patterns and, in doing so, make the extinction of other species in the community more likely? The existence of pattern may indicate persistence, but we do not know for certain. One way to resolve the problem is to look at the experimental introductions of species to communities and ask whether patterned communities are harder to invade. Ecologists have yet to do this.

11

Community Assembly;
or, Why Are There So Many
Kinds of Communities?

ASSEMBLY AND PERSISTENCE

In this chapter I return to an earlier result from models of community assembly. After assembly models reach their long-term average number of species, each successful species invasion makes the community harder to invade (chapter 8, figure 8.5). In other words, the assembly process increases persistence. This result is not inevitable, and I will introduce both models and real communities in which persistence does not increase. When communities do become more persistent over time, there are still important aspects of persistence to be discussed. For example, with any given set of species, will there be only one or many alternative persistent communities? Thus, do many persistent communities persist side by side, or does the diversity of communities indicate that one is slowly replacing the rest? With the aim of providing the terminology to describe the features of more complex models and of real communities, I shall begin by developing some simple models of the assembly process.

MODELS OF COMMUNITY ASSEMBLY

Topological Possibilities

One predator and one prey. Consider a model with one species of prey and one species of predator. The prey species can invade without the predator, whereas the predator can invade only when the prey is present. This is an obvious result, and I introduce it to develop a method to present results about assembly models in general. In figure 11.1 I show the four community states that are the combinations of predator present or absent and prey present or absent, as well as arrows that indicate the directions in which the community can move. This figure is a graph, in the sense in which I defined a graph in chapter 10, and I call it the *community-transition graph*. The objects are the various states of the community, and the relationships between the objects are

Community has no
species in it

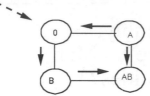

Figure 11.1. Community transition graph
for predator A and prey B. The four corners are the combinations of prey present
and absent and of predator present and
absent. The arrows indicate the possible
transitions from one state to another.
From a community with no species (0),
first B and then A will invade. A cannot
invade before B.

the transitions from one community state to another. I make the assumption
that the community cannot make two moves simultaneously. There are no di-
agonals that would represent the simultaneous invasions or extinctions of two
species.

Two competitors. Simple models of competition allow for both species to
coexist, or for one species to replace the other, or for the outcome to depend
on which species has the initial advantage. Graphs of all these possibilities are
shown in figure 11.2. There is a new phenomenon in these models and graphs:
alternative persistent states. The community may end up with one or the other
species depending on the transitions.

Two predators and one prey. Increasing the complexity of the model by one
species, consider two predators and one prey. Models of such a system suggest
several possibilities. Suppose we make the simplification that both predators
cannot coexist on the one species of prey. The two possibilities are (1) that one
predator always replaces the other or (2) that the outcome depends on which

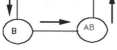

Figure 11.2. Community-transition graphs for models
of competition between two species A and B. Three
possibilities for these graphs are shown: A and B coexist (a), one species replaces the other (b) (A replaces B
in this example), and the outcome depends on which
species arrives first (c).

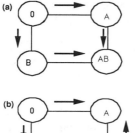

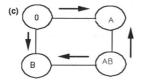

species of predator arrived in the community first. When the predators are present simultaneously, they are competing, so it is hardly surprising that the outcomes of this model are comparable to those of the two species competition described previously. The associated graphs are shown in figure 11.3; notice that with three species the transitions lie on a cube. The corners of that cube denote the presence and absence of each of the three species.

Three competitors. Next suppose we have three competitors. There are a variety of arrangements. For example, all three species may coexist, one species may exclude the other two, or two species may exclude the third. There is a plausible kind of behavior possible for such a model that I have not recognized before. There can be *cycles;* that is, there can be cyclical changes in

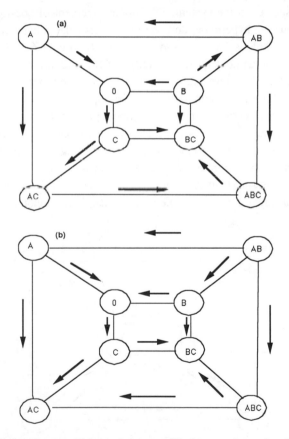

Figure 11.3. Community-transition graph for a model of two predators A and B and one prey C. The combinations of A, B, and C present and absent create a cube for the graph. The two possibilities shown are for one predator to always exclude the other (a) and for the outcome to depend on which predator arrived first (b).

composition, with species A replacing species B, which replaces species C, which replaces species A, and so on (figure 11.4). The obvious consequence of cycles is that the community composition does not persist. At some local scale, the community constantly changes in its composition.

Catalysts and humpty-dumpty effects. Another possibility for three species is that either one species or all three species may persist. In figure 11.5, I have created a model with four possible final states, either only A, or B, or C or all three species. In this particular arrangement, the final state with all three species is unreachable; from only A, B, or C present, it is not possible to reach any of the intermediate states with A and B present, A and C present, or B and C present. This graph represents the conventional folk wisdom that "there are some places to which you cannot get from here." Now let us suppose that there is a fourth species in the system, D. The arrangement of possibilities is now not a cube but is a hypercube, and one way to visualize a hypercube is to have two cubes one below the other and connected (figure 11.6). The top cube is for combinations without D; it is exactly the same as for figure 11.5, except that from each of the combinations there is a connection to the lower cube. The lower cube is the same combinations as the upper cube, except that now D is present. The arrows coming out of this cube represent the connections coming from the top cube. It is now possible to reach the three-species state of just A, B, and C. The pathways always involve D invading and then always being eliminated from the community. Indeed, in the way that I have drawn figure 11.6, if D can invade, then the state with A, B, and C will always be reached, and D will always be lost from the community. If D cannot invade, then the

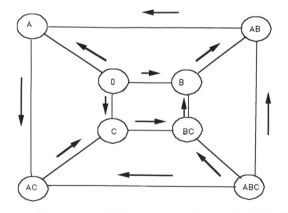

Figure 11.4. Community-transition graph for three competitors without a persistent community state. Species A can replace species B, which can replace species C, which can replace species A.

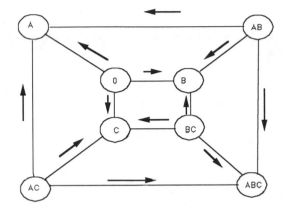

Figure 11.5. Community-transition graph illustrating the "you cannot reach there from here" principle. Starting from no species, it is possible to reach either A or B or C, but all the two-species states repel. The three-species state cannot be reached.

state with A, B, and C cannot be reached. This example contains what I have called a *humpty-dumpty effect* (Pimm 1986a). It is impossible to put the community of A, B, and C back together again from its pieces, that is, its constituent species. The absent species D is a catalyst, for it is essential to the community's construction (Drake 1985).

IMPLICATIONS: POSSIBLE TOPOLOGIES FOR COMMUNITY ASSEMBLY

Drawing out the possible transition graphs suggests a number of options for those transitions. The transitions may be strictly hierarchical (figures 11.1–11.3, 11.5, and 11.6), by which I mean that there are no cycles anywhere in the graphs. The other extreme is for there to be no persistent states at all. Figure 11.4, for example, has each community state with at least one arrow leading away from it. Such communities perpetually cycle in their composition. I shall present a field example of such a community later. An intermediate pattern is for there to be some cycles but for there always to be some ways of escaping them and reaching one or more of the persistent states. Suppose cycles are common and there are no persistent community states. If cycles are ubiquitous, then, at some spatial scale, communities do not persist, for the species constantly replace each other. I say "at some spatial scale" because obviously all the species persist on some large-enough scale—the planet, for example. Communities will not become increasingly persistent over time if

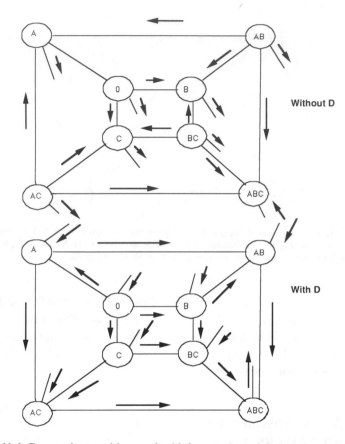

Figure 11.6. Community-transition graph with four species and that can be represented as two connected cubes, the top one without species D and the bottom one with species D. In this example, the community-transition graph without D is identical to that of figure 11.5. The addition of D is always possible, except from the community with A–C. Therefore, the final persistent state will have only A–C and not D, a species which is essential to create the final community. This community illustrates the humpty-dumpty effect: the community cannot be assembled from its pieces alone.

there are cycles. However, this constant cycling will lead to a diversity of communities and may be one explanation for why there are so many kinds of communities.

Mistakenly, we might conclude that there are alternative persistent states in nature, when what we are observing are the many transitional states through which the community moves. Thus Diamond's (1975) example of the checkerboard distribution of fruit doves (figure 9.6) may represent a snapshot of a process that involves one species replacing the other, then the other replacing the one, on each of many islands. For this example, I think this explanation is unlikely, and I shall elaborate on this presently. Roughgarden (1989) has suggested that there might be cyclical changes in species composition for species of *Anolis* lizards on small Caribbean islands.

On a real landscape we cannot have Escher's famous visual paradox of water flowing apparently continually downhill around the same track—the metaphorical equivalent of cycles. In community-transition graphs, if there are no cycles at all, then the various community states now can be arranged on abstract landscape. If there are no cycles, then this means that there is some abstraction, akin to height, that provides some overall measure of the community's ability to resist invasion. What the abstraction "akin to height" is, I do not know. I fully realize that attributing some overall measure to the community (comparable, say, to an organism's fitness) brings to mind arguments about communities behaving as superorganisms. I did not say communities are superorganisms—only that, *if cycles are absent,* then there might be an overall community statistic that measures the community's persistence. The assembly process pushes the community higher on the landscape—meaning the community is harder to invade, the longer the process of assembly has been running. Sometimes communities may reach the peaks quickly; at other times, the assembly process will be drawn out.

The possibilities of the last two sections prompt these questions about persistence. First, do communities persist, or do they cycle in their composition? Second, if there are no cycles, how many alternative persistent communities are there? In the landscape metaphor, why are there many peaks and not just one? Why do so many communities persist in nature, and why is there not one grand, global community? I shall tackle these problems by exploring theoretically why cycles may be rare and, if they are, what determines how rugged is the community landscape. Then, I shall look at multispecies models to see what they suggest about more complex patterns of community assembly. Finally, I shall review empirical studies of species assembly.

Why Might Cycles be Rare?

If there are no cycles, is this because they are statistically rare, in the sense that, given the various ways in which we could create the community-transition graphs, cyclical changes are not very likely? Or is it because there are ecological processes that make cycles rare? It is easy to show that community-transition graphs can have cycles; they are possible even in the simplest graph with two species (figure 11.1). There are two ways of picking each transition. The simplest graph has four transitions and therefore has $2^4 = 16$ possible configurations. Two of these are cyclical. With three species, there are twelve transitions, and so there are $2^{12} = 4,096$ combinations. There are some very simple cycles, around the faces of the cube, that are of the same form as those for two species. There are other, more complex cycles, around two or more faces of the cube. With more and more species, models with at least some cycles appear to be statistically inevitable, so why should they be rare ecologically?

Although theoretically possible, the very simple cycles do not seem ecologically plausible. The cycles are of the following form: A invades the empty community, B then invades A, A becomes extinct, and then B becomes extinct. When there are more species, this pattern becomes the following: A invades a community with several other species present, B then invades A, A becomes extinct, and then B becomes extinct. This does not seem very plausible either. In short, plausible community-transition graphs ought not to have cycles around the faces of their hypercubes.

What about the more complex cycles that involve three or more species— how likely are they? There are other reasons why more complex cycles should be rare. For example, a species might be able to invade a community by replacing a species that has the same species of prey. Species occupying a particular position in the food web would only be replaced by another species more efficient at exploiting their common prey. It becomes harder for other species to successfully invade, because they must be even more efficient. The characteristics that enable a species to invade are likely to be those that make it more difficult for another species to displace it. Post and Pimm (1983) provide examples of this mechanism. Cycles are rare, therefore, because the pattern of "A beats B, B beats C, and C beats A" is rare. When cycles are found, as in the examples I will discuss below, they occur because the species are not simply competing by exploiting a resource for which there is a simple ranking of competitive abilities.

The process of assembly must be considerably more complicated than this gradual improvement of efficiency. If species simply replace each other, then it ought to be possible to follow the changes in interaction parameters and to observe the improvements as "better" species replace "poorer" ones. Post and I tried to do this in our studies of community assembly (Post and Pimm 1983; also see chapter 8 above). Over the course of the assembly, we did not find trends in the parameters, no matter what parameters we examined. We found that, even after the number of species had become approximately fixed, an invading species did not generally occur in the same position in the food web as did a resident. For example, the addition of one species might cause the replacement of a monophagous predator, yet the next species added might be a polyphagus predator at a different trophic level. Progressive "improvements" in the species were impossible to follow, because successive invasions tended to be staged by trophically different kinds of species. Indeed, the fact that the communities became progressively harder to invade is not easy to understand, because the changes in food web, after the addition of a species, can occur in so many different ways.

Finally, cycles may not be rare in general—most community-transition graphs may contain them, even those in which we eliminate the obviously implausible cycles around the hypercube faces. What may be rare, however, are persistent cycles or persistent changes in composition such that the communities wander endlessly around a set of states. Thus a community may initially go through cyclical changes but eventually reach a sequence of transitions that lead it to a persistent community composition.

These various possibilities suggest a number of obvious questions about community-transition graphs. For graphs involving more than a very small number of species, how rare are graphs with not even one persistent state? We also need to know how common are graphs with no cycles at all, for these have an analogy in landscapes. The remaining class of graphs have both cycles and persistent states. We also need to know what proportion of these graphs will lead to cycles and what proportion may cycle but eventually lead to persistent states. There is a technical literature on comparable graphs in other disciplines (see, for example, Kauffman 1992), and some of these questions may have been answered already. All these questions need to be answered for ecologically sensible graphs which have their own special constraints and which may lead to somewhat different answers.

To summarize, we should not expect cycles for the very simplest models, and so there will be obvious persistent states ("peaks") in the community-transition graphs. Even in less simple models, one predator may replace another predator because it is more efficient at exploiting their prey. Consequently, there will be a single persistent state to which the community moves as progressively more efficient predators invade. For more complex models, however, it is not obvious why one invasion should make the next more difficult. It is far from clear how common are cycles in more complex models and, particularly, how common are models in which composition changes endlessly without reaching a single persistent community composition.

What Happens If Persistent Cycles Are Absent: The Metaphors of Adaptive Landscapes and Community Landscapes

Evolutionary biologists often metaphorically envision species as sitting on adaptive peaks. The invasion of new genotypes allows evolution to progressively alter the population's phenotypes so that species become better adapted to their environment, so pushing the population upward. Eventually species reach adaptive peaks that are phenotypes on which it is hard to improve. On peaks, the existing phenotype will persist, as no new genotypes will invade. There may be a single adaptive peak, or the adaptive landscape may be topologically very complex, with many peaks, and with selection pushing different populations onto different peaks. To understand the process of adaptation, we need to see how species fit their environment, but this is not sufficient. We must also know how that fit was derived, for the process itself will decide the particular peak on which a species lands. Similarly, although we may gain an important understanding of communities by looking at their structural patterns, this understanding will inevitably be incomplete if we do not know how the communities assembled.

Perhaps there is a parallel between these ideas on the invasion of new genotypes and the invasions of new species. Post and Pimm's models show that the process of community assembly makes communities more persistent because each successful invasion makes the community progressively harder to invade. This is true even after the connectance and number of species in the model communities have reached their long-term average values. In metaphorical terms, assembly pushes communities upward in the community landscape, much as evolution pushes populations up adaptive landscapes. When the community is moving upward, species are invading; but when it reaches the peak, the community is persistent, in just the same way that phenotypic change ceases when phenotypes reach adaptive peaks.

This metaphor of the community landscape immediately suggests that we ask, how many peaks are there? That is, what makes a landscape rugged or smooth? How many alternative, persistent community types are there, and how does this number vary with species number or connectance? Whatever factors make landscapes rugged will create a diversity of communities and so, inevitably, a diversity of species. If the community landscape has a complex surface with many peaks, which peak is achieved is going to depend on the sequence of species invasions.

The most obvious question, however, is whether the metaphor of the community landscape is sensible. What does height mean? In the adaptive-landscape metaphor, it is the organism's fitness. Do communities have a fitness? Obviously not, although their individual species do. In an adaptive landscape, if phenotype A is fitter than phenotype B and if phenotype B is fitter than C, then A is inevitably fitter than C. By contrast, in a community, it is at least possible for species A to replace B, B to replace C, and then C to replace A, even though this kind of cycling may be rare.

What Decides the Number of Peaks?

Kauffman (1992) has developed a comprehensive theory about the topology of adaptive landscapes. If community-transition graphs have no cycles, then their various states can be arranged on landscapes, and his theory makes some interesting predictions. Kauffman's simplest model supposes that there are a number of genes, that each comes in two types (or alleles), that each allele contributes to fitness, and that no gene modifies the fitness contribution of any other gene. For this model, the fitness of the organism is the sum of the individual fitness contributions of each gene present. Consequently, the increase in fitness when one allele is replaced by another does not depend on which other alleles are present. The adaptive landscape is a very simple one with one peak. Through evolution, the genotype moves progressively toward the combination of alleles that gives the organism its greatest fitness.

The models become more complicated when there are epistatic effects—that is, the presence of one allele modifies the relative fitness contributions of one or more other alleles. The relative fitness contribution of one allele relative to another now depends on not just the relative merits of the two alleles but on how each allele affects the fitness contributions of all the other alleles that it effects. As the effects of genes become increasingly interconnected, the landscape becomes increasingly rugged, with more and more adaptive peaks. The more rugged the landscape, the more likely evolution will reach a local peak and not the peak that gives the organism its greatest possible fitness. Returning to communities, when we consider only a simple replacement of one species by another, what determines whether it will succeed will be one or a few factors, such as how efficient a particular predator is at exploiting a particular species of prey. Successive replacements may be quite easy to understand, and models of this process may progress smoothly until the most efficient predator has invaded. There is a single peak in the community landscape. In more complex food webs, the total effect of one species on any other is very hard to predict (chapter 13). For example, in a complex food web, the effect of a predator directly reducing the density of its prey may be small relative to the effects that the predator has on the prey, competitors, predators, and mutualists of that prey—and on their prey, competitors, predators, prey, and mutualists, and so on. In highly connected food webs, as in genotypes, predicting whether one state will succeed another will depend on which other species (or alleles) are present. The more connected the components, the more local peaks there will be, and the more likely the assembly process will find a relatively low local peak and not the globally highest peak.

Summary

I have introduced a number of possibilities for community assembly. There may be either only one or many alternative persistent states for any given set of species. If there are many states, which will be obtained will depend on the chance sequence of invasions. There may be no final, persistent state; rather, the communities may repeatedly cycle through a simple or complex sequence of species. Species absent in the final state may be essential to reach the final state: it may be impossible to reach that state otherwise. If I have translated correctly Kauffman's models of adaptive landscapes to models of community assembly, then the existence of many alternative persistent community states is an inevitable consequence of the species interconnectedness that the food web documents.

MULTISPECIES SIMULATIONS

All the models considered so far have been very simple topological ones. Now that we have some terminology, we can examine species-assembly models to see what kinds of patterns of assembly emerge from much more complex models, those that incorporate dynamics. Are cycles common or rare in these models? And are there many alternative persistent states? While the final answer to these questions must come from nature, multispecies models give us a chance to see what to expect.

An obvious limitation in the Post and Pimm (1983) study of species assembly (chapter 8) is that only two hundred invasions were attempted, and each species was allowed only one attempt. In nature, species may have the potential to reach a community throughout the community's assembly, and a species might be able to invade when one set of species is present but not another set is. To investigate the effect of multiple invasion attempts, Drake (1985, 1988, 1990) created a program that keeps track of all the species and all their potential interactions. He created one hundred species—twenty-five species each of plants, herbivores, carnivores, and top-carnivores. At each trophic level, some species were specialized in their choice of animal or plant species used as food, and others were polyphagous. Drake's model randomly chose, as potential colonists, plant and animal species not present in the community. Species were added sequentially, for a total of twenty-five hundred attempted introductions. On average, each species had twenty-five attempts, spread across the entire community-assembly process, at entering the community.

In many ways Drake's results were similar to those of Post and Pimm. The
number of species in the community rose from the initial three species. Initially, species were added to and lost from the community. After a time, the
species composition became fixed; that is, the community became totally persistent. In one of the examples of figure 11.7, the persistent community had
fifteen species. None of the other eighty-five species could invade this community, and this situation persisted for over a thousand attempted invasions.
This is enough time to ensure that each of the eighty-five species not present
in the community had at least one chance to invade. This result is related to

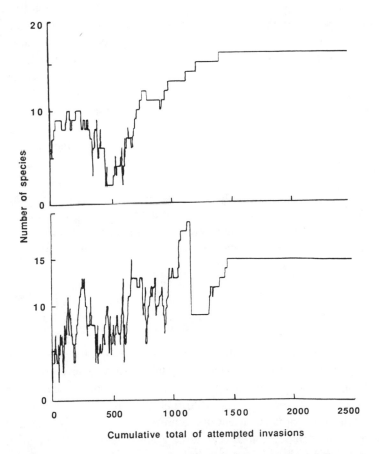

**Figure 11.7. Number of species present in a community, versus the number of attempted
introductions in Drake's simulations. (See figure 8.5 for the comparable figure from simulations from Post and Pimm 1983). In contrast to Post and Pimm, Drake used a finite species-
pool, and, in these two examples, the communities eventually become totally resistant to
invasions. (Redrawn from Drake 1991.)**

that of Post and Pimm, who found that the number of attempts required to get into a community increased after each successful invasion. With a finite pool of species available for invasion, we might expect that this number of attempts would exceed the number of species available. Drake's models, by keeping track of the individual species, show details of persistence hidden in Post and Pimm's models. When Drake repeated his simulations with the same one hundred species, the final species compositions were not the same (figure 11.7). There were alternative, persistent communities, and which particular one developed depended on the species sequence throughout assembly. Comparisons of pairs of persistent communities showed that they shared 17%–60% of their total species in common. Each species got a chance to invade the final, persistent communities, yet some species were present in some of these communities but were unable to invade others. Consequently, these alternative communities would be able to coexist side by side, with some species present in both communities, but other species were present in only some of the adjacent communities (Drake 1990).

Only once did Drake find a sequence of communities that was not persistent but which repeatedly cycled. We do not know whether Drake's models go through many cyclical changes in composition before reaching their persistent states, because his interests were different, and he did not record such information.

A final feature of these simulations is that there was usually a humpty-dumpty effect. Drake attempted to assemble the final, persistent communities from just the species present in them. He was unable to find ways to do this, suggesting that species that eventually become extinct play an essential role as catalysts in creating the communities.

EMPIRICAL STUDIES

There are four features of multispecies assembly models. First, communities become progressively harder to invade as assembly proceeds, even when they are not changing their average numbers of species or connectance. Second, with a fixed set of species, those species can be assembled into different combinations each of which is persistent in the presence of the others. Third, cyclical changes in composition appear rare. Fourth, persistent communities sometimes cannot be created from just their constituent species: other species, which become extinct in the process, may be needed. The first three aspects of these results on community assembly can be tested against what we know about natural communities. Are long-assembled communities more persistent than recently assembled communities? Are cyclical changes in composition indeed rare, and do alternative persistent states exist? The existence of humpty-dumpty effects may be tested in the future when we try to restore communities that we have destroyed. It may not be possible to restore communities once they have disappeared from the wild, even if we have all the species in botanic gardens and zoos.

Are Assembled Communities More Persistent?

Both the Post and Pimm simulations and those of Drake show that community persistence differs greatly, depending on how long the community assembly process has been running. Even after the number of species and the connectance have assumed their long-term average values, each successive species invasion makes the community harder to invade. Is this result reasonable?

What "Disturbance" Means.   I think the idea that recently assembled communities are special is not new. It is conventional wisdom that "disturbed" communities seem easy to invade (Crawley 1987). The problem is that ecologists use "disturb" in three separate ways. The first way describes communities that are clearly short of individuals. There might be an abundance of bare ground in which plants can germinate, for example. (The significance of this is not apparent in the assembly models, because they only consider invasions after all the other species have had time to reach their equilibrium densities.) The second meaning describes communities that are short of species, and both models and data (chapter 9) show such communities are easy to invade. The third meaning is used to describe often highly human-modified habitats. Crawley (1987) shows that such habitats have a high proportion of alien species in them, suggesting that these habitats may be readily invaded (although these habitats may also have few native species and may be close to sources of the aliens). These habitats, however, are not *inevitably* short of either individuals or species, but they are always recently assembled communities composed of species that have been recently introduced by humans. So perhaps, at least in part, they are easy to invade because they are recently assembled.

The idea that recently assembled communities are easy to invade is not an easy one to test, because it is hard to control for the many other variables that affect persistence. Suggestive evidence comes from Moulton's studies of the Hawaiian avifauna (chapters 8 and 9).

The Hawaiian Avifauna Revisited.   The lowland habitats on the main Hawaiian Islands contain almost exclusively introduced species of birds (chapters 8 and 9). The upland forests are temperate in climate and contain the remnants of a once diverse native avifauna. To what extent do the lowland and the upland faunas mix? The lowland faunas are relatively diverse, with ten or more introduced species; the upland faunas almost always have fewer than ten species. Introduced birds always became established (and usually became abundant subsequently) if there were fewer than ten introduced species on an island (chapter 9, figure 9.8). Yet, few of these diverse introduced species, often originally temperate in origins (chapter 8), have invaded the upland forests. When the native forest contains seven or more native species, about 30% of the individuals seen in the native forest are of introduced species (Scott et al. 1986). Even this percentage is probably an overestimate, because Scott et al.'s data included forests with differing degrees of disturbance and from wide altitudinal ranges. My unpublished census data show that in the interior of native forests the percentage of individuals that are of introduced species is about 10%.

It would seem that a small number of native species is much more effective at keeping abundances of introduced species at low levels than is perhaps twice the number of introduced species. The old, native, upland communities may be more persistent than the new, lowland communities. The fact that the introduced species of temperate origins survive in the lowland tropical habitats but rarely penetrate the upland temperate habitats seems to rule out an explanation for their absence in the uplands due to lack of appropriate adaptations.

Finally, a point about conservation. The world's ecosystems will soon resemble those of Hawai'i's, with small fragments of native habitat surrounded by large areas of human-modified habitats. We are not surprised that the human-modified habitats are full of introduced species. Human-modified habits may be disturbed in all the aforementioned meanings of the word. What we need to understand is how long native communities will persist once they are juxtaposed with communities composed of introduced species.

Are Cycles Common or Rare?

Species-change diagrams such as those in figures 11.1–11.4 suggest that cyclical changes in species composition can occur in a large number of different ways. Yet discussion of such cycles is almost absent in the ecological literature. Perhaps this is because cycles are overlooked, or perhaps it is because they are rare. The one study involving more than a very small number of species is that by Buss and Jackson (1979), on competitive relationships in cryptic coral reef environments.

The Cryptic Coral Reef Community.   Most of the hard surface area of coral reefs lies within crevices, caves, and other cavities of the reef. The most com-

mon of these cryptic surfaces on Caribbean reefs are the exposed skeletal undersurfaces of filiaceous corals. These surfaces hold a diverse encrusting fauna
of some three hundred species of sponges, ectoprocts, colonial ascidians, and
coralline algae, as well as foraminiferans, ahermatypic corals, serpulids, bivalves, and brachiopods. Unoccupied areas are scarce, and competition for
space is intense (Jackson 1977a, 1977b). Buss and Jackson collected colonies
of foliaceous corals from depths of 15–30 m. They determined the competitive
abilities of twenty common cryptic species inhabiting the undersurfaces of
these corals, by examining which species could overgrow with other species.
The twenty species included eight sponges, one colonial ascidian, foramini-
fera, coralline alga, and colonial ahermatypic coral.

Two types of interactions were found: either (1) one colony overgrew the
other or (2) growth had apparently stopped along the margin of contact. Observations of the interactions between two species allowed Buss and Jackson
to determine which member of the two-species set was the dominant. They
compiled these observations into a contact matrix of all possible pairwise interactions. Examination of the contact matrix allowed them to see whether
cyclical competitive networks exist. In all, Buss and Jackson observed 152 of
the 190 possible pairwise interactions among the twenty species (figure 11.8).
In figure 11.8 notice that every species can be overgrown by at least one other
species. These patterns of overgrowth are for single points in time, and the
data do not include long-term observations of cyclical changes in species composition. The competitive relationships are consistent with cycles, however.
There should be no persistent community state. On the underside of a particular
coral, we might, over a long enough period of time, see each of the twenty or
so species in Buss and Jackson's study. All the twenty species may persist over
a larger spatial scale—the entire reef, for example. (Of course, even this may
not be case, for there may be cyclical changes in the set of species that occupy
a particular reef: some of the current set of twenty species may be replaced
next century, only to return next millenium.)

What this study shows is that there can be communities without a persistent
state, at least on some spatial scales. There are other studies comparable to this
one; Buss (1986) reviews other examples of hard surfaces, finding some communities that have cycles and others that do not. In contrast, Quinn (1982),
found generally hierarchically arranged benthic species, and Hallett (1982, and
in preparation) found hierarchically arranged competition among sets of three
or four small mammal species in communities ranging from boreal forests to
tropical forests.

Obviously there are patterns suggestive of cycles, but there are far too few
data to decide whether cycles are rare or common. Cycles seem to require

Code	E2	E3	E5	E6	E8	E9	E10	S1	S2	S3	S4	S5	S6	S7	S8	CL	F	C	A		Code	Species	major group
E1	R	R	R	R	R	RT	R	RT	RT	RT	RT	T	R	R	RT	RT	RT	RT	T		E1	Steginoporella spp.	Ectoproct
E2	.	RT	R	RT	T	T	RT		T	T	T	T	T	T	RT	T	RT	T	T		E2	Stylopoma spongites	Ectoproct
E3		.	RT	RT	T	R	T	T	T	T	T	T	RT		T	T	RT	RT	T		E3	Retadeonella violacea	Ectoproct
E5			.					RT	T				R				R	T			E5	Trematppceia aviculifera	Ectoproct
E6				.	T	RT	R		T	T	T	RT	T			RT	T		T		E6	Disporella spp.	Ectoproct
E8					.	T		T	T	T	T	R	T	T	T				T		E8	Smittipora levesensi	Ectoproct
E9						.	R		T		T	R	RT	R	R	R	T		T		E9		Ectoproct
E10							.	T	T	T	R	T					RT	T	T		E10		Ectoproct
S1								.	T	T	T	T	T			T	R				S1		Sponge
S2									.	R	R	T	R		R	R	R	T	T		S2	? Toxemnma spp.	Sponge
S3										.	T	R	R	R	T	R	T	R	T		S3	? Tenaciella spp.	Sponge
S4											.	R	R		R	RT	R	R			S4		Sponge
S5												.	R	T	T	R	T	T	T		S5		Sponge
S6													.	T	T	T	T	R	T		S6		Sponge
S7														.	T	T	T		T		S7		Sponge
S8															.	R	T	R	T		S8	Ectyoplasia ferox	Sponge
CL																.	T	T			CL		Coral
F																	.	R	T		F	Gypsina spp.	Foram
C																		.	T		C		Coraline alga
A																			.		A	Diemnum spp.	Ascidian

Figure 11.8. Competitive relationships between species growing on the undersurfaces of corals. The letters indicate which of any pair of species beats (that is, overgrows) the other. Thus, R means that the species at the right beats the species at the top of the table, and T means that the species at the top of the table beats the species at the right; no letter indicates that there are no data, and both letters in conjunction means that both outcomes are observed. Every species beats at least one other species and is beaten by at least one other species. Thus, while *Steginoporella* beats many other species, it is beaten both by the sponge S5 and the ascidian *Dilemnun*. (Redrawn from Buss and Jackson 1979.)

special conditions and, as a consequence, may only occur rarely. Cycles have been found where the competition is for space and where the mechanism is direct interference. It may be impossible to arrange for, say, A to be more efficient at exploiting seeds than is B, B to be more efficient than is C, and C to be more efficient than is A. Yet it is possible for A to poison only B, B to poison only C, and C to poison only A. The mechanisms of interference competition may allow cycles, where the mechanisms of exploitive competition may not.

The facts that the communities in which cycles are found do appear special, that the processes that generate them are also special ones, that multispecies models find cycles to be rare, and that long-assembled communities appear more persistent that recent ones lead me to conclude that many communities probably do not have cycles. This conclusion is a very tentative one, but it is worth considering its implications for persistence.

Alternative Persistent States

An Overview.   The great diversity of natural communities is not in itself evidence that there are alternative persistent community states. Another explanation for this diversity is that each community experiences a different set of environmental conditions and that physiological adaptation to these conditions decides which species occur where. This may be so, but it is not a priori any more likely an explanation for the diversity of communities than are hypotheses based on species interaction. The relative roles of physiological adaptation and interspecific interactions in deciding species ranges are rarely compared, and, when they are, the interactions can be more important (chapter 8). In consequence, my view is that we should accept tentatively the evidence of alternative persistent communities, even if we cannot prove the impossible— that is, that the communities are in identical environments. Large differences in the composition of communities with similar environmental conditions are compelling evidence for alternative communities, because most species can occur over environmental conditions wider than species interactions permit.

With this viewpoint, there are many studies which show alternative persistent states. Diamond's "checkerboard distributions" (figure 9.7) are examples. Others have found similar examples in which the sequence of assembly produces alternative states in terrestrial, marine, and aquatic communities, in plants, invertebrates, and vertebrates, in temperate regions and tropical regions. Drake (1989) has discussed many of these examples, and table 11.1 summarizes them.

The only certain way to ensure that alternative persistent communities develop in identical environments is to create those environments experimentally. To explore the effect of assembly on community composition, a number of ecologists have constructed experimental communities under controlled laboratory conditions by using different orders of species invasion. Examples include experiments by Robinson and Dickerson (1987), Robinson and Edgemon (1988), and Drake (1991). I shall now discuss Drake's study in detail.

A Microcosm Study

Drake chose a set of species and then permuted the sequence of invasions, allowing each community to be colonized by all species in the species pool. He used two types of aquatic systems. The small-volume systems held a volume of 250 ml, and the large-volume system held 40 liters. The large-volume systems were used to explore the dynamics of community assembly by using ten different temporal sequences of invasion. The small-volume systems were

TABLE 11.1   Cases in Which Assembly Phenomena Have Been Documented

System	Mechanism or Effect	Source(s)
Algae	Multiple stable states	Drake (1990)
Trees	Alternative states and multiple assembly pathways	McCune and Allen (1985), Connell (1978)
Insects	Historical effects	Brown (1985), Kneidel (1983), McBrien et al. (1983)
Ants	Alternative states	Cole (1983)
Diptera	Alternative states and multiple domains of attraction, priority effects	Gilpin et al. (1986)
Aquatic microcosms	Alternative states which were sequence driven	Drake (1990), Robinson and Dickerson (1987), Robinson and Edgemon (1988)
Marine invertebrates	Alternative states	Sutherland (1974), Barkai and McQuaid (1988), Buss and Jackson (1979)
Reef fishes	Competitive lottery	Sale (1978)
Amphibians	Priority effects, predator reversal of competitive effects	Alford and Wilbur (1985)
Lizards	Species coocurrence	Roughgarden (1989)
Birds	Forbidden combinations in species coocurrences, character displacement	Diamond (1975)

Note.—Most of these studies illustrate rules for only a portion of the communities which contain them, though in several of the laboratory studies, species composition was completely known. (From Drake 1989.)

used to explore competitive interactions and sequence effects among species found to be key components of the large-volume systems. Each of the large-volume treatments was replicated four times, while treatments were replicated three times in the small-volume systems. If a consumer species became extinct or suffered substantial mortality within the initial 24-hour period after it was introduced, Drake reintroduced it. This reduced the probability of extinction due to chance factors.

Algae Assembly. *Small-volume results.* Drake first established the patterns of population growth for three of the algal species—*Scenedesmus quadracauda, Selenastrum bibrium,* and *Ankistrodesmus falcatus*—growing in isolation. Both *Selenastrum* and *Ankistrodesmus* achieved larger populations than did *Scenedesmus.* Yet, despite this, *Scenedesmus* always became numberically dominant in treatments with more than one species and so depressed the abundance of both *Selenastrum* and *Ankistrodesmus.* This result eventually occurred regardless of when *Scenedesmus* was introduced in the assembly sequence.

In contrast to that of *Scenedesmus,* the population growth rates of both *Ankistrodesmus* and *Selenastrum* were strongly influenced by the order in which the species were introduced. Drake found that both species developed substantial populations only when they were introduced first, before *Scenedesmus* was introduced. When introduced after *Scenedesmus,* neither of these species was able to increase in abundance much above introduction levels. Long-term coexistence of two or more species was possible only when *Scenedesmus* colonized at a stage later during the assembly sequence.

In sum, the effect of *Scenedesmus* on the abundance of the other plants was strong in all the assembly treatments. Nevertheless, two alternative community states developed as a function of assembly sequence: communities containing only *Scenedesmus* and communities dominated by *Scenedesmus* but with small persistent populations of the other two species.

Large-volume results. The order of species introduction had even more obvious effects on invasion success—and, ultimately, on the relative abundance of the species—in the large-volume systems. Indeed, Drake found that the effect of invasion sequence was so strong that order was the primary cause of the relative abundance of the algae. In contrast to the small-volume systems, the overwhelming competitive advantage shown by *Scenedesmus* was not observed in the large-volume systems. This contrast provides a neat example of how the spatial scale of a study can alter the results.

In all but sixteen of forty replicates, the dominant algal species at the end of the experiment was the first species to invade (table 11.2). In half of the ten treatments (treatments 1, 4, 5, 7, and 10; table 11.2), the dominant species was consistently the same in all four of the replicates. Each of the remaining three algal species was second in numerical abundance in one or more of the remaining replicates. The remaining treatments (treatments 2, 3, 6, 8, and 9) had some variation in the dominant algal species. In two treatments (treatments 4 and 10), the initial algal colonist was unable to maintain numerical dominance in any of the replicates. In this situation, the initial colonist was rapidly overtaken by the next algal to invade. In one case, *Chlamydomonas* colonized

TABLE 11.2   Relative Dominance of Algal Species in Drake's (1991) Large-Volume Systems after Completion of the Assembly Sequences, in Day 195 of the Experiment

Treatment/ Replicates	Sequence of Plant Invasions	Order of Relative Dominances[a]	Proximity of Invasions[b] (days)	Proportion of Time Dominant[c]
1/1	AN-SE-CH-SC	AN CH SE.	120	0.98
1/2		AN CH..		
1/3		AN CH SE.		
1/4		AN CH SE.		
2/1	SC-SE-CH-AN	SC CH SE AN	60	0.94
2/2		SC CH SE AN		
2/3		SC CH SE AN		
2/4		AN SC SE CH		
3/1	SE-CH-AN-SC	SE AN SC.	105	0.95
3/2[d]		. . .		
3/3		SE AN SC CH		
3/4		AN SE SC CH		
4/1	CH-AN-SC-SE	AN CH SE SC	105	0.29
4/2		AN SC SE CH		
4/3		AN SE SE CH		
4/4		AN SE SC CH		
5/1	SC-CH-AN-SE	SC AN CH SE	135	0.88
5/2		SC AN CH SE		
5/3		SC AN CH SE		
5/4		SC AN CH SE		
6/1	SE-CH-SC-AN	SE AN SC.	150	0.98
6/2		SE CH SC AN		
6/3		SE AN SC CH		
6/4		AN SC SE.		
7/1	AN-SC-CH-SE	AN .CH SC SE	150	1.00
7/2		AN CH SC SE		
7/3		AN CH SC.		
7/4		AN CH SC.		
8/1	CH-AN-SE-SC	AN SE..	105	0.40
8/2		AN SC CH.		
8/3		CH SC AN.		
8/4		SE SC AN.		
9/1	SC-AN-CH-SE	AN SC SE CH	60	0.71
9/2		SC AN..		
9/3		SC AN..		
9/4		AN SC..		
10/1	AN-SE-CH-SC	SE SC AN.	60	0.15
10/2		SE SC AN.		
10/3		SE SC AN CH		
10/4		SE SC AN CH		

Note.—AN = *Ankistrodesmus falcatus;* SE = *Selenastrum bibrium;* CH = *Chlamydomonas reinhardtii;* SC = *Scenedesmus quadracauda.*

[a]Between-species population sizes that were statistically indistinguishable between replicates ($P < 0.05$) are underlined. Species which became extinct are missing from the species list.

[b]Time elapsed between the first and last plant introduction.

[c]Proportion of time that the first plant colonist was significantly more abundant than any other species.

[d]Ended because of contamination.

first and was rapidly replaced by *Ankistrodesmus*. In the other case, *Ankistrodesmus* colonized first and was replaced by *Selenastrum*. The two treatments in which *Ankistrodesmus* colonized first illustrate the dramatic effect that timing can have on community assembly. *Ankistrodesmus* quickly lost dominance when the system was next invaded by *Selenastrum* fifteen days later (treatment 10; table 11.2). In the other treatment (treatment 1) in which *Ankistrodesmus* invaded first and in which *Selenastrum* again was introduced second, but thirty days later, *Ankistrodesmus* maintained its dominance.

Consumer Assembly.   Drake found that the consumers varied considerably in whether they could invade the large-volume systems (table 11.3). Some treatments were readily invaded and supported large populations of many consumer species. Other treatments were very difficult to invade and were essentially resistant to invasion by all consumer species. In addition, although some species were able to colonize a treatment and persist for a period of time, they were incapable of reproduction. Among consumers, there were rarely differences between replicates within a treatment. Either the species successfully invaded all replicates, or it failed in all replicates. *Daphnia magna, Cyclops vernalis, Cypris* sp., and *Gammarus lacustris* were the most successful invaders. These species successfully invaded at least half of the treatments, and *D. magna* colonized all treatments. Both *D. magna* and *Cyclops* are filter feeders, although *Cyclops* is predatory on very small *Simocephalus*. *Cypris* sp. and *Gammarus* were grazers foraging on the bottom or sides of the ecosystems. *D.*

TABLE 11.3   **Mean Persistence Time (Days) of Consumer Species (from Drake's Large-Volume Systems)**

	Mean Persistence Time (in days), by Treatment Number[a]									
Species Name	1	2	3	4	5	6	7	8	9	10
Daphnia magna	30	23	53	23	23	23	30	71	15	38
Cypris species	0	75	0	0	0	38	53	128	0	98
Gammarus lacustris	0	23	38	0	0	23	0	68	0	53
Euglena gracilis	0	0	0	0	0	0	0	0	0	0
Paramecium multimicronucleatun	0	0	0	0	0	0	0	0	0	0
Cyclops vernalis	23	23	68	23	0	11	23	15	90	68
Pleuroxus truncatus	0	0	0	0	0	0	0	23	0	0
Daphnia pulex	0	0	0	0	0	0	0	49	0	0
Simocephalus vetulus	0	0	0	0	0	0	0	19	0	68
Total no. of days	53	144	159	46	23	95	106	373	105	325

[a]Treatment numbers (i.e., numbers used as column heads) are as in table 11.2. Data in body of table are number of days of treatment.

pulex and *Simocephalus* were both short-term residents in some of the treatments, though they were unsuccessful in most of them. *D. pulex* is a filter feeder, and *Simocephalus* grazed on the ecosystem walls. *Pleuroxus truncatus* colonized only a single treatment and persisted for a short period of time. Neither of the protozoans was a successful colonist.

The difficulty with detecting alternative persistent states among the consumers is that their environments differed from treatment to treatment because the algal species showed so many different states. However, one can compare similar algal communities and look at which consumers predominated. Table 11.3 shows the persistence times of the various consumer species. For example, in treatments 1, 4, and 7, *Ankistrodesmus* was the dominant algal species; in all three treatments, *D. magna* and *Cyclops vernalis* persisted, but only in treatment 7 did *Cypris* persist.

Drake concluded that "the sequence or order of invasions determines the set of rules under which a given ensemble of species assembles. Different sequences showed different sets of assembly rules even when the species pool is held constant . . . To fully understand communities, one must not only have a knowledge of the extant community, but must also understand the historical processes which created that community. Sequences and historical effects which produce divergent communities do so because the actions of these effects are influenced by chance and timing."

SUMMARY AND CONCLUSIONS

This chapter shows that persistence depends not just on community's present composition but also on the sequence of events that lead to that composition. Model communities become progressively harder to invade as assembly proceeds, even when they are not changing their average numbers of species or connectance. Assembly increases community persistence, so "old" communities should be harder to invade than "young" ones. This may be a familiar idea, but it is an extremely difficult one to test. Ecologists readily accept that disturbed communities are more readily invaded than undisturbed ones. Disturbed means several things, but one of them is that the communities are recent assemblages of introduced species. In Hawai'i, relatively species-poor upland native communities are harder to invade than the relatively species-rich lowland communities composed of recently introduced species.

In models with a fixed set of species, those species can very easily be assembled into different combinations each of which is persistent in the presence of the others. There are many examples of possible alternative persistent states in nature, and in laboratory experiments these alternative communities seem very easy to obtain. In contrast, cyclical changes in composition are rare in models. Cycles certainly occur in nature, and the patterns of competition between species growing on coral reefs show how complex they can be; however, cycles may be rare in nature because the conditions needed to generate them are special ones.

In models, persistent communities may not be created from just their constituent species: other species, which may become extinct in the process, may be needed. If this is true in nature, then, when we destroy communities, we may not be able to reassemble them, even if we have kept all the species alive in botanical gardens and zoos.

Implications for Community Diversity

Alternative persistent communities are probably common in nature, and they are certainly easy to obtain in the laboratory. Drake, for example, was not deliberately adjusting the conditions of his experiment to obtain them. The community landscape is likely to be a rugged one, with many peaks. The great diversity of communities we see in nature may well be a consequence of the abundance of alternative persistent states and of the ability of these states to coexist over a landscape (Drake 1990). There are also two other possible explanations for the diversity of communities we observe.

The first possibility is that the diversity of communities reflects the diversity of environmental conditions. This must surely be the case; undoubtedly, most species of mountaintops or deserts would not flourish in marshlands, for they lack the appropriate adaptations. What concerns me is that this alternative, which provides a population-level explanation for community diversity, is sometimes considered both sufficient and as enjoying a kind of logical supremacy. Given the species-composition changes that we observe over small ranges of environmental conditions, and given the wide range of physiological tolerances that species enjoy, it is far from clear that this alternative provides a *sufficient* explanation for community diversity. Nor should this alternative enjoy the status of a hypothesis we must reject before any others are considered. Community-level models predict the existence of alternative persistent communities by using the simplest of assumptions. It is equally reasonable to ask that those who favor population-level explanations for community diversity be certain that they have rejected alternative persistent states as an alternative.

The second possibility is that what we think are alternative persistent states are the intermediate transitions of constantly cycling communities. If this is true, then the cycling must be very slow—or we would see the changes. Obviously, the only way to test this possibility is to have very long-term data on community composition. Such data are unlikely to be available in the near future.'

A Note on Resistance

How much do "things" change when we permanently alter the environment in "some way"? Are these "things" resistant to the alteration or sensitive to it? Quite what these "things" are, I shall explain presently. Please note that resistance is a measure of *permanent* change—we are comparing "before" with "after," rather than the transient behavior in between. Therefore, resistance is a very different way of measuring "stability" than those already discussed.

Ecologists may estimate resistance simply as the ratio of the value of some variable after the change divided by the value of that variable before the change. The more the values change, the less the resistance; and, the closer the ratio is to unity, the more the resistance. Therefore, resistance is measured by quantities that are dimensionless, unlike resilience, which is measured by a rate (the change per unit time), and persistence, which is measured as a time (how long things last). Bender et al. (1984) introduced two terms relevant to this discussion. They called experiments yielding information about permanent changes "press" experiments. These are contrasted with "pulse" experiments, which look at transient changes and thus yield information on resilience. Much of my discussion here will center on the question of whether ecologists have waited long enough for changes to be permanent—and whether it is indeed *possible* to wait long enough.

There are many ways in which the environment can be altered and many features of species, communities, and ecosystems that can change as a consequence. Moreover, the alterations may occur on many different spatial scales. Discussing resistance, therefore, is a matter of asking, Resistance of what, to what, over what spatial scale? Perhaps more than any other of the various meanings of "stability," resistance is the one which immediately brings a large number of possibilities to mind.

Organizational Scale

What is changed and what is affected as a consequence can be events at very different levels of ecological organization. Ecosystem alterations are one example. Through pollution, humans are substantially altering the climate, the chemistry of the soil, the atmosphere, rivers, lakes, oceans, and global nutrient cycles. How much do gene frequencies, population densities, or the species compositions of communities change when we alter the patterns of nutrient and energy flow? There are alterations within a population. How much do gene frequencies, population densities, the species composition of the community, and the flows of energy and nutrients change when we alter the genetic structure of populations, perhaps by fragmenting them or by introducing individuals with novel genotypes?

What Is Altered, and What is Changed as a Consequence? My discussion will focus on the consequences of altering species densities. Even when we restrict what changes, we are still faced with the question of what has been altered as a consequence. For example, species introductions obviously involve a change in density. One consequence of this change is that introductions can alter the genetic structure of native populations. Hybridization can occur when an exotic species that is a member of the same species group as is a native plant or animal is introduced into a community. Hybrids are often infertile and die out each generation, so reducing the reproductive rate of the native species. Sometimes the introduction of novel genes can have significant population effects. One widely cited but anecdotal study is from the Tatras, in Czechoslovakia, where Neubian ibex were introduced. These animals then interbred with the native ibex. Because they originated in warmer climates, the exotic goats bred earlier in the season. This trait was passed on to the hybrids, whose young were born two months earlier, at a time when cold in the mountains was intense, and most of the young froze to death (Turcek 1951).

At the opposite end of the organizational spectrum, the introduction of species can have profound effects on ecosystem processes. Investigating these effects has been a special interest of Peter Vitousek and his colleagues. In Hawai'i, the introduced nitrogen-fixing tree *Myrica faya* invades sites of recent lava flows where the growth of native plants is limited by a lack of nitrogen. *Myrica* increases, by a factor of four, the amount of nitrogen entering some sites and thus greatly modifies the nature of ecosystem development after volcanic eruptions (Vitousek 1986). These genetic and ecosystem effects are both interesting and important, but I want to look here at a more restricted area—namely, changes at the intermediate level of ecological organization. Of the many questions one could ask about resistance, I shall concentrate on how changes in the densities of some species affect the densities of other species.

The Problems of Spatial Scale

Over what spatial scale do we perform the press experiments that look for the effects that changing the density of one species has on the density of other species? A press experiment usually is performed by removing individuals, and the size of the removal experiment is set by the size of the enclosure or area from which individuals are removed or excluded. The enclosure may be quite small, say, a fraction of a square meter in some experiments in intertidal habitats. Similar experiments in terrestrial habitats may be much larger, perhaps on the order of hectares. It is obviously difficult to build enclosures around areas much larger, but, if we introduce or exterminate species, we are performing an experiment on a scale that may affect thousands of square kilometers. Ideas about resistance need to be considered across this spectrum of spatial scales, for there is no guarantee that the same results that hold true for a small experiment will also hold true for a large one.

Resistance to Species Additions and Removals

There has been considerable discussion about whether species abundances *are* altered by either the removals of species in enclosure experiments or the introductions of species to new areas. Does the removal of species increase the abundance of their putative competitors? That is, were the species really competing (chapter 9)? Similarly, do predators regulate the density of their prey, or are these prey regulated by other factors, such as availability of *their* prey? To answer these questions, I will turn first to several excellent compilations of research on predator and competitor removals. While there is complete agreement on the broad range of effects that can be documented, it is difficult to draw conclusions about the extent of these effects. These compilations have an obvious limitation. For instance, if 60% of studies of removals of predators show increases in prey density, does this mean that predators generally regulate their prey? The obvious bias will be for ecologists to report those experiments that yield "significant" results—that is, those that find effects. Moreover, the experiments may have been undertaken only because observations of natural history changes, such as local variations of predator density, suggested that these predators would have an effect on their prey. We may not be able to answer precise questions about what percentage of randomly chosen species removals have effects.

If altering the density of some species affects the densities of some species but not those of others, can we predict generally *which* species will be affected—and, roughly, by how much? There are a number of theories that answer this question with a yes. These theories predict differing degrees of resistance, on the basis of such ecosystem characteristics as the level of physical disturbance or primary productivity. These theories and the reviews of small-scale removal experiments used to evaluate them are also the subjects of chapter 12.

Chapter 13 considers a food-web theory of resistance. This community-level theory looks at how food-web complexities lead to indirect pathways that confound any simple attempts to predict the affect that one species has on another. This theory replies with an emphatic no to the question posed in the previous paragraph, for it argues that we cannot even predict whether species increase or decrease after the removal of any given species. This theory seems inconsistent with the data cited in the reviews of removal experiments; I shall argue that it is the data that need reinterpretation.

The topic of chapter 14 is how much small-scale introductions or removals alter either the overall species composition of the community or the total species abundance of some interesting *group* of species in the community. The changes in total density or biomass show patterns not appreciated by looking at either individual species or community composition. Among other things, I shall discuss the paradox that, after the loss or addition of a species, relatively large changes in species composition are generally accompanied by relatively small changes in the total density of the group of species.

Chapter 15 discusses the consequences of species introductions and extinctions; these are changes in species abundance on a large spatial scale. With species introductions and extinctions, we will encounter a bias that is the reverse of that found in species removals. In the case of many introduced species, it is not known whether their introduction has any effect, and this lack of knowledge is sometimes equated with the introduced species not having an effect. Do introduced species generally have an impact on the density of the native species, or do the introduced species merely fill "vacant" niches? As with small-scale species removals, quantitative assessments of the effects can only be done with extreme caution. This being the case, how can we begin to understand and predict effects of species introductions or extinctions?

The "experiments" of species introductions or extinctions do not usually tell us about the changes in the density of particular species. What these experiments *do* yield is data on changes in species composition—typically, which species went extinct after the introduction or extinction of another species. With these data, is it possible to predict these changes in the species composition of the community? For example, are there species whose removal or introduction will cause great changes in the composition of the remaining species? (Ecologists call these the "keystone" species.) Clearly, these questions of resistance are different than those discussed in chapters 12–14. Again, there are some theories about how we might interpret the empirical data on species introductions and extinctions, and I shall present both the theories and the data.

Notice that, unlike my procedure in previous major sections, I have omitted all discussion of the effects that a species' natural history has on resistance. This theory exists, of course. One example is the degree to which trophically specialized insect parasitoids depress the densities of their hosts (typically, pests). It is a subject which has attracted considerable attention, because introductions of highly specific parasitoids to control often accidentally introduced hosts is a topic of considerable economic importance. Beddington et al. (1978) documented the results of how much densities of insect herbivores were reduced by insect parasitoids, both in the laboratory and in the field. In the field, densities were reduced down to a few percent (sometimes less) of the host densities that existed without the parasitoids. In the laboratory, such reductions were 15%–50%. It is not just the contrast between laboratory and field studies that is interesting. Mathematical models also predict how much a predator can reduce the density of its prey, and, for most simple models, prey are quite resistant to predators. One feature missing from both the simple models and the laboratory studies was spatial patchiness. The models assume that the environment is homogeneous, and laboratory studies generally create such environments. If the environment is patchy, there may be refuges where the prey are relatively invulnerable, while in other places the predators may quickly eliminate them. In such patchy environments, efficient predators can reduce their prey to the number the refuges can hold. The refuges may be simply areas of very low prey densities if predators aggregate and feed preferentially in areas of relatively high prey densities. These simple ideas have been supported (Heads and Lawton 1983), rejected (Murdoch et al. 1985), extended, reviewed, and argued over (Strong 1988, 1989; May and Hassell 1989). Irrespective of the details of these arguments, it is the prey's dispersal and the predator's searching behavior that are thought to determine the degree to which prey density is reduced by the predator.

The reason why I give these population-level explanations for resistance little space is twofold. First, I think the discussions of resistance are long and complex enough without a review of these population-level arguments. Second, I think I no longer need to reiterate the fact that explanations for the various meanings of stability span the range of organizational scales. A general theme running through the next four chapters should now be a familiar one: even the simple question of how changes in abundance of one species

affect the abundances of other species leads to a complex of theories and experimental studies. The complexity arises from the different spatial and organization scales over which the simple question can be asked. I begin with the
most obvious way of posing the question—that is, how the removal of a
competitor or predator affects another specified species over spatial scales so
small that the question can be answered by direct experiment.

12

Small-Scale Experimental Removals of Species

DO SPECIES REMOVALS HAVE EFFECTS?

Do predators affect the densities of the species they prey on? The answer is not obviously yes, for prey may be limited by factors other than their predators. Predators may feed mainly on individuals—the weak or the sick, for example—whose deaths are almost certainly due to other causes. In such cases, predators are taking individuals doomed for other reasons, and removing these predators may not affect the density of prey at all. If prey densities do not change when predators are removed, then prey are resistant to predator removals. Are there patterns in which prey species are resistant to predators? Similarly, are species resistant to the removal of their competitors? If some species compete and some do not, can we predict which? Competition, unlike predation, may act either directly, between a pair of species, or indirectly, by the species' sharing of resources. Direct competition is often called "interference competition," and it occurs when one or both species harm the other directly. Indirect competition is often called "exploitive competition," and it occurs when one of the competitors alters the abundance of something important to the other. This "something" may be the space in which the competitors live, or it may be the densities of one or more species of prey on which both species feed. As well as through their prey, species may harm each other indirectly through their predators, by apparent competitive interactions through shared predators or diseases (chapters 9 and 10).

Finally, should the increase in the density of a prey species X increase the density of its predator Y? The answer is yes only if that predator Y has no predators feeding on it. If the predator Y has predators Z, then the latter may "take up the slack," killing all the additional predators Y produced when species X increases in density. I shall discuss all these possibilities in turn.

The Removal of Predators

Do predators alter the density of their prey, and, if so, which predators produce such effects? These are the questions asked by Sih et al. (1985) in an important survey of predation experiments. They compiled over one hundred experimental studies on predation. Most of these studies appeared in the three years prior to their study, and most had been done in temperate regions. Marine experiments primarily removed nonarthropod invertebrate herbivores, terrestrial experiments generally removed arthropod herbivores, and freshwater work concentrated on vertebrate carnivores. Manipulations of herbivore densities had rarely been done in freshwater; on land and in freshwater, few studies looked at the effects that nonarthropod invertebrates have on their prey.

Almost all of the studies showed some significant effects, meaning that the densities of some prey species increased by a statistically significant amount when predators were removed. The great majority of the studies showed some "large" effects. "Large" sometimes meant that a prey species was at least twice as abundant when the predator was absent. (Sometimes, the change was in the opposite direction—when "large" meant that the prey was at least twice as abundant when the predator was present! I shall return to such unexpected effects later.) Clearly, few communities remain unaltered in the face of predator removals. Sih et al. asked whether the effects of predation removal appeared regardless of latitude, type of community, the predator's trophic level, or other variables.

One of the most consistent patterns in communities is a latitudinal trend in species diversity—there are more species in the tropics (Pianka 1966). One popular explanation for high levels of diversity in the tropics is that increased predation increases species diversity (chapters 8 and 9) and that predation is more intense in the tropics (Janzen 1970; Connell 1978). Few studies have specifically attempted to compare the relative intensity of predation in temperate versus tropical regions. Sih et al.'s survey indicated a slight trend toward both more frequent and larger effects of predation in tropical latitudes. This trend was not statistically significant, however, perhaps because there were relatively few experimental studies on predation in the tropics, making this a difficult comparison to make (table 12.1).

Is predation more important in certain types of habitat? The importance of predators in setting the abundance of their prey has been stressed by ecologists working in marine communities, especially those of rocky intertidal habitats. The studies of Robert Paine (for example, Paine 1980) are well known to every ecologist and are staple material in every introductory ecology text. In freshwater communities, most commonly in lakes, there is also a tradition of dem-

TABLE 12.1   The Percentage of Studies of Species Removals or Additions that Yielded
Various Results (After Sih et al. 1985)

How Data Were	Percentage of Studies		
Classified	Significant[a]	Large[b]	Unexpected[c]
Region:			
Temperate or polar	95	85	44
Tropical	100	100	33
Habitat:			
Intertidal	94	91	40
Other marine	100	75	32
Streams, rivers, and lakes	100	82	46
Terrestrial	90	76	22
Types of predator:			
Vertebrate	91	70	43
Arthropod	98	80	31
Other invertebrate	94	88	37

[a]Studies that showed any effects significant at the 5% level, as a percentage of all studies.
[b]Studies that showed at least a twofold change in species population density, as a percentage of studies with any significant effects.
[c]Studies that showed that prey species decreased when predators were removed, as a percentage of studies with any significant effects.

onstrating the effects of predators, as Zaret's influential book *Predation and Freshwater Communities* (1980) clearly demonstrates. In contrast, the importance of predation occurring on land has not been emphasized as often. Sih et al. asked whether this difference between terrestrial and aquatic habitats appeared in their data; it did (table 12.2). In addition, experiments in the intertidal and lake habitats showed larger effects of predator manipulations than did experiments in other communities. Sih et al. cautioned that the difference, in both frequency and magnitude of effects, between various communities was not striking.

Why might predation be more important in the intertidal and lake habitats than elsewhere? Sih et al. suggested that the answer might lie in the lower structural heterogeneity typically found in intertidal and lake habitats. Compared with other habitats, the open-water habitat of lakes and the rocky surfaces of the intertidal system are generally thought to be structurally simple; in structurally complex habitats, prey have more places to hide from predators than they do in simple habitats; consequently, prey densities may be more resistant to predation.

Field Experiments in Competition

Connell (1983) and Schoener (1983a) published, independently, two excellent reviews of competitor-removal experiments. Some of their results are sim-

TABLE 12.2   The Percentage of Studies of Species Removals or Additions Showing any Significant Effects or Large Effects (after Sih et al. 1985)

Habitat	Percentage of Studies Showing Significant Effects			Percentage of Studies Showing Large Effects[a]		
	Herbivore	Predator	Top-predator	Herbivore	Predator	Top-predator
Intertidal	68	58		84	70	
Other marine	59	41		95	68	
Lakes, etc.		67	49		73	75
Streams and rivers		73	64		61	55
Terrestrial	63	53	29	74	61	

[a]Effects showing at least a twofold change in density.

ilar, but they disagree on a number of important issues. Schoener assembled 164 experimental studies of competitor removals involving 390 species or groups of species that might be subjected to competition. Connell's study focused on seventy-two papers, reporting studies of a total of 215 species and 527 experiments. Most of the papers Schoener studied were done in the few years prior to his review, and the papers Connell studied were published in six journals during the nine years prior to his survey.

Schoener's experiments were concerned with terrestrial plants, terrestrial animals, and marine organisms and with a smaller number of studies of freshwater organisms. He found that the great majority of competitor removals effected change in the species for which population densities were measured. There were no differences among types of community in this respect: 91% of freshwater, 94% of marine, and 89% of terrestrial studies showed some competition. Nearly all the studies were done in temperate regions. Schoener noted that few studies were available for folivorous insects. No one who has heard John Lawton or Donald Strong give seminars would have failed to learn that folivorous insects are taxonomically extremely diverse (about 25% of the planet's known species) and very abundant. These ecologists argue that many folivores do not usually compete for plant resources (Strong et al. 1984); therefore, percentages such as those Schoener found need to be weighted taxonomically.

Connell had an alternative and quantitative approach to estimating how widespread competition might be. He counted the number of studies in which competitive effects had been found and related this to the number of pairs of species interactions studied. In studies in which one experiment was done on only one species at one time and at one place, fourteen (93%) of fifteen studies had found competition. Interestingly, when more studies were reported by the experimenters, the percentage dropped. When there were more than one species studied by a given experimenter, only 48% of the studies found evidence of competition. When there were four or more experiments reported by each experimenter, only 22% of the studies found evidence of competition. Perhaps at least one statistically significant result is required before an experiment can be published; a *single* nonsignificant experiment, as opposed to a cluster of experiments performed simultaneously, may not even be reported.

FOUR HYPOTHESES AS TO WHICH SPECIES CAN BE REMOVED

Making sense of what happens when we add or subtract species from communities is clearly difficult. Predators can affect their prey, and putative competitors often compete, but exactly how extensive are these effects? The syntheses of the effects of press experiments involving predators (Sih et al. 1985) and competitors (Connell 1983; Schoener 1983a) show that there are many effects. The very nature of predator and competitor-removal studies makes generalizations about them difficult, however, for the species removed are hardly a random selection. We must be cautious in interpreting the patterns observed in these syntheses, especially those we find by subjecting the combined data to large numbers of statistical analyses, because some differences may well be spurious.

An alternative approach to viewing these data is to ask what the relevant theory might tell us. Can we predict how a species' density changes when other species are added or removed from the community? Do these surveys provide support for the theories? There are four theories that use a knowledge of ecosystem features to predict general patterns of resistance. A fifth, food-web-based theory about resistance will be a topic of the next chapter.

The World is Green

In one of the most widely cited ecological papers, Hairston et al. (1960) made predictions about the species that should be resistant to removals of their predators or competitors: carnivores should compete because they have no predators holding their densities down; herbivore populations are held in check by their predators and so should not compete; and, because they are not limited by their herbivores, plants should compete for resources such as nutrients, space, or light. Competition and predation should alternate in importance. This basic idea was clarified later (Slobodkin et al. 1967) with the argument that herbivores are to be defined as folivores, not as frugivores, granivores, or nectarivores. Herbivores other than folivores would be expected to compete. The implications for resistance to predator removals follow simply (figure

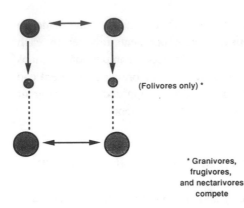

(Folivores only) *

* Granivores,
frugivores,
and nectarivores
compete

Figure 12.1. Sketch of "the world is green" hypothesis of Hairston et al. (1960). Solid lines indicate strong interactions, and dashed lines indicate weak interactions. Predators compete and limit herbivores, which then do not limit plants, and, in consequence, the plants do compete. Granivores, frugivores, and nectarivores do compete, however, because seeds, fruits, and nectar are in limited supply.

12.1): folivore removals should have relatively little effect on plant densities; carnivore removals should have a strong effect on herbivore densities; and plants and carnivores should compete but herbivores should not.

The World is Prickly and Tastes Bad

In a response to Hairston et al., Murdoch (1966) proposed an alternative explanation of why plants are relatively abundant and herbivores relatively scarce. Plant defenses may make life very difficult for folivores—foliage may be protected by physical barriers such as spines, may contain few nutrients, and may be laden with toxins. Herbivores may be scarce, but they might still compete intensively for resources of very limited quality, and predators might also compete for the scarce herbivores. Predator removals or herbivore removals would have little effect on the species on which they feed (figure 12.2).

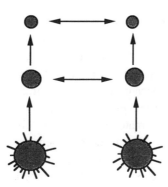

Figure 12.2 Sketch of "the world is prickly and tastes bad" hypothesis of Murdoch (1966). Plants are difficult to eat, and so the availability of digestible, accessible plants limits herbivores (which compete), and these, in turn, limit carnivores (which also compete).

Life at the Top and Life at the Bottom

Certain terrestrial systems may be green, but the observation seems much less compelling for aquatic habitats. Menge and Sutherland (1976) developed a hypothesis that does indeed differ from that of Hairston et al.; they predicted that predation should be more important at lower trophic levels, and they argued that the effects of predation should decrease as one moves from herbivores to carnivores to top-carnivores. This hypothesis hinges on the presence of species that feed at more than one trophic level. Species low in the food chain are likely to be the prey of more predators. The higher the species is in the food chain, the more it is likely to be competing with other species.

There is one addendum to this idea. Species composition at the top of a food web may be altered by the degree of physical disturbance in the habitat. For example, where there is much wave action, barnacles and mussels may be the top species, while, in less disturbed environments, their predators, such as the snail *Thais,* may be the top-predators in the community (figure 12.3).

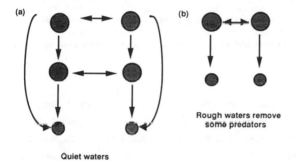

Figure 12.3. Sketch of Menge and Sutherland's (1976) hypothesis. (a), Predators that feed extensively on various trophic levels beneath them, so that competition is more important at higher trophic levels than at lower trophic levels and so that species at the lowest trophic levels are more likely to be limited by predation than by competition. (b), How the physical environment—unusually rough water, for example—may remove species at some of the top trophic levels.

The World is White, Yellow, and Green

As Oksanen writes, "it makes no sense to ask why the world is green while standing in the middle of the Atacama desert or the northern shores of Greenland" (1988, 429). A biogeographical extension of the hypothesis of Hairston et al. was proposed by Fretwell (1977) and was expanded by Oksanen et al. in 1981. They suggested that the terrestrial species' resistance to the removal of their predators or competitors varies along a gradient of primary productivity (figure 12.4). In extremely unproductive or "white" ecosystems, natural graz-

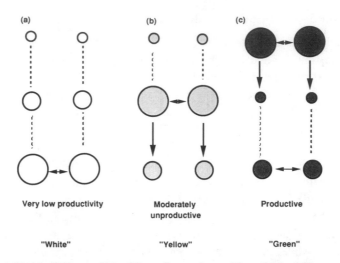

Figure 12.4. Sketch of "the world is white, yellow and green" hypothesis of Oksanen et al. (1981). Solid lines indicate strong interactions, and dashed lines indicate weak interactions. (a), Habitats of very low productivity, where herbivores cannot limit plants, which then compete. (b), Moderately unproductive habitats, where plants are limited by the herbivores and where the herbivores then compete but the plants do not. (c), Most productive habitats, where predators can limit the herbivores and where "the world is green" hypothesis operates (see figure 12.1).

ing pressure will be very light, because there is not enough forage to support effective grazers. At the highest levels of plant productivity, or in "green" eco-systems, there will also be only light grazing, because the grazers will be limited by predators. Between these two extremes, there will be a class of "yellow" ecosystems, where the plants are limited by their herbivores, because there are insufficient herbivores to support large numbers of predators.

These ideas should seem immediately discordant with my repeated claims that food-chain lengths are more or less constant and that they do not vary with primary productivity (chapter 10). In fact, Oksanen and I agree far more than would seem possible. Oksanen recognizes that ecosystems with low primary productivities often have both herbivores and carnivores. He argues that the presence of higher trophic levels is often sporadic in both space and time and that, when these higher levels are present continuously, their impact on lower trophic levels may be minimal. I agree that, in unproductive ecosystems, the herbivores that are present may have little effect on the plants and that their predators may have little effect on them: in published food webs, there is usu-ally no indication of the strength of the interspecific interactions (Pimm and Kitching 1987). In short, Oksanen and I are counting trophic levels differently; I am merely recording their presence, whereas he is concerned with their functional importance. (The special issue of *Oikos* [volume 50, number 3, 1987] provides an extended discussion of this topic.)

The same ideas make more complicated predictions about resistance than do the previous hypotheses. Plants should compete in both the least productive and the most productive ecosystems but not in ecosystems of intermediate productivity. Herbivore removals should have effects on ecosystems of intermediate productivity but not in either the least and the most productive ecosystems. The most productive habitats are green but would not be if predators were removed. Herbivores compete in ecosystems of intermediate productivity, whereas predators compete in ecosystems of greatest productivity.

Which Hypothesis do the Surveys Support?

The four hypothesis make different predictions about the relative effects that the removal of plants, herbivores, and carnivores has both on their competitors (plants, herbivores, and carnivores, respectively) and on their prey (that is, the effect that herbivore removal has on plants and the effect that carnivore removal has on herbivores).

Results of Predator Removals.   Sih et al. (1985) argued that their survey of predation research provided strong support for the predictions of Menge and Sutherland. Because of the confounding effects of habitat, they broke their data into a two-way table, by habitat and trophic level (table 12.2). Their survey found that herbivore removals had significant effects on plant densities more commonly than carnivore removals had significant effects on herbivore densities. When Sih et al. looked at the magnitude of these effects, they found that the herbivores also had generally larger effects on plant communities than carnivores had on herbivore communities. In short, the lower the species in the food chain, the less resistant the species is to predator removals. Sih's et al. survey showed that the trend predicted by Menge and Sutherland held for both terrestrial and aquatic habitats. Plants were *not* resistant to herbivore removals, and this result rejects both "the world is green" and "the world is prickly and tastes bad" hypotheses.

Implications for Competitor Removals.   The ideas of Hairston et al. have implications for competition as well as for predation—namely, that carnivores and plants should compete but that folivores should not, because folivores are limited by their predators. Schoener's (1983a) analyses involved a number of different comparisons and ways of grouping data. His conclusions were that freshwater species probably supported the predictions suggested by Hairston et al.: competition is less important among herbivores than among plants and carnivores. Marine species showed "substantially weaker support" for the predictions. Finally, terrestrial species were "highly supportive" of the predictions, with one caveat: the extent of competition between carnivores was relatively low, in part because four of five experiments on spiders failed to find evidence for competition. Connell, reviewing both his own survey and Schoener's, concluded that "taken together the evidence is still insufficient and too inconsistent to either support or reject" (1983, 671) the ideas of Hairston et al.

The Role of Primary Productivity. "The world is white, yellow, and green" is the most difficult hypothesis to test. The importance of competition and predation alternates between trophic levels, as it does in "the world is green" hypothesis. In addition, the trophic levels at which competition or predation are important change along gradients of productivity. Oksanen (1988) reviews the evidence from the competition and predation surveys and finds much data to support his ideas. Of course, he is not an impartial spectator, and much of the evidence he assembles is anecdotal; but three of his arguments are worth considering in detail.

First, the prevalence of significant effects when herbivores are removed means that Murdoch's "the world is prickly and tastes bad" hypothesis cannot be generally true. Second, the prevalence of significant increases after herbivore removals does not refute Oksanen's ideas. Of course, herbivore removals should not have an effect if they are folivores in either the least or most productive terrestrial habitats. However, experimental removals of herbivores have usually been done in aquatic habitats or in "yellow" terrestrial habitats, such as grasslands, where the hypothesis predicts that there should be effects. Oksanen has a rather ingenious argument to explain why nobody has removed herbivores from "green" habitats. Where these habitats are concerned, "the prevailing attitude . . . is that herbivory is usually unimportant" (1988, 426). In other words, experimenters only remove species when they expect effects.

As Donald McQueen has pointed out (McQueen et al. 1986, and personal communication), there is an interesting complication in freshwater ecosystems. In relatively productive habitats, manipulations of herbivores or planktivores often have little effect on the algae. In these habitats the algae are usually blue-green algae, which are less edible than green algae, and it is because of their unpalatability that the blue-greens are little affected by herbivore removals. This supports the "the world is prickly and tastes bad" hypothesis—but only for these systems.

Oksanen's most convincing evidence comes from an explicit test of his ideas about plant competition. Plants should compete in habitats of either low or high productivity but not in intermediately productive habitats. Reanalyzing the plant-competition experiments in Schoener's survey neatly confirms this prediction. Thirty-one experiments in "green worlds" such as forests, meadows, and tall-grass prairies clearly demonstrated competition, and six experiments did not. In moderately productive habitats with scanty vegetation (grasslands and desert grasslands), four experiments demonstrated competition, but twenty-five did not. In the least productive systems (deserts), four experiments demonstrated competition, and none failed to do so.

Which Hypothesis is Best?   The data in the surveys unequivocally reject both "the world is green" and "the world is prickly and tastes bad" hypotheses. No herbivore removals should have effects on the vegetation if these ideas are correct. Herbivore removals *do* have effects, and, what is more, introduced herbivores often have quite devastating effects on the vegetation (chapter 15). *Any* effect at all rejects the simple application of these hypotheses, yet, surely it is too much to expect that not one herbivore removal would ever have an effect Plants quite generally may not be effected by herbivores, yet ecologists may have carefully chosen those few exceptions as subjects for their removal experiments. Deciding between "the world is generally green" and "the world is white, yellow, and green" hypotheses may be very difficult.

Quite simply, I doubt whether the surveys of species removals are really appropriate for testing the four hypotheses. The theories and the data are at different organizational scales. For predation, the theories deal with the *average response* that *all* the species at one trophic level have to the removal of the average of *all* the species at the level above. For competition, the theories deal with the removal of an average species and with that removal's effect on the average species at the same trophic level. For convenience, I shall say that the theories deal with the *average one-on-one* effect.

The experimental data usually report the effects that the removal of one particular species has on the abundance of another particular species. Such experiments deal with *particular one-on-one* effects. Sometimes, the experiments involve the effect that one species has on a small number of other species: *one on many* effects. Only rarely do the experiments involve the effect that removal of all or most of the species at one trophic level has on all or most of the species at a different trophic level. This latter category is the total of the average one-on-one effects. What the four hypotheses do is make predictions about such experiments, and I shall examine these experiments presently.

The four hypotheses, however, do *not* make convincing predictions about the particular one-on-one effects, for at least two reasons. As already noted, the species removed may have been carefully selected to be those perhaps very few species whose removal should have effects. Second, in addition to the

trophic level of the species removed and the productivity of the habitat, there
are many other factors likely to affect resistance. These factors are the subjects
of the next three chapters. Unless trophic level and productivity are of overwhelming importance, we may miss their influence unless we have very large
numbers of randomly chosen removal experiments.

What is needed are experiments in which the species removed are those
chosen to test the four hypotheses directly. The closest we can come to this
ideal are those studies in which many or all of the species at one trophic level
are removed and in which the experimenters record the effect on all the species
at the trophic levels beneath them. It is to these experiments that I now turn.

McQueen's Review of Trophic-Level Removals

McQueen et al. (1989) have reviewed the extensive literature on freshwater
pelagic community structure. He wrote that there is "substantial agreement
among limnologists" that fish have a major impact on the biomass of the zooplankton on which they feed. This is a discovery about the effect of one trophic
level on the trophic level immediately beneath it, and aquatic ecologists have
begun to expand the results across three or more trophic levels. Shapiro and
Wright (1984) suggested that reductions in fish (trophic level three) might lead
to increases in zooplankton (trophic level two) and so to decreases in phytoplankton (trophic level one). Carpenter et al. (1985) expanded this idea to
include the effects that could cascade downward from the fourth trophic level—
fish which eat other fish. Carpenter et al. (1987) tested this idea by adding fish-
eating fish to lakes and showing an increase in zooplankton and a reduction in
algae.

These results are consistent with all but "the world is prickly and tastes bad"
hypothesis, for all three of the alternatives postulate the importance of predators in controlling species at lower trophic levels. The critical difference between the hypotheses involves the effect of herbivores on plants—that is, zooplankton and phytoplankton, in the case of lakes. In reviewing studies of
ponds, McQueen et al. found that the results of zooplankton manipulations
were not clear-cut. They wrote that "some exclosure work (Lynch and Shapiro
1981; Levitan et al. 1985; Hambright et al. 1986; Langeland et al. 1987; and
Vanni 1987a, 1987b) found a direct relationship between . . . decreased abundance of all zooplankton (or large zooplankton) and increased phytoplankton
biomass . . . Other enclosure studies (did not find) decreases in overall phytoplankton biomass."

He reviewed ten whole-lake studies of the impact of zooplankton on phytoplankton and found that four reported that decreases in zooplankton increased
algal biomass and that one reported the comparable result that increases in

zooplankton decreased algal biomass. The other half of the studies reported
only short-term changes. McQueen et al. suggested that the differences lay in
the productivity of the lakes, with zooplankton having stronger effects on phytoplankton in lakes with low productivity than in lakes with high productivity
(McQueen et al. 1990). At least one reason why herbivores have relatively
little impact in unproductive lakes, however, is because such lakes contain
blue-green algae, which are less edible (McQueen et al. 1991). In short,
McQueen et al. find that the limited data may support the more complex theory
of Oksanen et al. but perhaps also the local importance of unpalatable plants.
Effects cascade downward, but where the regulation occurs depends on the
productivity of the system.

LOOKING UPWARD: RESISTANCE TO CHANGES IN LOWER TROPHIC LEVELS

We have reviews of the effects of predators on their prey (and sometimes on
the prey's prey), and we have reviews of the effects of species on their putative
competitors. However, there are no extensive reviews of mutualistic interactions, probably because there are few manipulative studies of mutualisms.
What is perhaps more surprising is that there are no reviews of the effect that
altering the abundance of prey species has on their predators. In other words,
most manipulations look downward—not upward—through the food web.
Why should this be the case? Predators are generally larger than their prey, and
competitors are of the same general size. The larger the animal, the larger its
home range (Southwood 1978). Consequently, removing predators leads to
changes in the densities of their prey, on a spatial scale smaller than that of the
prey, and the removal of competitors leads to changes in competitors, on the
same spatial scale as that of the prey. In contrast, removing prey leads to
changes in their predators, on spatial scales much larger than that of the experiment.

There are two kinds of studies that look "upward." There are experiments
in lakes and ponds where ecologists have added nutrients and measured
whether there are increases in the phytoplankton and higher trophic levels.
There are also nonexperimental studies that look at how densities of herbivores
and carnivores change along natural gradients of primary productivity.

Freshwater systems. I have examined the relationship between natural plant
productivity and carnivore biomass by using previously published data (Pimm
1982). Increased productivity leads to increased carnivore biomasses. Mc-
Queen et al. (1989) reviewed a wider range of aquatic studies, including those

of systems in which productivity had been experimentally enhanced with nutrient inputs. He found the same result. Interestingly, increases in nutrients in lakes have effects that are progressively weaker on phytoplankton than on zooplankton than on planktivores.

Terrestrial systems. McNaughton et al. (1989) find that, along natural gradients of plant productivity, a two-orders-of-magnitude increase in productivity causes a four-orders-of-magnitude increase in herbivore biomass. I also found that, along gradients of increased primary productivity, carnivore biomasses increased too (Pimm 1982). Arditi and Ginzburg (1989), using data originally compiled by Whittaker (1975), found that plant biomass and animal biomass increased along gradients of increased primary productivity, and, in a result which parallels those for aquatic systems, the increase in plant biomass is stronger than the increase in animal biomass. (The animals were probably mostly herbivores, because the latter are generally more abundant than carnivores.)

These increases in the biomass of one trophic level that occur with an increase in either primary productivity or the biomass of a lower trophic level are not necessarily what theory predicts. Changes ought to be more complex than a simple increase of one trophic level when there is an increase in the trophic level beneath it. The difficulty is that, with the exception of "the world is white, yellow, and green" hypothesis (to which I shall return), no hypothesis appears to deal explicitly with the changes to be expected.

Let me propose some possible changes, by considering a three-trophic-level system where predators control herbivores. When a simple three-species, three-trophic-level Lotka-Volterra model is used, some easy algebra tells us that increases in plant productivity ought to lead to changes in the densities of the plants and the predators but not to changes in the densities of the herbivores, since the predators would consume the increased production of the herbivores. Consequently, changes in plant productivity ought to cause *smaller* changes in herbivore biomass than in predator biomass. This prediction is at odds with the observations. Arditi and Ginzburg (1989) and McQueen et al. (1989) observed, along gradients of increasing primary productivity, larger changes in herbivores than in plants. McNaughton et al. provide no comparisons, but they do find that herbivore biomass increases faster than primary productivity.

There are at least two possibilities to explain these discrepancies. The Lotka-Volterra models may be wrong. This is Arditi and Ginzburg's conclusion, and they develop an alternative model that correctly predicts increasing herbivore biomass with increasing primary productivity. (I shall return to this model in chapter 16.) It may be that it is the *simple* Lotka-Volterra model that is wrong. If we build a Lotka-Volterra model of a reasonably complex food web—that is, one that has more than one species per trophic level—then it is far from obvious that changes in primary productivity will lead only to changes in the abundance of the top-predators. Quite how the changes should be distributed has not been investigated.

Even if the models are all wrong, they force us to ask the interesting question of why an increase in primary productivity does not benefit the top-predators alone. Why do the top-predators not take up the increased production of lower trophic levels (as they do in the simplest Lotka-Volterra models)?

The second possibility is that, as productivity increases, the regulatory role of the top-predators may be changing as well. Return again to simple Lotka-Volterra models that have one species per trophic level, but now compare how models with one, two, and three trophic levels respond to changes in primary productivity. A simple way to alter primary productivity is to increase the carrying-capacity term for the species at the base of the food chain. (This means reducing the term in a_{11} of equation [6.4], for it determines how much a species' density reduces its own growth rate.) Obviously, for a one-species model (of the plant), this increase in primary production causes an increase in the equilibrium density of the plant. For a two-species model, the plant density remains the same, but the herbivore density increases, and, for the three-species model, plant and carnivore biomass increase, and the herbivore biomass remains the same.

Over gradients of primary productivity large enough to support one-, two-, and three-trophic-level systems, we would indeed see relationships, in all pairwise combinations, between plant biomass, and herbivore biomass, and carnivore biomass, and primary productivity. The relationships would be nonlinear; herbivore biomass, for example, should increase quite rapidly up to the point where the productivity is sufficient to support carnivores, and it would remain constant thereafter. This argument, of course, is a restatement of Oksanen et al.'s "the world is white, yellow, and green" hypothesis.

I have already discussed Oksanen's argument that, even in systems that *functionally* have one or two trophic levels, species at higher trophic levels may be present but may feed locally or sporadically, perhaps on prey that, for other reasons, are doomed to die. Such species may be quite rare, but may increase in biomass rapidly when the productivity of the system is sufficient to support viable populations. An explicit prediction of Oksanan et al.'s model is that plant biomass should not change linearly over gradients of plant productivity, and Oksanen (1983) assembled data to support this prediction.

Nonlinear changes are not obvious in the studies I discussed, though Neill (1988) shows how complex the responses to nutrient enrichment of lakes can be. The ranges of productivities may not be appropriate, however—they may all fall within the range of "yellow" or "green" ecosystems or across the boundaries between these ecosystems. McNaughton et al.'s finding that herbivore biomass increased much faster than primary productivity is consistent with a nonlinear response, however.

In sum, I find that the observed and theoretical patterns of resistance to changes in higher trophic levels after alterations in lower trophic levels present a considerable challenge. The empirical patterns may well support the hypothesis that changing productivity changes the regulatory role of the herbivores and predators, but a full range of alternatives has not been explored.

SUMMARY

Resistance is the degree to which something is unchanged when something else is permanently changed. This chapter has focused on the extent to which species densities are changed when densities of other species in the community are altered by either additions or removals. Even this restricted topic is complex, both because additions or removals may be at very different spatial scales and because their consequences may be examined at different levels of biological organization.

It is not obvious that removing one species should have effects on the density of another, even when the latter species is its prey or putative competitor. If predator removals do cause an increase in the density of the predator's prey, or if competitor removals produce an increase in the density of competing species, can we predict which species will increase—and by how much?

Several surveys have reviewed studies on predator and competitor removals. The great majority of these removals produces the expected increases in the species counted; species were not resistant to the removals. Such studies are biased in many ways, however. Studies that find nonsignificant results are less likely to be published than those that find significant results. Studies that do find significant results are probably suggested by simple natural history observations that indicate the effect that one species has on another. Major classes of species and communities are greatly underrepresented in the surveys.

Perhaps the most famous hypothesis that tries to predict the species that are most resistant to removals is that of Hairston et al. (1960): predators should limit the density of their herbivorous prey, and, in turn, these herbivores should not be able to limit the density of the plants on which they feed; the world, in consequence, should be green. The evidence from removals of both competitors and predators usually does not support this hypothesis. The data often reject the alternative explanation that "the world is green" because plants are so well protected by their physical and chemical defenses. An intriguing extension of "the world is green" hypothesis was proposed by Fretwell (1977) and Oksanen et al. (1981). They argue that, at the lowest levels of plant productivity, there will not be enough food for effective herbivores; that, at higher levels of plant productivity, effective herbivory will be possible and will greatly reduce the density of the plants; and that, at yet higher levels, predators will begin to control the herbivores. Most predator and competitor removal studies have not been examined in the light of this hypothesis, but some data on plant competition support it.

The data available may not be appropriate for testing the hypotheses. All four of these hypotheses predict changes of all the species at one trophic level when all the species at another trophic level are removed. In contrast, the data thus far accumulated address the effects that removing one species will have on the densities of (at most) a small number of other species. This is a startling mismatch of scale; any one of the theories could be correct, yet studies of small numbers of species might fail to verify it. Worst of all, the four hypotheses make their most distinctive predictions about where removals should *not* have effects, and ecologists are likely not to report negative studies—or even to undertake studies that are likely to find such results.

Only in aquatic systems have ecologists manipulated entire trophic levels. There is evidence that top-predators affect predators and that predators affect herbivores; the effects that herbivores have on phytoplankton are mixed. Some studies find such effects; others do not. There is some suggestion that the difference is due to the level of plant productivity, so supporting the hypothesis of Oksanen et al. In lakes with high productivity, herbivores may have little impact on the blue-green algae that dominate such systems, because these latter species are unpalatable. This result supports "the world is prickly and tastes bad" hypothesis.

13

Food Webs and Resistance

A FOOD-WEB THEORY

When we alter the density of a species and look for consequent changes in the density of other species, these changes involve both direct and indirect effects. Indirect effects might occur when species A affects species B through a chain of intermediate species: species A affects species X, which affects species Y, which affects other species, until one or more of these species affects species B. Purely exploitive competition is an indirect effect. While we may be aware that indirect effects are possible in principle, do we regard these effects (other than exploitive competition) as being of greater magnitude than the direct effects? Ecologists generally have looked at the direct effects. Indirect effects can be quite different from direct ones: a plant species might *benefit* from its herbivores because they feed on its competitors more than they feed on it. If direct effects predominate, then we can ask simple questions about resistance: How much does a prey species increase in density when we reduce the density of its predator? If indirect effects predominate, we need to ask, *Does* a prey species increase in density when we remove its predator? Uncertainty about the direction of the change in density might make it unlikely that we could produce a simple theory about resistance.

If indirect effects are important, then interpreting removal experiments will not be simple. The last chapter suggested that, in order to be able to predict the experiment's outcome, we may need to know both the productivity of the habitat in which the experiment was carried out and the trophic level of the species removed. If indirect effects are important, then we may also need to know the structure of the community's food web (Bradley 1983). To explore these possibilities, Yodzis (1988) simulated press experiments for sixteen observed food webs. He concluded that indirect effects are of overwhelming importance. It was even more surprising that the outcomes of his simulated press experiments—that is, investigations into whether species increased or decreased—exhibited a high degree of sensitivity to the values of the parameters that describe the interactions between the species. In practical terms, given the difficulty of measuring these parameters and their considerable inherent variability, this means that the outcomes of press experiments may be unpredictable. Yodzis has produced an uncompromisingly negative theory: not only may we be unable to say anything about the resistance to change of a species' density, but we may be unable even to say anything about whether it increases or decreases.

Yodzis' theory may seem both surprising and unlikely, in view of the results of real press experiments discussed in chapter 12. Results of these press experiments seem both to be predictable and to involve changes in the expected direction. I think Yodzis' models are probably giving the right answer, not just because we both use the same modeling recipe, but because Yodzis' models point to rarely appreciated features of real press experiments. Chapter 16 provides the details of how to go from a known food-web structure to a dynamical model of the species' densities in the corresponding food web, but to understand the results I must now present the outline of the modeling recipe.

Details of the Model

In the first step, one must assume that the dynamics of each species in each of several communities with known food webs can be described by a particular set of equations. Second, one must assign values to the parameters of these equations, for each pair of interacting species. The known food web can provide the *signs* of these interactions—that is, whether a species increases, decreases, or leaves unchanged the growth rate of other species in the community. The next step in getting the dynamics involves making further assumptions about the *magnitudes* of the interaction parameters. To arrive at these magnitudes one makes plausible guesses, based on the nature of the particular organisms involved, about the relative magnitudes of the parameters. For example, predators typically have a greater effect on the growth rates of their prey than the prey have on the growth rates of the predator (Pimm and Lawton 1978). For each prey killed, only a small number of baby predators may be born, the remaining energy going to support the maintenance costs of the predator. Finally, the actual *parameters* are chosen randomly over the ecologically reasonable magnitudes dictated by the ecology of the interactions.

Given the model and its parameters, some simple algebra yields the *community matrix*. Each term in this matrix specifies how the density of one species alters the growth rate of itself or another species when all the species are at their equilibrium densities. Obviously, in a model with n species, there are n^2 such terms. The community matrices formed by Yodzis were generally the simplest ones that could plausibly be associated with the sixteen real food webs, in that they had the smallest possible number of direct interactions. An example of one of these sixteen webs is shown in Figure 13.1, and the elements of the community matrix are shown in graphic form in figure 13.2. This process of randomly selecting parameters over sensible ranges is repeated many times to produce a set of models for each real food web. Yodzis performed this process one hundred times for each food web.

The community matrix is not the final stage, for what we require to know is the effect that changing the density of one species has on the *density* of another species, not the effect that it has on that other species' *growth rate*. For any community matrix, one can obtain another matrix, called the *inverse* of the community matrix, by a straightforward if computationally tedious process. This inverse provides the required information and can be thought of as a table that summarizes the outcomes of all n^2 possible press experiments. The signs of the inverse of the community matrix indicate the directions for the outcomes of press experiments—that is, whether adding members of species A will cause an increase or a decrease in the equilibrium density of species B. Figure

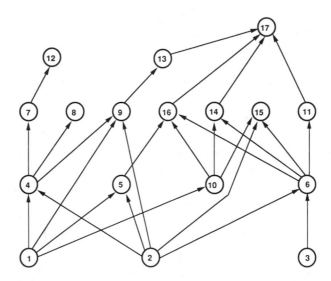

Figure 13.1. Food web for Naragansset Bay, from an original food web presented by Kremmer and Nixon (1978). Arrows indicate which species are the prey of which other species. This web is in the form published by Yodzis (1988) who grouped species (those indicated with an asterisk in the list that follows) if they shared exactly the same prey and predators. Designations are: 1, flagellates and diatoms; 2, particulate detritus; 3, macroalgae and eelgrass; 4,* Acartia, other copepods, and sponges; 5, benthic macrofauna; 6, clams; 7, ctenophores; 8, meroplankton and fish larvae; 9, Pacific menhaden; 10, bivalves; 11, crabs and lobsters; 12, butterfish; 13*, striped bass, bluefish, and mackerel; 14, demersal species; 15, starfish; 16, flounder; 17, man.

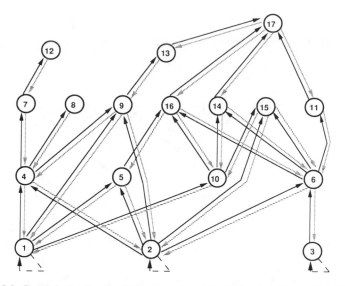

Figure 13.2. Graph indicating the likely dynamics of the food web in figure 13.1. The solid arrows indicate the positive effect that the prey's density has on the growth rate of its predator, and the light arrows indicate the negative effect that the predator's density has on the growth rate of its prey. Species at the base of the food web are likely to be limited by resources (such as space or light) and to have negative feedbacks acting on them as a consequence. Designations are as in figure 13.1. (Redrawn from Yodzis 1988.)

13.3 shows a graphic representation of one possible inverse of the food web I have been using as an example.

The Results

We expect that if we construct a hundred different plausible community matrices with different parameters then the inverses of the community matrices will be different as well. It is possible that even the *signs* of those matrix elements will vary somewhat, because the different models will place different emphasis on different indirect pathways through the food web. If indirect effects are important, even our most elementary expectations for press experiments cannot be taken for granted. For example, removing predators will not necessarily result in an increase in density of one of its prey species, because that prey species and another prey species may be strongly competing. Whether that one prey species increases or decreases when the predator is removed will depend on the strengths of the various interactions between predators and prey and between the prey and the resources for which they compete.

Yodzis computed the inverse matrix for each of the one hundred community matrices corresponding to each of his sixteen food webs. He then defined each

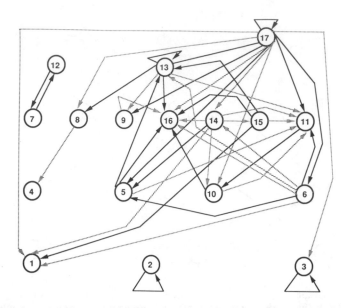

Figure 13.3. Possible graph of the effects that small changes of equilibrium species densities have on the equilibrium densities of other species in the food web of figure 13.2, when both indirect and direct effects are considered. (In figure 13.2, the arrows show how changes in equilibrium densities affect the growth rates of other species.) The patterns of these changes are extremely sensitive to small changes in the parameters that describe the interactions between species. Small changes in the parameters often cause even the direction of the changes to alter. Solid arrows indicate where an increase in the density of one species will have a positive affect on the density of another, and light arrows indicate negative effects. For example, although species 17 feeds on species 11, increases in 17 may actually benefit species 11, because 17 also feeds on a competitor (species 14) of species 11. Species 14 and 11 compete because they share species 6. Designations are as in figure 13.1. (Redrawn from Yodzis 1988.)

effect of a particular species on another particular species to be *directionally undetermined* if *less than* ninety five of the resultant elements of the inverse of the community matrix had the same sign.

Table 13.1 summarizes Yodzis' results, classifying the effects as belonging to several classes. In what follows, whenever I speak of the *effect* of species A on species B, I will mean it in the context of a *removal* experiment. The "predator-on-prey" effect involves removing predator individuals and observing the effect on prey density: we expect the prey density to increase. The prey-on-predator effect corresponds to removing prey individuals and observing the effect on predator density: we expect the predator density to decrease. Yodzis found that more than half of these effects are directionally undetermined, and that, in some cases where the effect is determined, the effect is the opposite of what we would expect (table 13.1)!

As an example, Yodzis considers a food web for Narragansett Bay (figure 13.1). In this figure, species 4 has a direct effect on species 1, because species 4 eats species 1. Species 4 also has an indirect effect on species 1, because species 4 feeds on species 9, which in turn feeds on species 1. Species 4 also affects species 1, through the path from species 4 to species 9 to species 13 to species 17 to species 16 to species 5 to species 1. Another indirect effect runs from species 4 to species 9 to species 2 to species 5 to species 1. You can find other indirect pathways from species 4 to species 1.

The net effect of a predatory-prey interaction is the sum of the direct effect plus all the individual indirect effects. It is this net effect that we observe in real press experiments and that I have been calling simply "the effect." The indeterminacy and the "reverse determinacy" in these net effects can come about only if the direct component is swamped by a stronger indirect component.

Finally, Yodzis analyzed the effects that are purely indirect. In the Narragansett Bay community, species 4 has no direct effect on species 15, because there is no link from species 4 to species 15. There is an indirect effect of species 4 on species 15, through a number of pathways such as 4—2—15, 4—9—13—17—14—10—15, and so on. Here too, Yodzis finds considerable indeterminacy (table 13.1). Of particular interest is that subset of indirect interactions for which we might expect competitive effects because the two species involved share a prey. The expected effect is that increasing one species will

TABLE 13.1   The Degree to which Density Changes Will Lead to Other Density Changes that are Predictable in Direction (after Yodzis 1988.)

Type of Effect	Percentage Directionally Determined[a]	Percentage not in Expected Direction[b]
All effects:		
Predator on prey	52	11
Prey on predator	54	7
Indirect:	50	(directions are not predictable)
All Competitive	58	29
Important effects:		
Predator on prey	55	10
Prey on predator	58	2
Indirect:	53	(directions are not predictable)
All Competitive	64	31

[a]To be directionally determined, at least 95% of all the interactions between a given pair of species should be in the same direction. All percentages were based on 200–3,400 pairs of species interactions from sixteen food webs; see text for details.

[b]Expected directions are (a) for predator removals to lead to increases in the density of their prey, (b) for prey removals to lead to decreases in their predators, and (c) for competitor removals to lead to increases in the densities of their competitors. For example, 29% of competitor removals actually lead to decreases in the density of putative competitors.

result in a decreased density of its competitor. For those competitive effects that are determined, 29% show the reverse of the expected effect. Thus 29% of putative competitors are indirect mutualists.

Some of the many effects in these systems will be stronger than others. Yodzis continued by considering the possibility that his results mix large effects and small effects indiscriminately. Possibly the largest effects are highly determined (and in the expected direction), with most of the indeterminism confined to effects that are so weak as to be uninteresting. Where are the effects that indicate points at which the community is least resistant to change? Yodzis also presented data on what he defined as the important effects (meaning, in a certain sense, the largest effects). He showed that even the most important effects have still a high degree of directional indeterminacy.

What these results do not obviously tell us is how the magnitude of effects varies with the length of the pathway between species. First, are direct effects (pathways of length one) generally more important than indirect effects (longer pathways)? Comparison of figures 13.2 and 13.3 shows that the direct effects are typically the important ones and that indirect effects are typically the less important ones. There are some exceptions: in the example of figure 13.3, species 1 and 6 have important effects on each other, even though they do not interact directly. Second, how do the magnitudes of effects vary with path length in general? This is a topic that, to my knowledge, has not yet been explored, though Thomas Schoener (personal communication) is beginning to do so. His suggestion is that effects become less important, the more indirect they are. My reply is to agree but to notice that there are some kinds of indirect effects that may be very large. We shall encounter these in the next chapter.

Are Yodzis' Results Sensible?

Yodzis analyzed sixteen food webs from a wide variety of habitats, geographical locations, and researchers. All the models suggest that indirect effects make it difficult to discern whether the increase in one species' density will increase or decrease the density of another species. We seem to be a long way from being able to predict how large or how small these changes would be—and so a long way from being able to build a simple theory of resistance.

One problem with Yodzis' work stands out. At least at face value, his results seem totally inconsistent with what we know about real press experiments. The data discussed in chapter 12 seem to indicate that predators reduce the density of their prey and that putative competitors usually compete and are not indirect mutualists. In other words, the real world seems far more determined than Yodzis' models would predict. What is wrong with Yodzis' results? While the models have many failings, it would be a mistake to dismiss them. Making the models more biologically realistic would probably increase rather than diminish the degree of indeterminacy. Perhaps we should consider how to conduct and interpret press experiments in the field rather than consider how we might modify the theory.

INDIRECT EFFECTS IN REMOVAL EXPERIMENTS

Yodzis obtains his results because the effects of one species on another that are mediated indirectly through other species are more important than the direct effects. It is the complexity of food webs that leads to the very large number of possible pathways along which indirect effects propagate. There are two kinds of evidence for indirect effects. First, some removal experiments show initial effects on one group of species and then we observe directly the secondary effects—that is, the influence of these species on other species. Second, after some species' removals, some species' densities behave in ways unexpected on the basis of direct effects alone; indirect effects are invoked to explain the anomalous results.

A Summary of Food-Web Studies

I assembled data from nineteen studies of predator removals (Pimm 1982), based on several criteria. First, I wanted either research that contained explicit food-web descriptions or research that was directly comparable to that of another study in which this information was available. My objectives were different than those of Sih et al. (1985), for I wanted to look at more than the direct effects that a predator X has on its immediate prey Y. The significance of looking at studies in which the food web was known was that it enabled me to consider indirect effects of X on Y, of Y on Z, and hence of X on Z. At the time, this criterion of requiring food-web data excluded all terrestrial studies; therefore the nineteen studies I chose consisted of data on freshwater ponds and rocky intertidal habitats.

Data in two of the nineteen studies failed to show that significant changes in either species density or species composition occurred after predator removal, a proportion of significant results entirely consistent with the results of the survey by Sih et al. (1985).

In those seventeen studies that did show effects of the removal of a predator, most removals caused multiple species effects. Typical is a study that removed fish from a pond in which their prey were principally larger zooplankton (Hall et al. 1970). After the fish were removed, a large species of cladocera, *Ceriodaphnia* sp., became the predominant zooplankton, and how dense it became depended on the nutrient levels in the pond. Smaller species of cladoceran and other herbivores (rotifers and copepods) were eliminated through competition. In another study, Lynch (1979) removed fish and observed similar results. When the fish were present, smaller species of zooplankton such as *Bosmina* sp., *Cyclops* sp., and rotifers predominated. The rotifers supported a predator, *Asplancha*. Predator removal resulted in an increase both in larger species such as species of *Daphnia* and *Ceriodaphnia*, and in predators such as *Chaoborus*. Rotifers were reduced in numbers and were not sufficiently abundant to support *Asplancha*. Post and McQueen (1987) found the effects of their fish removals to be comparable. The fish removed the larger cladocerans, and the smaller species increased. Changes in fish density did not alter the density of phytoplankton, however.

The conclusion I reach from this admittedly small set of studies is that indirect effects are the rule rather than the exception. Since I completed this survey, many studies have been published showing the importance of indirect effects. These include Kneib's (1988) study of fish and shrimp in salt marshes; Power et al.'s (1985) study of algae, herbivorous, and predatory fish; and Morin et al.'s (1983) and Morin's (1986) studies of how predators mediate competition in aquatic systems.

Unexpected Effects

That we routinely observe the effects of press experiments but have difficulty predicting their exact outcomes a priori is further substantiated by looking at what Sih et al. (1985) called "unexpected effects." The prey of some species not only fail to increase significantly in density when predators are removed, but some of the prey species significantly *decrease*. Sih et al. (1985) called a result in which removals of predators seemed detrimental to the prey species an "unexpected effect." Predator removals might be detrimental by causing a decrease in their prey's feeding or growth rate, survivorship, reproduction, population size (of one prey species or of all prey species summed) or species richness.

Unexpected effects arise when the *direct* effects that predator removals have on prey density are less important than effects that occur through *indirect* pathways. For example, a predator may reduce the density of an aggressively competitive prey species to the point at which less aggressive competitors benefit. Remove the predator, and some of its prey may suffer declines or even local extinctions at the hand of the superior competitor. Another example involves a three-trophic-level system. When a top-predator consumes a middle-level predator, it may release from middle-level predation the prey at the lowest trophic level.

In their survey, Sih et al. (1985) found that 30%–50% of all studies yielded some unexpected effects (table 12.1). Unexpected effects were often "large" ones, by the factor-of-two change-in-density criterion. Of studies showing significant effects, 76% of expected effects were large, while 60% of unexpected effects were large. Unexpected effects—and so, presumably, indirect effects—appear to be common.

Unexpected effects were more common when carnivores were removed than when top-carnivores were removed. Removals of herbivores had the fewest unexpected effects on their prey. Removals of vertebrates caused more unexpected effects than did removals of invertebrates. These differences may occur because, at higher trophic levels, removals of predators can be beneficial to their prey's prey.

Connell (1983) but not Schoener (1983a) reported experiments in which the removal of a species *decreased* the density of a putative competitor. This result is "unexpected" in the sense in which Sih et al. (1985) used the term. The species might have been directly mutualists, or the result could come about in a variety of ways that involve indirect effects. For instance, removing a species might increase the density of a competitor, an increase that might cause another competitor to decline. Such unexpected changes were documented in four studies of plants, seven studies of herbivores, and three studies of omnivores or carnivores. Connell (1983) noticed that these responses involved only one member of a pair, except in two cases where two species showed mutual effects. In some cases, the results might be statistical artifacts, stemming from the large number of experiments performed.

UNCERTAINTY IN REMOVAL EXPERIMENTS

There is much evidence for indirect effects in species-removal experiments. This makes Yodzis' ideas possible, for they depend on the importance of indirect effects. But there remains the problem that I have just discussed: the conflict between the unpredictable modeling results and the generally predictable outcomes of field experiments. In fact, the certainty in outcomes of competitor and predator removals in the field is far more apparent than real. There are many biases that lead to this apparent certainty. To uncover these biases, let us consider, first, how the field experiments are performed and, second, examine what happens when field experiments are repeated. It is a crucial prediction of Yodzis' results that, when the experiments are repeated, they should have highly variable outcomes because the parameters that describe interactions between species are probably constantly changing.

The Nature of Field Experiments

Experimenters do not pick just any pair of species and try to predict the effect that removal or addition of one species has on the other. Instead they examine the effects that removal of a carefully chosen species has on either another carefully chosen species or several carefully chosen other species. In the latter case, when a predator is removed, for example, we might expect that at least one of its several species of its prey would increase in density. That is not a surprising result, but then we are not trying to predict *which* of those several species of prey increases in density. Yodzis' models are trying to predict which species will increase in this circumstance, not whether *any* of them do.

What do I mean by experimenters carefully choosing species? Experimenters may often see that the occasional accidental, local, and natural removal of, say, a species of predator causes a particular prey species to increase. Carefully choosing that predator as the one to remove makes an expected outcome much more likely: we may already know the outcome before the experiment is undertaken; in short, the ways in which experiments are performed makes expected outcomes much more likely. Given that the odds seem stacked in favor of finding an expected outcome, the actual experimental results seem much less predictable than they are on first encounter.

First, consider the studies of predator removals. Sih et al. (1985) found that studies looking for density changes in a particular species did not find significant effects as often as did studies that looked for changes across several different species in the community. They had two explanations for this result. First, significant effects could arise when only one or a few of the many prey species responded strongly. Second, consistent but nonsignificant changes in many species may well be statistically significant when combined. Sih et al. did not make much of this difference between studies with one and studies with many species, but I find this difference very interesting. The first explanation Sih et al. provide makes sense only if the original experimenters had little idea which prey species should increase in density after the predator removals. Therefore, the more species of prey the experimenters measured, the more likely they were to find one that increased in density.

Second, consider the competitor-removal studies. Recall Connell's (1983) results on the proportion of studies that showed competition. He found that, as more species were included in each study, the proportion of experiments in which a significant effect of a removal was detected dropped. Perhaps, by including a wide variety of species in their analyses, the experimenters are seeking to reduce the bias created by choosing the species most likely to be competing. Again, I would guess that experimenters use their knowledge of the natural history of their organisms to predict which species should compete. I would also guess that these species would often be the species chosen to be counted. If I am right, then both the predation and competition experiments have a similarity. The odds are stacked in favor of the experienced field experimenter predicting which species should show responses, yet the experimenters only get the desired significant results when they cast their nets widely. Experimenters can predict that some species will change when they remove others—but not which species will change.

The Variability of Interspecific Interactions

Schoener (1983a) examined variation in the effects caused by removing competitors, though his reasons for doing so were different from those of Sih et al. (1985), who looked at the variation in effects of predation. When reports mentioned competition, or when it could be determined from the data, year-to-year variation in competition definitely occurred in eleven cases and possibly occurred in three others, but it was not detected in twelve cases. In only two or possibly three of the eleven cases did competition appear in some years but not in others. Connell (1983) addressed the same question and found Schoener's half-filled glass to be half empty instead, coming to a different conclusion with very similar results. Thirty-six of the seventy-two studies Connell reviewed gave data on interspecific competition between the same species at different places or times. In about half of these studies no competition was found. Of those species that showed some competition, 59% showed year-to-year variation, and 31% showed differences between locations. When data on annual variation were available, competition occurred in 35% of the years and in 46% of the locations. Connell concluded that "apparently it is not rare for competition to vary in either time or space" (1983, 682).

Summary

Removal experiments are performed in ways that should greatly increase the chance of finding expected outcomes. Which species are removed and which species are counted are both often chosen on the basis of prior information that suggests likely outcomes. Alternatively, experimenters look for the effects that removal has on many species, rather than predict the species whose density should change most. In other words, the odds are stacked in favor of finding at least one expected outcome. Yet, many outcomes of removal experiments have been unexpected in both direction and statistical significance, and the outcomes have varied when the experiments have been repeated. Only by undertaking a large set of removal experiments are we likely to understand the full degree of uncertainty that indirect interactions create. I shall presently address one study that does this.

THE TIME SCALE OF REMOVAL EXPERIMENTS

There is another way in which Yodzis' results can be reconciled with field experiments, and it points to what I consider to be the principal limitation of field experiments. The effects considered by Yodzis contrast equilibrium densities present before a press experiment has taken place with the densities attained after the press experiment has been operating long enough to allow the system to reach a new equilibrium. That is, these are *long-term* effects. The intervening, "transient" behavior—and, especially, the short-term behavior—can easily be different. Indirect effects take time to appear.

Yodzis considers short-term changes too and looks at the *transient* changes in density that occur immediately after the alteration of a species density. He shows that theoretically there is no indeterminacy in this short-term behavior. The transient changes are exactly what we expect on elementary intuitive grounds: predators reduce the density of their prey, and competitors compete. Only over the long term do we encounter unexpected changes. How long is "long term"—how long need we wait to be reasonably confident that the system has settled near its new equilibrium? Yodzis shows that this length of time will depend on the length of the food-web pathways from the species whose density we alter to any species whose response is of interest (figure 13.4). Most field press experiments are relatively short-term, so they appear to contain more results that are expected in both direction and magnitude. Only long-term studies will pick up indirect effects.

In addition, the time needed for indirect effects to appear will depend on the generation times of all species that occur in the pathway. For species with populations that can respond quickly, indirect effects will appear quickly too; but other species may be very long-lived, with populations that respond slowly. Indirect effects through such long-lived species may take a very long time to appear.

To my knowledge, there are few press experiments that expressly look at how effects change over very different time scales and at how long it takes for indirect effects to appear. Quite simply, press experiments usually fit neatly into my classification of most ecology being on scales of one to three years, one to ten species, and less than 10 hectares (figure 1.1). At such scales, the transient effects which emphasize direct species interactions are those that pre-

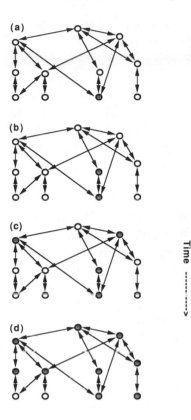

Figure 13.4. Species density changes. When a species' density is permanently changed (a; the shaded species which is the second plant from the right in the food web), the first effects are on those species with which the species interacts (b; its herbivore). These effects then effect other species, and the changes ripple through the community (c, d). While the short-term effects tend to be predictable—increases in predators decrease densities of their prey, for example—the eventual long-term effects are much more difficult to predict.

dominate. Indirect effects will appear on different scales; and, they become progressively more important, the longer we observe the consequences of our initial species removal. There is one fascinating example of the temporal development of indirect effects: the studies by Brown, Davidson, and their colleagues that consider seed-eating birds, mammals, and ants in the deserts of Arizona (Davidson et al. 1985; Brown et al. 1986).

Brown et al.'s Experimental Studies of Desert Granivores

The System and the Experiments. The system of seeds and granivores is an important one in desert communities. Annual plants, because they can complete the vegetative part of their lives after brief and unpredictable rains, account for a large part of the primary production and for as much as 95% of the total seed production in Arizona deserts. These seeds form the primary food for three groups of granivores. The rodents are nocturnal and include such seed specialists as kangaroo rats (*Dipodomys* sp.) and pocket mice (*Perognathus* sp.), as well as the more omnivorous deer mice (*Peromyscus* sp.). These rodents are resident throughout the year, storing large numbers of seeds in underground caches to tide them over periods of food shortages.

The second group of granivores are harvester ants, including specialized granivores such as the genus *Pogonomyrmex* as well as more omnivorous species. The ants live in colonies and forage only during limited periods of warm temperatures and high humidity. They also store seeds underground. Third, there are various species of birds, which arrive in flocks during the winter. They cannot store seeds, and their numbers are highly variable over both space and time, as the flocks seek out local concentrations of seeds (Raitt and Pimm 1976).

As much as possible, Brown et al. picked areas of desert that were uniform, divided them into replicated plots, and assigned treatments at random. These treatments were selective species removals in the tradition of the press experiments I have been discussing. Brown et al. (1986) review two sets of experiments. One lasted from 1973 until 1977 and involved eight circular plots, each 0.1 hectares, in the Sonoran Desert near Tucson, Arizona. Rodents were excluded from two of the plots by fencing; ants were removed from two plots by poisoning; both ants and rodents were removed from two of the plots; and two of the plots were control plots. The second, more ambitious study was begun in 1977 and is still continuing. Brown and his group established twenty-four fenced plots, each 0.25 hectares, in the Chihuahuan Desert very close to the border of New Mexico and approximately 100 km north of the U.S.-Mexican border. These plots permitted different press experiments. Some or all of the rodent species could be removed by trapping, while other species could be allowed free movement by placing holes of various sizes in the fences. In some experiments, the group removed ants by poisoning them; in another class of experiment, they added millet seeds. In all the experiments, the densities and distributions of the plants and animals were measured at regular intervals. The results of three groups of press experiments addressed competition among rodent species, competition between rodents and other granivores, and the effects that granivores had on annual plants.

Competition among Rodent Species.   Brown et al. conducted two tests of the effects of competition between selected species of rodents. In the first, they explored the effect of removing the three larger species of kangaroo rats by trapping and removing them from four plots. Each plot was surrounded by a fence that allowed only the smaller rodent species to move through it. There were four control plots. Once the larger kangaroo rats were removed, four of the five species of granivorous rodents showed statistically significant increases in density, and the other showed a nonsignificant increase. Two species of insectivorous grasshopper mice in the genus *Onychomys* did not show increases in density.

Such results are what we would expect. The increase in densities of the smaller rodents did not appear until after "a lag of approximately nine months" (Brown et al. 1986, 45). Indeed, the data (figure 13.5) suggest that the increase in density of the smaller species continued five years after the kangaroo rats had been removed. Competition is an indirect effect, and the food web–based theory would lead us to expect such delays.

Brown et al. also noted delays in their second press experiment that excluded only the largest rodent, *D. spectabilis*. This experiment was begun in 1980 on two plots, only one of which produced statistically significant changes. When the largest rodent was excluded, five of the remaining seven granivorous rodents changed their foraging strategies to exploit plot areas they had not used previously. Such a habitat shift would be expected if the large kangaroo rat had behaviorally excluded the smaller species from certain areas before the kangaroo rat itself was removed. When Brown et al. published in 1986, this change in feeding had not yet led to changes in the density of the smaller species.

In addition to commenting on the time it took for rodent species to increase in density, Brown et al. also looked at the magnitude of changes in density. While these increases were statistically significant, they were "nevertheless much less than would be expected if those food resources made available by [our] manipulations were completely utilized by those rodent species that potentially had access to them" (Brown et al. 1986, 48). Why this should be so was not clear. Brown et al. raised the possibility that the small increase in the density of the smaller rodent species was because they were partly limited by predators. In other words, indirect effects would seem to be playing an important role in determining the degree of resistance to change.

Competition Between Rodents and Other Granivores.   Brown et al.'s first experiment to test possible competition between rodents and ants was at the Sonoran Desert site, and they found the expected effect. The number of ant colonies almost doubled when rodents were removed, and rodent densities increased by 20% or more when ants were removed. In the Chihuahuan Desert experiments, experimental results were not so clear-cut, and the density of only one species of ant, the smallest of the granivorous species, increased consistently. The density of another species actually declined in plots from which rodents had been removed.

In the experiment at the Sonoran Desert site, the numbers of birds increased in the plots from which both ants and rodents had been removed. The numbers

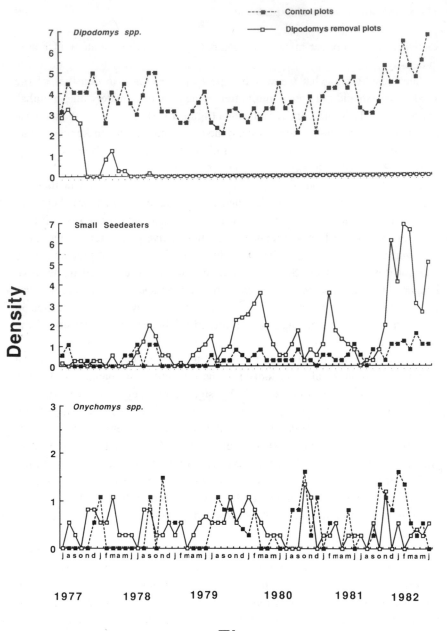

Figure 13.5. Changes in the densities of three groups of rodents on four experimental plots in the Chihuahuan Desert, following the removal of all three *Dipodomys* species in October 1977 (open squares), compared with the densities of four control plots (solid squares). *Dipodomys* were indeed completely removed and none were caught on the removal plots after early 1978 (top). There were compensatory changes in five species of small seed-eating rodents (middle), but there were no effects on the insectivorous *Onychomys* (bottom). (Redrawn from Brown et al. 1986.)

of birds also increased in the Chihuahuan Desert sites to which millet seeds had been added. But, in the Chihuahuan Desert site, birds were not counted until five years after the treatments; then, birds foraged *less* in the plots from which rodents had been removed. Again, these results suggest the importance of indirect effects, for the direction of the changes is not consistent from site to site and some species changed density in unexpected ways.

Predation on Annual Plants.   In the case of indirect mutualism between ants and rodents, Brown et al. were able to suggest an explanatory mechanism that involves indirect interactions with annual plants. The densities of both seeds and adults of annual plants increased significantly when either rodents or ants were excluded—and did so especially when both were removed. The effects of time delays were again evident. Shortly after the initiation of the experiments, there were no differences between control and experimental plots. Differences appeared *after* about three years, and the plant communities were reported as "still changing" when Brown et al. published their review paper in 1986. (Incidentally, three years is the length of a grant in the United States and, as a consequence is typically the length of an experimental study.) Rodent removals consistently altered the composition of annual plant communities, with the densities of species with relatively large seeds increasing and the densities of plants with smaller seeds tending to decrease (figure 13.6). Brown et al. undertook thinning experiments, which showed that large-seeded annuals inhibited the numbers of smaller-seeded species.

These competitive interactions among annuals might provide an explanation for why the ants generally did not increase in density when the rodents were removed. Decreasing rodent densities would mean increasing densities of large-seeded annuals, decreasing densities of small-seeded annuals, and uncertain changes in the ants species that feed on different proportions of large- and small-seeded species. The density of ants that feed on small seeds might be expected to decline after rodent removals. Indeed, the densities of such ant species did decline eventually, though their initial response was to increase. Ants that feed on large seeds should increase and whether the density of ants increased or decreased would depend on the details of their diets.

The final indirect interaction occurred when the larger-seeded species did increase on the plots from which rodents had been removed. One of the large-seeded plants became progressively more infected with a specific pathogen (figure 13.7).

From these detailed studies, Brown et al. draw conclusions that are in remarkably close agreement in with food web–based theories: short-term changes are often predictable, but many changes take time to develop; and

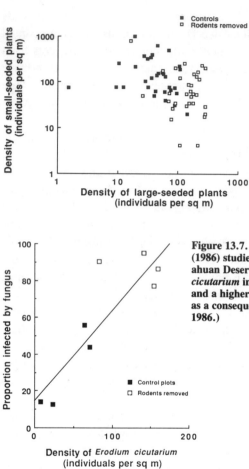

Figure 13.6. Removal of rodents increasing the density of the large-seeded annuals at the expense of the small-seeded species. The data are the densities of large- and small-seeded annual plants on plots from which rodents had (open squares) and had not (solid squares) been removed. (From Brown et al.'s [1986] studies of species removals in the Chihuahuan Desert.)

Figure 13.7. Another result from Brown et al.'s (1986) studies of species removals in the Chihuahuan Desert. Density of the annual plant *Erodium cicutarium* increased when rodents were removed, and a higher proportion were infected with fungus as a consequence. (Redrawn from Brown et al. 1986.)

indirect effects are important and, when they eventually take place, often involve changes that would not be expected on the basis of simple pairwise interactions. How long it took for indirect effects to develop depended both on just how indirect they were and on the generation times of the organisms involved.

SUMMARY

Food web–based ideas suggest that it may be difficult to make any statements about resistance. For, while species A should increase if its predator B is removed, this interaction is complicated when there are indirect effects: B may also feed on species C, which competes with species A. Perhaps, with a knowledge of the food web, we can predict what the effects of removing any given species will be. Yodzis (1988) explores this possibility and produces an uncompromisingly negative theory: even when the food web is known, slight changes in the parameters that describe interactions between species are amplified by complex pathways that describe all the indirect effects between any pair of species. The result is a high degree of uncertainty—we cannot be certain that removing a predator will generally cause an increase in the density of one of its prey species!

This theory may seem unlikely in view of the surveys of field experiments discussed in chapter 12. Yet it is the conclusions of the field experiments—and not those of the theory—that need reinterpretation. There is ample evidence of indirect effects in field studies, which also show a considerable degree of uncertainty in their outcomes. The same removal repeated several times often does not yield the same result. Perhaps more significant is that, although field experimenters probably remove species known from natural history information and natural experiments to affect the species that the experimenters then choose to measure, effects are most often reported when the densities of large numbers of species are measured. Thus, it may be possible to say that a predator reduces the densities of *some* of its prey species, but it is not possible to say *which* species. Indeed, the prey species that increase may differ from experiment to experiment.

Finally, the directional uncertainty inherent in more complex food webs only applies to the final change in a species density. When we examine the transient changes—for example, whether a prey species will increase in density when its predator is removed—there is no uncertainty. Therefore it is possible that we see expected changes in density only because of the short duration of the field study. The importance both of indirect effects and of how long it takes for the effects of a species removal to be manifested are best illustrated by studies of the removals of ants and rodents from desert communities. Ants and rodents may be indirect mutualists rather than competitors, because they take seeds from different species of plants that compete with one another. The pathway of effects from rodent to plant to plant resource to competing plant and so to ant is a long one; it is not surprising that decreases in rodent numbers when ants are removed take some time to appear.

Can There Be a Theory of Resistance?

Predicting the change of one species' density when another is removed may be simple in the short-term. In the long-term, the complexity of natural food webs will make predictions impossible, except, perhaps, in some special circumstances. What if we seek a more general theory by asking how much species composition changes when we add or subtract species? We may be unable to predict which individual species will change, but we may be able to predict how many species will become extinct. We know that removing predators will cause both increased densities of some of their prey and decreased densities of others. Can we predict how much the total density of those prey will be changed? It is these questions, which arise when we consider higher levels of organization, that I shall address next.

14

Changes in Total Density and Species Composition

RESISTANCE AT THE COMMUNITY LEVEL

Indirect effects propagating through complex food webs make it difficult to understand what happens when we add or subtract species from communities. The extent to which press experiments are interpreted simply—by saying, for example, that the removal of a particular predator will always increase the density of a particular species of prey—may be a reflection of the very short time interval over which the experiment was undertaken. When really long-term press experiments are performed, indirect effects become more apparent. Density changes take time to be manifested, and, when they appear, they are rarely predictable in direction, let alone magnitude.

I have been discussing press experiments on only one organizational scale: the effects of one species on another. Equally sensibly, we could ask how resistant is the species composition of the community or the total density of some group of species to our adding or subtracting species. Many press experiments record such information, but is there any point in looking at the resistance of either species composition or total density? Do we have any reason to expect that changes either in species composition in a community or in its total density will be predictable when changes in individual species densities are not? I shall argue that the effects that species removals have on changes in species composition and total density do appear to be predictable. At these different levels of organization, it may be possible to develop a theory of resistance.

A THEORY FOR CHANGES IN SPECIES COMPOSITION

Recall both MacArthur's (1955) argument that increased food-web complexity dampens the effect of abnormal densities and Watt's (1964) argument for the opposite effect (chapters 1 and 4). These ideas appeared in the context of population variability, but they have implications for resistance, too. MacArthur's

315

idea was that, when species are removed completely, other species densities should change less if the food web is a complex one. MacArthur's ideas are incomplete (chapter 4), for they deal with altering the density of species at the base of the food chain, but they were the stimulus for more extensive studies of resistance.

Resistance to Species Removals

I modeled the effects that food-web structure had on the resistance of community composition when species at many different positions in the food web were removed (Pimm 1979a, 1980b). My procedures first produced a set of food-web models that satisfied certain conditions sufficient for their species to persist, and then they removed species from these models. (These models are discussed at length in chapter 16.) Suppose there were n species in the model. I removed a species from a model by the simple expedient of setting its density to zero and keeping it there. Would the model community with $n - 1$ species also satisfy the same conditions that are sufficient for its species to persist? If so, the model would be resistant to the species' deletion from it: I call such models *species-deletion stable* (Pimm 1979a, 1980b, 1982). Or would the model progressively collapse, reducing to $n - 2, n - 3, \ldots$ species? Such a model would not be resistant to the loss of the first species. My models comprised from one to four three-species food chains, each chain consisting of a plant, a herbivore, and a carnivore. As many as five predator-prey interactions were added to increase the complexity of the models. These interactions could be within food chains (which involved species feeding on more than one tropic level) or between food chains. Altering the number of interactions also affects the model's *connectance* (p. 185).

These models produced two major results. First, webs of differing numbers of species and of different degrees of food-web complexity showed varying levels of resistance to species removals. Second, resistance to species removals depended on which species were removed. On average, more highly connected food webs are less resistant to species removals than are more loosely connected food webs (figure 14.1). For a given connectance, food webs with more species are also less resistant to species removals.

The results for six-species models show that there are marked differences in resistance, depending on which species are removed from the food web (figure 14.2). For species at the base of the food web, increasing connectance *does* increase resistance to species removal, but increasing connectance will rapidly decrease the community's resistance to the removal of its top-predators. MacArthur's intuition about stability was correct—but not complete. In his argument, the effects of a press experiment "from below" were considered. His

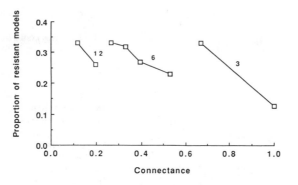

Figure 14.1. Proportion of three-, six-, and twelve-species food-web models from which species could be removed without resulting in the loss of more species. This proportion of resistant models decreases as food-web connectance increases. The proportion also decreases with increasing numbers of species. This is harder to see; try extrapolating each of the three lines to the same values of connectance. Only a limited portion of the range of connectances can be shown for each number of species. It is not possible to explore a complete set of combinations of connectances and numbers of species. The limits are imposed by two factors. First, the models are three-, six-, and twelve-species communities organized into simple linear chains of three species per chain—for example, a plant, its herbivore, and a carnivore. The simplest models involve interactions only within food chains, and the corresponding connectances are 2/3, 4/15, and 8/33, respectively. Second, as connectance increases, it becomes progressively more difficult to find models that satisfy the essential constraints discussed in the text. (From Pimm 1980b.)

argument is correct for such experiments, but the results are reversed when press experiments "from above" are included. These results parallel the results on variability that have been considered in chapter 4, where polyphagous species are argued to be less variable than monophagous species to changes in the densities of their prey species—but to be more variable in response to changes in the densities of their predators.

Which species can be removed without further loss of species from the web, and which species are "keystone" species, whose losses will cause extinctions to cascade throughout the community? To determine which species were the keystone species, I investigated ten food webs, each with twelve species and with five additional predator-prey interactions added to the four three-species chains. The additional predator-prey interactions were placed randomly. Figure 14.3 shows an example with each species coded according to the proportions of removals that caused further species losses. Figure 14.4 summarizes the three basic results for all ten models:

1. Removals of plant species cause the fewest subsequent species losses when the herbivore feeds on a variety of other species—that is, when it is polyphagous.

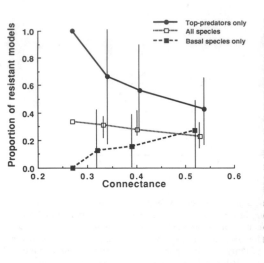

Figure 14.2. Proportion of six-species food-web models from which species could be removed without resulting in the loss of more species changes with increasing connectance. The mean values for all species are the same as in figure 14.1. Notice, however, that this mean is determined in part by removing species from the top of the food web (where resistance to loss progressively decreases with increasing connectance) and in part by removing species from the bottom of the food web (where resistance to loss progressively increases). Lines indicate the ranges of the proportions in different models. (From Pimm 1980b.)

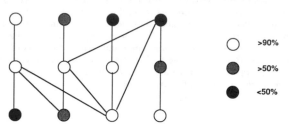

Figure 14.3. Example of a twelve-species food web coded by which species effect greatest changes when removed. Solid black circles indicate species whose removal results in further losses in less than half of the models that were analyzed with this structure but with randomly chosen parameters. Open circles indicate species whose removal results in species losses in more than 90% of the models that were analyzed: these might be considered keystone species. Notice that keystone species occur at all trophic levels. (From Pimm 1980b.)

2. Removal of herbivores produces a comparable pattern. The more polyphagous the herbivore's predators, the less the effect caused by removing the herbivore.

3. Predator removal shows that further species losses are *more common* when the herbivores on which the predators feed are polyphagous.

Notice that effect 3 is contrary to those seen after plant and herbivore removals (effects 1 and 2). When plant and herbivore species are removed, the more complex the food web, the more resistant to species removals it will be. The effect of removing predators, however, is sufficiently strong to counter these effects, producing a pattern of resistance to species removals that, when averaged over all the species in the food web, decreases with increased food-web complexity.

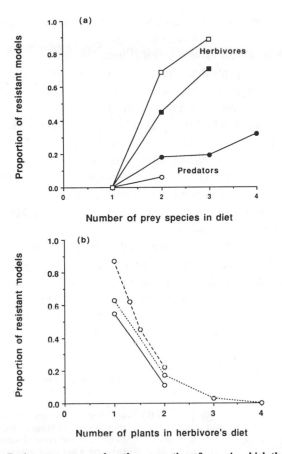

Figure 14.4. (a), Resistance, measured as the proportion of cases in which the removal of a species of prey will not cause further species losses, increases with increasing numbers of species of prey in the predators' diet. For predators feeding on only one herbivore, the herbivore's loss will always cause the loss of the predator, while predators with more generalized diets are more resistant to the loss of a prey species. Closed symbols indicate a herbivore that has one species of predator feeding on it, and open symbols indicate a herbivore that has two species of predator. The results for the herbivores are the same as those for the predators, but now it is the interaction between herbivores and the plant species that they exploit. (b), Resistance, measured as the proportion of cases in which the removal of a species of predator will not cause further species losses, decreases with increasing numbers of plant species exploited by the herbivore on which the predator feeds. The lowest line indicates where the predator feeds on three species of herbivore, the middle line indicates where it feeds on two species of herbivore, and the top line indicates where it feeds on one species of herbivore. For the lower two lines, the number of plant species is the mean number used by the two or three species of herbivore. (Redrawn from Pimm 1980b.)

Some factors appear to have no effect on resistance to species removals. In particular, whether a predator is monophagous or polyphagous seems to have no bearing on the effects of its removal (Pimm 1980b).

Given these very general results, can we predict which species become extinct after the removal of another species? We might expect it to be very difficult to do this. The complex indirect pathways mean that food-web models cannot predict simply whether a particular species' density increases or decreases when the density of another species is altered (Yodzis 1988; also see chapter 12 above). It may indeed be impossible to predict *which* species will become extinct in these models, after another species has been removed. What is interesting, however, is that it is possible to say from *which group* of species the extinctions will occur, and it may be possible to predict whether these secondary extinctions are few or many.

When plants or herbivores are removed, it is the species in the same food chain (that is, those species that feed on the species removed—or the predators of species that feed on the removed species) that are lost. Obviously, trophically specialized species are at greatest risk. Thus the models predict that plants that support specialized herbivores and herbivores that support specialized carnivores will be keystone species (figure 14.5). The patterns that follow remov-

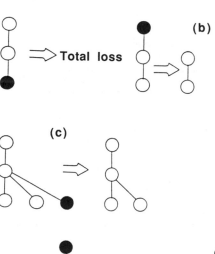

(a)

Total loss

(b)

(c)

(d)

Figure 14.5. Summary of previous figures. The effects of species removals depends on the complexity of the food web and from where the species is removed. A plant removed from a simple food chain may cause the entire food chain to be lost (a), and a predator removed from the same food chain may have no effect (b). In contrast, a plant lost from a more complex web may have little effect (c), but a predator lost from the same food web may release its herbivore and so lead to the loss of several species of plant (d). (From Pimm 1987a.)

als of carnivores are less obvious, for the removal of a carnivore often causes species losses within the plant trophic level. Why should the predator of a polyphagous herbivore be a keystone species whose removal will have such dramatic effects? There are three reasons. First, in the absence of the carnivore, the densities of herbivores on which the carnivore preyed will increase, and the plants' densities depend, in part, on the herbivore's density: higher herbivore density causes lower plant density. Therefore, after the removal of the carnivore, a polyphagous herbivore will increase in density more than will a monophagous herbivore, a finding that recapitulates Watt's (1964) argument discussed in chapter 4. Each plant species has a greater chance of extinction after an increase in herbivore density, if the herbivore is more polyphagous. Second, even if that chance were to remain constant, the chance of at least one plant species becoming extinct would increase with increasing numbers of plant species. The more species of plants, the more there are to lose. Finally, a plant does not need to be lost from the food web in order for other species losses to occur. Herbivores feeding on the plants may become extinct because the plant densities become too low to support them. All three reasons operate together. Consequently, in complex food webs, the removal of a carnivore species can cause very extensive extinctions (figure 14.5).

A THEORY FOR CHANGES IN TOTAL DENSITY

What happens to the total density of some set of species when we add or subtract a species from a community? This was the question asked by McNaughton in his stinging criticism of the existing theory about ecological stability (McNaughton 1977; also see chapter 1 above). His paper specifically discussed removals of large grazing herbivores from grasslands—and the subsequent changes in the composition and biomass of plant communities. (Changes in total biomass should follow much the same patterns as do changes in total density; so, while the theory predicts changes in densities, sometimes I shall use changes in total biomass to test the theory.) We have already glimpsed McNaughton's results, in chapter 1. Rather than discuss them here immediately, I shall present a theoretical exploration of his question first and shall return to the results later. Although the models to be presented next were designed to investigate the effects of additions or removals of large herbivores from grasslands, they are equally applicable to other cases, such as the removal of invertebrate predators from intertidal habitats. Data to test the models come from more studies than just McNaughton's. For the sake of convenience, however, I shall continue to call the species in the models "plants" and "grazers."

The Models

Anthony King and I established a set of structured food-web models composed of one grazer and a variable number of plant species (King and Pimm 1983). We followed the familiar food-web recipe of creating structured models with parameters chosen at random over ecologically sensible intervals (chapter 16). All the models retained for analysis had to satisfy the sufficient requirements for species persistence, and from models meeting these requirements we removed the grazer. We wanted to follow the resultant changes in plant densities until a new stable equilibrium was achieved, and these changes might involve either no extinctions of plant species or extinctions of several species. To follow the density changes, we simulated the equations until the species densities reached an equilibrium.

How appropriate are these models as descriptions of grazing systems? Including the interspecific competition between the plants and including the grazer-plant interactions means that the plants' responses to reduced grazing will be composites of individual plant responses to the removal of the grazer and community-wide responses involving shifts in species composition. These features are in general agreement with McNaughton's (1985) interpretation of his experimental studies, in which he also recognized both species-specific and community-wide responses. In the models that King and I established, one feature certainly did not match reality: in our models, maximum plant-population growth rate occurs when levels of grazing reach the point at which the grazer has reduced the plant population to half the density that would occur in the grazer's absence. McNaughton showed that his plants remained highly productive despite very high levels of grazing that reduced their levels well below that expected in its absence. Our models do not contain this refinement, and its effect on our overall results remains to be uncovered by subsequent analyses.

The relationship between resistance and food-web complexity was explored by using three measures of complexity. First, was the number of species in the system; our models contained from two to twelve species of plant. Second, we varied the number of competitive interactions between plants, which determines the food-web connectance once the number of species is known. Third, ecologists often report one or both of two statistics that describe relative density. The first of these indices is *evenness*, which is maximum and equal to unity if all the species have equal densities. Evenness approaches zero as one species dominates over all the others. The second index is *diversity*, and to explain diversity indices requires a brief digression.

The many diversity indices have been very popular with community ecologists, and a fine recent book-length review of them is provided by Magurran (1988). Each index contains both a component that measures the number of species in the sample and an evenness component, and different indices weight the components differently. I have great reservations about these indices. They are rather like having one index to describe the contents of a fruit bowl—an index that arbitrarily gives one apple the value x and one orange the value y, so that the value of the index is x times the number of apples plus y times the number of oranges. Conventional wisdom dictates that we cannot equate apples with oranges, and that, if we do, the index value we obtain will vary on those arbitrary values of x and y as well as on the contents of the fruit bowl. Returning to ecology, notice that, if we fix one component—for example, the number of species in the sample—then any changes in diversity must reflect differences in evenness. Most of my discussion will be about evenness and not about diversity. Indeed, King and I included diversity values only to facilitate comparisons with McNaughton's (1977) results. We calculated diversities in both the presence and absence of the grazer, using plant densities generated by the simulations.

Results

The models generate a large number of individual simulations. The different numbers of species and the numbers of competitive interactions between them were determined by King and me directly. Different species densities arise from the randomly chosen interaction coefficients, and these varying densities lead to plant communities that have different values of evenness. The change in densities after the removal of the grazer must be interpreted in terms of all three of these features.

Interaction between Plant Species. In figure 14.6a, the ratio of densities (without herbivore divided by with herbivore) falls as the number of competitive interactions increases—that is, increased connectance, obtained by adding competitive interactions between plant species, results in increased resistance. This should strike you as odd. Does it not contradict the results on species composition that were discussed earlier in this chapter, where increased connectance resulted in greater changes (less resistance) in species composition when species were removed from the top of a model food web? The changes in species composition that are associated with the grazer removals are shown in figure 14.6b. Comparing the two parts of figure 14.6 shows that plant communities that change the least in total density are those that change the *most* in species composition, when grazers are removed.

These superficially paradoxical results suggest a resolution to Mc-

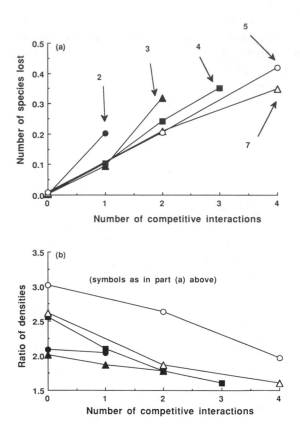

Figure 14.6. Result when an herbivore is removed from a plant community. The *least* change in total plant density is accompanied by the *greatest* changes in species composition. In models with one herbivore feeding on two to seven species of plant, the greater the number of competitive interactions between plants, the greater the average number of plant species lost when the herbivore is removed (a). When the plants do not compete (that is, there are zero competitive interactions), there can be no losses. Where there are one or more competitive interactions, the herbivore's loss may cause the increase of a competitively superior plant which will then cause its competitor's demise. In contrast, the ratio of total plant densities after the herbivore is removed divided by when the herbivore is present, decreases with increasing number of competitive interactions (b) (Redrawn from Pimm 1984a.)

Naughton's concern over the conflict between the theory and the observation of complexity-stability relationships. In model grazing systems, increased complexity can increase one kind of stability and simultaneously decrease another. While increased connectance between plant species reduces the changes in total density, it can increase the changes in species composition.

Diversity and Evenness. For each of the twenty-three combinations of number of species and number of competitive interactions between species, there were many simulations. Each simulation had a different set of plant densities, both with and without the grazer, and each had an associated ratio of densities. Measures of resistance are ratios, and ratios often tend to be skewed toward high values. Consequently, King and I analyzed the logarithms of the total plant densities without the grazers divided by total plant densities with the grazers. Henceforth, I will call these logarithms of the proportional change in densities the *log density ratios*. A log density ratio of 0 means total resistance, while a log density ratio of 1.0 means that the plant densities increased tenfold when the grazers were removed. To best summarize the large number of observations, we calculated the regression lines through the log density ratios versus the corresponding plant-species diversities, for the many individual simulations for each of the twenty-three food-web structures.

Figure 14.7 shows the regression lines for the log density ratios plotted against the diversity of plant species without the grazer. Within each model, the number of plant species is fixed; thus these changes in resistance are due only to differences in evenness. The obvious result is that, the more even the plant densities, the greater the resistance of the total plant density. The proportional changes in density decrease as plant diversity increases.

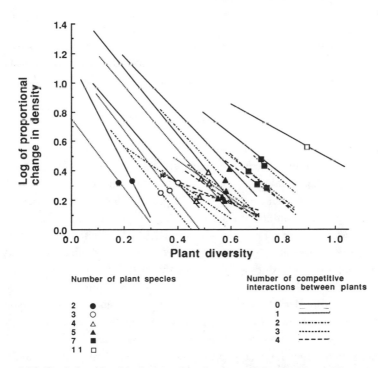

Figure 14.7. Total plant-density changes after the removal of an herbivore. These changes are least when the plant species are relatively equal in their densities. The lines are the regression lines through large numbers of points generated by the simulations (see text for details). These points are the logarithms of the change in total density when the herbivore is removed and the corresponding diversity of the plant densities when the herbivore is absent. Each line is for a fixed number of species, so the only component of diversity that is changing is the evenness. Symbols on the lines indicate the number of plant species, and the style of line used indicates the number of competitive interactions between those plant species. The extent of the line indicates the range of diversity in each subset of the data. (From King and Pimm 1983.)

Less obvious in figure 14.7 are the effects of the competitive interaction between the plants. These are harder to see because there are so many lines, so just consider the two of them, at the extreme left of the figure, which have solid circles on them. These regression lines are for models with one grazer and two plant species: the upper, solid line is for the model without competition between the plants, and the lower, dashed line is for the model with such competition. The effect of competition between the plants is to make the communities more resistant, for any given evenness—that is, with competition, the line is lower, and the plant communities change less and so are more resistant. Although the graph's complexity makes the fact difficult to see, this effect is true for all the other models in the figure; the greater the number of competitive interactions between the plant species, the lower the line and the greater the resistance.

Changes in evenness. The greatest changes in total density were associated with increases in evenness that were caused by the addition of grazers to the models (fig. 14.8). Total density changed less when the addition of the grazer caused a decrease in plant evenness.

Why Do the Models Produce These Results?

Models without Competition between the Plants. By definition, communities where one or a few species are unusually abundant are highly uneven and so have low evenness values. In such communities, grazers that always feed on the most abundant species will always produce increased evenness. The more the most abundant species are cropped, the greater the change in total density the grazer causes. The more a community is dominated by one or a few species, the more a relatively specialized grazer can both reduce the density of those dominant plants and, thereby, change the total densities. In contrast, if a grazer specializes on the rarer plant species, it may significantly reduce evenness, for the more dense species will become even more prevalent. Notice, however, that such a specialized grazer cannot reduce the total plant density by much, because it is feeding on plants that are already rare. In all the models, even those with no resource competition between the plants, large changes in density come from models in which the plant densities are uneven in the absence of the herbivore. In short, these results are consequence of the grazers feeding *selectively* on the more abundant species. The grazers do not feed selectively on the rare species, and, if the grazers were to feed *proportionately* on all the plant species, then, although total densities would change, there would be no changes in evenness.

The only problem with this simple explanation is that the models did not

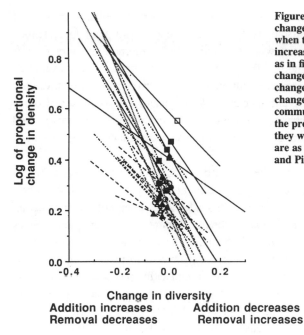

Figure 14.8. Total plant density changes. The change is greatest when the addition of an herbivore increases diversity. The situation is as in figure 14.7, but the density changes are plotted against the change in diversity. A negative change in diversity means that the communities were less diverse in the presence of the herbivore than they were in its absence. Symbols are as in figure 14.7. (From King and Pimm 1983.)

explicitly incorporate selective feeding by the grazers on the most dense species! The models forced grazers to feed on all the plant species, and the parameter chosen to describe the effect on a plant species was independent of that plant species' density. Why, therefore, are the grazers so selective, when we did not make them so? The answer comes from the constraint of the grazer and plants being able to coexist. Consider two plant species that do not compete for resources. *By the rules of the simulations,* the two plant species have to coexist in the presence of the grazer (though not in its absence). (Is this a reasonable rule? It would seem to be for the Serengeti grazing system described by McNaughton [1985], in which the plants and grazers have coexisted across geological time scales. In contrast, this rule would be totally inappropriate to the introduction of large herbivores to oceanic islands where the plant communities would have evolved without such species [chapter 15]. I will show later that this rule of coexistence is a very important one in understanding patterns of resistance.)

Put simply, the coexistence of plants and grazer imposes a limit both on the relative densities of the plants and on how the grazer feeds on them. If the grazer were nonselective, the more common plant species could not be too abundant, because, if it were, it would support so many grazers that the rare plant species would be extirpated. Therefore, there is *apparent competition* (chapters 12 and 13) among plants that is mediated by the plants' enemy. Of course, if the grazer feeds selectively on the more abundant plant species, then its effect on the rare species will be reduced; both common and rare species will coexist in the presence of the grazer. When the grazer is removed, the preferred plant may become very abundant. Nonselective grazers can have only relatively little effect on plant species *if those species are to coexist.* Consequently, the only major changes in plant density will occur when grazers feed highly selectively on the species that would be very abundant in their absence.

The Effects of Competition.   Now suppose that, in addition to this apparent grazer-mediated competition between the plants, the plant species also compete with each other for resources. This additional competition imposes an *additional* limit on the plants' relative density. The more abundant plant species cannot be too dense, or else it will eliminate the other species through competition for resources. Therefore, plant coexistence also places limits on how intensely the more abundant species can compete for resources and thus on how dense it can be. Without competition, the less dense plant species can only become rarer when the grazer is present. With competition, the less dense plant species may actually increase in density as the grazers crop the more dense plant species that through competition were reducing the other plant species' density. When grazers are added to a community of competing plant species, dense plant species may decrease as an immediate consequence and less dense plants may increase as a secondary consequence. Thus the total change in plant densities will not be as great as when the less dense plant species merely decreased.

Of course, when there is competition between plants species, the removal of the grazer can cause some plant species to eliminate other species through competition. This cannot happen, obviously, if the plants do not compete.

TESTING THE THEORIES

The theories developed in this chapter are complicated. Resistance to species removals depends on food-web complexity, from where in the food web the species is removed, and whether one considers species composition or total density. Tests of ideas concerning changes in community composition after removals of species from the top or bottom of simple or complex food webs are difficult to test. I shall use the literature on species introductions and extinctions to do this in chapter 15. Most removal experiments look at the effects on species' densities at trophic levels beneath the species removed. Such experiments provide limited tests of the density-change theories developed in this chapter, and it is to these tests that I now turn.

Serengeti Grazers

McNaughton's (1977) experiment was a simple one. In the Serengeti National Park of Tanzania, he chose two grassland areas that different in the diversity of their plant communities. Within each area, African buffalo were allowed to graze in some plots and were excluded from others. McNaughton measured biomass, and, strictly considered, the theory deals with total density. I shall assume that density is sufficiently closely correlated with biomass that we can use the terms interchangeably. In the diverse area, total plant biomass was only about 11% lower in grazed than in ungrazed plots, but the comparable figure for the less diverse area was 69%. The total plant biomass of the more diverse area was much more resistant to the grazer than was that of the less diverse area. This is exactly what the theory predicts (see above and figure 14.7). With the addition of the buffalo, diversity in the more diverse site decreased, while it increased in the less diverse site. These changes in diversity are not easy to interpret because diversity is a composite index of evenness and species richness. If we assume that there were similar numbers of plant species in grazed and ungrazed sites, then these changes in diversity would be caused by changes in evenness. The interpretation would be that, with the addition of the buffalo, the site in which total biomass changed less showed a decrease in evenness, whereas the site in which total biomass changed more showed an increase in evenness.

This interpretation supports the theory, but this is only one study. Fortunately, there is evidence from two other sources that supports the theory. The first source is a more detailed examination of the effects of grazers that was conducted by McNaughton on the same system, and the second source is a review of similar experiments in aquatic communities.

In 1985, McNaughton reported a more extensive study of the effects of zebra, buffalo, wildebeest, and Thompson's gazelle on grasslands in the Serengeti. He considered only biomass resistance, defined operationally as the proportion of the aboveground vegetation not eaten during either (*a*) a single passage of a herd (in the dry season) or (*b*) a twelve-day period (in the wet season). McNaughton's measure of resistance is one taken over a short period of time.

The grasslands differed both in the diversities of their plant composition and in whether they were grazed by one or more species. In the dry season, the three species that had relatively little impact (zebra, buffalo, and gazelles) had least impact on areas with greatest plant diversity. The same was true for areas grazed by wildebeest (figure 14.9). In the wet season, wildebeest did not graze much more than the other species, and, for all grazing species combined, the more diverse grasslands were again more resistant (figure 14.10). Again, these results support the theory. In contrast, in the dry season, in grasslands that were successively grazed by all four species, more diverse grasslands were not more resistant. Indeed, the grasslands on which all four species fed were very heavily grazed, with only 18% of the vegetation remaining.

McNaughton estimated the degree of dietary specialization of the grazers. The grazers were more selective in the more diverse grasslands than in the less diverse grasslands. So the more diverse grasslands were more resistant, in part, because a smaller proportion of their plants were being grazed (by their more

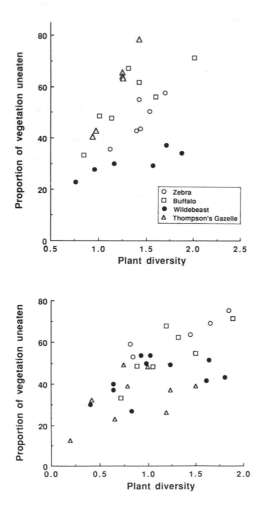

Figure 14.9. Resistance of vegetation to grazing (as the proportion of vegetation uneaten by four species of large ungulate). Resistance increases with the diversity of the plants on which the ungulates feed ($P < 0.05$). These are dry-season data. (Redrawn from McNaughton 1985.)

Figure 14.10. Resistance of vegetation to grazing (as the proportion of vegetation uneaten by four species of large ungulate). Resistance increases with the diversity of the plants on which the ungulates feed. Combined data for all species is significant at $P < 0.001$. These are wet-season data. (Redrawn from McNaughton 1985.)

selective grazers). The grazers' dietary differences could be factored out by using a statistical model. When this was done, the more diverse grasslands were still more resistant than the less diverse grasslands, a result that suggests that competitive processes acting within plant communities played a role in the communities resistance. Indeed, McNaughton had independent evidence of such processes.

I will draw some conclusions about these studies shortly. First, however, notice that there is nothing special about the models' predictions that restricts them to large herbivores in grasslands. The predictions ought to apply to other kinds of communities.

Aquatic Communities

Anthony King and I tested the models' predictions by using experimental removals or additions of predators or herbivores in aquatic systems, for that is where most press experiments have been performed (chapter 12). Our analyses appear here for the first time. We found four studies that allowed us to calculate both the evenness of species densities with and without the species' consumers and, simultaneously, the corresponding change in total density. These studies were Addicott's (1974) study of the effects that the predatory mosquito *Wyeomia* had on its protozoan and rotiferan prey in *Sarracenia* pitcher plants, Dayton's (1971) study of the effects that intertidal limpets had on marine algae, Nicotri's (1977) study of the effects that intertidal gastropods had on marine algae, and Virnstein's (1977) study of the effects that fishes and crabs had on marine benthic infauna. The data confirm the theoretical expectations. Total densities change least when consumers are added to communities that have relatively even prey densities (figure 14.11). The greatest changes in density occur when the predator substantially increases evenness (figure 14.12).

Grazing Experiments on Land and in the Water

Conventional wisdom in Australia records that it is always the "tall poppies" that society cuts down to size. The results of the terrestrial and aquatic grazing studies can be explained "simply" by the predators or herbivores feeding differentially on the more dense species—that is, the tall poppies. This *tall-poppy effect* may be no more than the consequence of the predators always taking the most dense species. If so, this effect begs a subtle question. Why should the most dense species always be the preferred species of the predator unless *only* density determines preference? I argue that, if predators preferred the less dense species or even fed on all species in equal proportions, then the less dense species would not survive in the presence of the predators. Consequently,

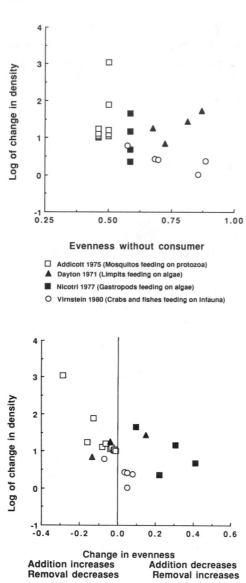

Figure 14.11. Changes in the total density when predators or herbivores are added or removed. Changes are least when the species on which they feed have more even densities. The data are combined from the four different sources indicated.

Figure 14.12. Changes in the total density. Changes are greatest when the predator species decrease the evenness of their prey's densities when the predators are removed (negative changes in evenness). In an equivalent manner, when predators are added, the greatest changes in total density are when they increase evenness. Sources are as in figure 14.10.

it is this requirement of coexistence that imposes the relationships observed in the data. The relationships are those we expect from the apparent competition of species that share predators. Were this requirement discarded, we could observe the opposite relationships between evenness and resistance—and between change in evenness and change in resistance.

In natural communities, we know the predator coexists with its less dense prey, just because those prey are there! In contrast, there is no guarantee that rare plants will survive the introduction of an alien herbivore or that rare animals will survive the introduction of alien predators. As we shall see in the next chapter, introduced species often cause extensive extinctions and effect great changes in total density. Introduced species do not have to obey the rules of coexistence.

SUMMARY

Previous chapters have considered the effects that one species has on another; this chapter has considered the effects that one species has on many species. There are two ways of summarizing such effects: change in species composition and changes in the total density or biomass. Theory predicts that the species composition of complex food webs will be most resistant to removal of species at their base and least resistant to the removal of their top-predators. For simple food webs, these results are reversed. These predictions are not easy to test; I shall try to do so in the next chapter.

Theory also predicts that, when grazers are added to a community, increased competition between the plants will lead to large changes in species composition but to relatively small changes in total density. Adding grazers to communities where some plant species are much more abundant than others will lead to larger changes in total density than occur when the plant species are more even in their densities. The greatest changes in total density will come when grazers increase evenness; the smallest changes will come when grazers decrease evenness. The result depends on the coexistence of the grazers and the plants. If the plants cannot coexist when the grazers are present, then the grazers may have a substantial impact on the total density of the plant communities.

Complex though these changes are, they are supported both by field studies of large mammalian herbivores in grasslands in east Africa and by a number of studies of aquatic systems involving both plant-animal and animal-animal interactions. Communities with more even densities change less when predators are added. In contrast, introduced species may be expected to effect great changes in both composition and total density.

15

The Consequences of
Introductions and Extinctions

RESISTANCE ON A LARGE SCALE

People have moved species around the world for millenia, sometimes by accident but often with considerable enthusiasm. Hunters have demanded exotic birds and mammals to shoot, fishermen want challenging fish, and gardeners want novel plants. Nineteenth-century New Yorkers wanted all the bird species in the works of Shakespeare, and they imported the starling and the house sparrow into North America as a consequence (Long 1981). Hawai'i's Hui Manu was an organization that collected coins from schoolchildren to help produce a lowland avifauna that is now totally exotic (Berger 1981). More passerine birds have been introduced into Hawai'i than anywhere else (Long 1981). English garden birds in New Zealand may merely be quaint curiosities introduced by settlers wanting familiar species from their former homes, but many introductions have been devastating. If the drive to introduce species has been reduced, faster international travel makes *accidental* introductions more likely. Some potential introductions are radically new, such as genetically engineered organisms, be they microbes, plants, or domestic animals. Evaluating the effects of introduced species is a topic of considerable practical importance. Do alien species have an effect on the communities into which humans introduce them, or are these communities resistant to such introductions?

A parallel question can be asked about species extinctions. Species are being driven to extinction in numbers paralleling those of the great extinction events in geological history (Ehrlich and Ehrlich 1981). There are many reasons for these extinctions, and one of them is *secondary extinction:* some species become extinct because other species, on whom their survival depends, become extinct. Secondary extinctions raise the question of whether communities are generally resistant or vulnerable to species losses.

These questions about the resistance of communities to introductions or extinctions are different in scale from those addressed in chapters 12–14. Species may be introduced into or exterminated from entire continents, and this is

334

a press experiment many orders of magnitude larger than that of an experimenter who removes organisms from an enclosure. There are differences in organizational scale as well. An experimenter will carefully measure the densities of one or more of the species in an enclosure and their responses to species removals, but we rarely have such information about species after the introduction or extinction of another species. What data are available tend to address species composition—that is, whether other species were lost after an introduction or extinction.

Do extinctions and introductions cause further changes in species composition? Given that some extinctions and introductions affect community change while others do not, are there particularly important species, so-called *keystone species,* whose loss will cause a cascade of subsequent extinctions? Are there some types of communities particularly sensitive to species extinctions or introductions? In chapter 14, I introduced a food web–based theory to answer these questions. Once I have presented a survey of the relevant data, I shall examine the question of whether the theory is able to predict the details of these data.

Do Extinctions and Introductions Cause Further Changes in Species Composition?

The data that may allow us to answer questions about community resistance are many, scattered, and of very uneven quality, so their interpretation is often problematical. I know of no reviews of secondary extinctions. However, there are reviews of the literature on species introduction which attempt to determine whether introduced species generally alter the species composition of a community or leave it unchanged.

Simberloff's Review

Simberloff (1981) argued that introduced species generally do not affect the communities they invade. He reviewed ten studies covering a total of 850 plant and animal introductions. He concluded that less than 10% of the introductions caused species extinctions, for he claimed that there were only seventy-one recorded extinctions. Of the ten studies, Greenway's (1967) book, which was devoted to avian extinctions, accounts for fifty-five of the seventy-one extinctions in Simberloff's analyses. Simberloff concluded that introductions apparently tend to add species to a community rather than to cause extinctions; in other words, Simberloff argued that species composition is resistant to the effects of introduced species.

Of Greenway's fifty-five extinctions, more than 90% occurred on islands. Over all the ten studies, and within Greenway's study, predation was the principal cause of extinction, accounting for fifty-one of the seventy-one extinctions in Simberloff's analysis and for forty-two of the fifty-five extinctions in Greenway's study. Habitat change accounted for eleven of seventy-one extinctions, whereas competition from the introduced species caused only three extinctions and so ranked a distant third.

Herbold and Moyle (1986) were sharply critical of Simberloff's paper. The first of their two principal criticisms was that Simberloff might have overlooked many of the effects of introduced species because he worked with other reviews of species introductions rather than with the original papers themselves. Moreover, they argued, in his restriction of "effect" to mean "extinction," Simberloff ignored very substantial density changes caused by introduced species. Their second criticism was that many introductions regarded as having no effect had been in habitats already extensively modified by human activity.

A Closer Look at the Effects of Introduced Species

Herbold and Moyle's examination of the three studies from which Simberloff had tallied 525 of his 854 cases showed that his determination of "no effect" generally conflicted with the conclusions in the original papers. From Elton's 1958 book on biological invasions (see chapter 1), Simberloff extracted 241 instances of species introductions and reported only four of them as showing "any effect at all." However, Elton's accounts included examinations of the effects of chestnut blight, spruce budworm, gypsy moth, Argentine fire ant, Norway rat, and black rat, all of which have had major impacts on communities. Indeed, Elton reported that only four species had no apparent effect and had "been able to edge in without producing any noticeable disturbances or making . . . species extinct" (1958, 124).

Several surveys present more detailed evaluations of the effects of introduced species. In looking at these surveys, my aim is to see whether a more careful reading of the original papers leads to the conclusion that introduced species generally do have impacts.

Mammals. By compiling data on nearly sixteen hundred mammal and bird introductions, Ebenhard (1988) has produced perhaps the best compilation of the effects of introduced vertebrate species. He identified those species affecting abundance or causing extinctions. For simplicity, I shall refer to both of these changes as "effects." Overall, effects are likely to be missed; those reported almost certainly involve major changes to the communities.

The ten most commonly introduced mammals are the rabbit, three species of rats, the house mouse, and domesticated pigs, cattle, cats, dogs, and goats. These species make up about half of the eight hundred mammal introductions which Ebenhard examined. Ebenhard analyzed the data according to (a) whether the mammals were strictly herbivorous, strictly predatory, or omnivorous, (b) type of herbivore or predator (see below), (c) taxonomic criteria and (d) the site of species introductions.

Herbivores. Herbivorous mammals introduced to continents were recorded as having effects in 26% of 86 cases. Yet, on oceanic islands, which probably lack large vertebrate predators, 49% of 188 introductions show effects. Almost all the herbivorous mammals have generalized diets, but, in five introductions of specialized herbivores, none was recorded as having an effect.

Ebenhard noted that it is difficult to predict the effects of introducing a herbivorous mammal that has a generalized diet into a community. For example, sheep are typically grazers, and it might seem improbable for them to compete with the endangered Hawaiian bird, the palila. The palila, however, depends on the seed pods of the mamane tree on which introduced sheep and other exotic ungulates feed whenever possible. Moreover, the browsing of sheep and other introduced ungulates severely impairs the regeneration of the mamane (Roots 1976; Scott et al. 1986). Thus, on Hawai'i, sheep compete with the palila through exploitation of a common resource.

Goats, sheep, and rabbits are capable of causing severe damage to habitats; figure 15.1 is a photograph of catastrophic rabbit and goat damage to Round Island, Mauritius. (Fortunately, these species have been removed, and the vegetation is recovering.) On oceanic islands where these herbivores have destroyed the native habitat, complete reptilian or avian communities have been affected. For example, a study of feral goats on Santa Catalina Island in the California Channel showed that they had destroyed much of the habitat, causing decreases in the number of both California quail and some species of rodents. Furthermore, these declines had additional effects on both mammalian predators and snakes, which also diminished in numbers (Coblentz 1978). On Laysan Island in the northwest Hawaiian chain, the Laysan honeycreeper became extinct after rabbits destroyed most of the vegetation, and a windstorm removed the rest (Berger 1981).

Predators and omnivores. Introduced mammalian predators or those species which are both herbivores and carnivores were found to have effects in 32% of the cases, this being an average of 9% of the introductions to continents but 42% of the introductions to islands. Strictly predatory species (52% of the cases) apparently had more effects than did omnivores (24% of the cases). Arboreal predators had more effects than did ground-living predators.

Some examples are needed to flesh out these bare statistics. Several extinc-

Figure 15.1. Devastation caused by introduced rabbits on Round Island. With thanks to Ian Atkinson for supplying this photograph.

tions on New Zealand have been caused by cats. The cats of a lighthouse keeper on Stephen Island were responsible for the extinction of the Stephen Island bush wren in 1894 (Greenway 1967). Cats alone apparently exterminated several species of land birds on Herekopare Island, off Stewart Island, before the New Zealand Wildlife Service exterminated the cats in 1970. Since the removal of the cats from Cuvier Island in 1964, the subsequent reintroduction of saddlebacks has been very successful (Gibb and Flux 1973). Ebenhard claims that the domestic cat is "the most dangerous predator ever introduced by man, with thirty-eight known or probable cases of abundance-shifting predation" (1988, 35).

Rats.   Atkinson (1977) has discussed the effects of rat introductions throughout the world. His careful and detailed review shows that serious effects and extinctions are legion. Three species of rats, which are mostly generalized predators, have had a greater negative effect on native faunas than is often realized. The brown rat and black rat have spread all over the world, and the Polynesian rat has spread over much of the Pacific Ocean, from New Zealand to Hawai'i (Ebenhard 1988). The black rat is the most damaging of the three because it is arboreal and can eat the young and eggs of many bird species, while the brown rat is mainly terrestrial and the Polynesian rat is partly terrestrial and partly arboreal (Ebenhard 1988). Several bird species became extinct after the introduction of black rats on Big South Cape Island off New Zealand (Bell 1977).

Birds. Ebenhard (1988) also compiled data on the effects of nearly eight hundred bird introductions, mainly from Long's (1981) book about bird introductions. The nine most commonly introduced birds are the rock dove, the house sparrow, the domestic fowl, the starling, the pheasant, the common mynah, the Java sparrow, the goldfinch, and the zebra dove. These nine species constitute about 30% of all the introductions.

Ebenhard found that only three species were reported to have induced changes in vegetation: the mynah in Hawai'i, where it spreads the introduced plant *Lantana;* the mute swan, which alters the density of aquatic vegetation; and the monk parakeet, which damages the buds of trees. The number of introduced predatory birds producing effects is likewise small—Ebenhard lists only three species, the most important of which is the weka rail, which has been transferred to islands surrounding New Zealand, where it preys on other rail species and on various nesting seabirds. These few cases are in contrast to the pronounced results of mammal introductions—a large percentage of introduced mammals cause effects, and a substantial number of such introductions caused species extinctions.

Fish. Herbold and Moyle argue that introduced fishes have been the cause of the decline and near or actual extinction of native fish populations throughout North America. Unlike the bird and mammal introductions, in which the evidence of competition is often equivocal, many of the fish introductions are reported to have caused extinctions through competition. For example, Herbold and Moyle report that, throughout the Great Basin, cutthroat trout have been replaced by the more aggressive brown trout and rainbow trout. Mosquito fish have replaced the Gila topminnow in the Southwest, and arroyo chub have replaced the native Mohave tui chub in the Mohave River (Moyle et al. 1986).

Species introductions also eliminate species through predation. Zaret and Paine (1973) reported that the disappearance of native fishes in Central America was due to the introduction and establishment of the peacock cichlid in Lake Gatun in Panama. Similarly, the Nile perch is drastically altering fish communities in Lake Victoria in East Africa (Miller 1989). Aquatic plant communities are affected as well. For example, when grass carp are introduced into lakes or ponds at high densities, major shifts in the community structure occur (Ware and Gasaway 1976).

The Story of One Introduced Snake Species.   Guam is one of the Mariana Islands, a north-south chain of about fifteen islands about halfway between Japan and New Guinea. The fate of Guam's birds has been described by Savidge (1987). Until recently, the resident avifauna consisted of eighteen native bird species (twelve land-bird species, three breeding-seabird species, two wetland-bird species, and one reef heron species) and seven introduced species. Historically, most species were common or abundant throughout the islands, but, over the past twenty years, populations of the species have declined dramatically. Seven species are probably now extinct, and four others are so rare that their survival seems unlikely. In contrast, bird populations on the small island of Cocos, off Guam's southern coast, and on the other islands in the Marianas have remained relatively constant.

Savidge reported that all ten species of forest birds followed an unusually similar pattern of decline. The birds disappeared from southern ravine forests in the 1960s, and then the populations declined progressively to the north. By early 1983, all the forest species occurred together only in the 160 hectares of mature forest beneath the cliff line at the northern tip of Guam. By 1986, the species were gone from this site as well (figure 15.2).

Other native and introduced species suffered a similar pattern of decline. The native white tern was common over the whole island in 1961 but by 1980 was found only along the northern coastline. This species is still abundant on Cocos (personal observation, 1988). It is very different ecologically from the forest birds, as it feeds on fish from the ocean, but it nests in trees. Certain introduced species have also declined dramatically. Only three species have not declined—the native yellow bittern and two introduced game-bird species. These species nest either in marshy habitats or in open areas away from trees.

Savidge assembled various types of evidence to demonstrate that an introduced snake, *Boiga irregularis,* was responsible for the decline. The range expansion of the snake correlated well with the contraction of the avifauna (figure 15.2). Exceptionally high predation by snakes on bird-baited traps occurred where the bird populations had declined, but no snake predation occurred where the bird populations were intact. The diversity of species affected would seem to rule out other causes such as pollution of terrestrial habitats; the tern, for example, is marine.

Boiga's nocturnal and arboreal habits make roosting and nesting birds, as well as their eggs and young, particularly vulnerable. In addition to birds, *Boiga* feeds on small mammals and lizards. Savidge argued that, by including abundant small lizards in its diet, *Boiga* has maintained high densities while exterminating its more vulnerable prey. *Boiga's* efficiency is remarkable: Guam's forests have no birds in them.

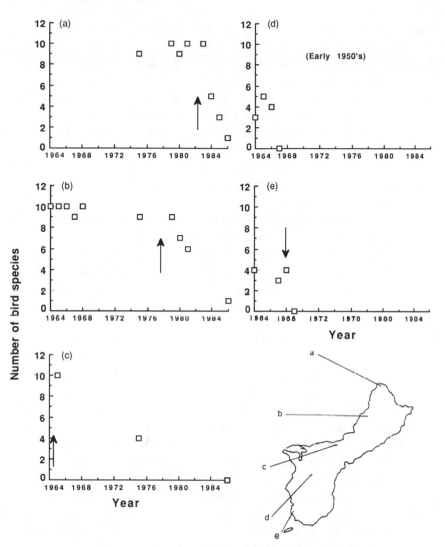

Figure 15.2. Numbers of species of birds in forest on Guam. The number of species dropped to zero after the snake *Boiga* was first found in those forests. Arrows indicate when the snake was first seen, except near the center of the island (d), where it appeared before the counts started. The snake proceeded southward (e), then northward (b, c) and reached the northern tip of the island (a) last. (Redrawn from Pimm 1987b, after data in Savidge 1987.)

Disease and Parasites.   Humans are perhaps the most mobile species, and we have taken our diseases with us, often with devastating effects. For example, May (1984) has argued that the prehistorical densities of native Amazonian peoples were probably much higher than at present. Some groups were wiped out by influenza within hours of their contact with the outside world. In contrast, there is little information on the effects that introduced diseases and parasites have on endemic faunas; the importance of such introductions may be underestimated, for the cases we have tend to be highly anecdotal. For example, on Christmas Island, the native bulldog rat disappeared soon after the appearance of black rats, and sick and dying bulldog rats were observed in large numbers (Day 1981).

One of the most widely cited instances of disease causing extinctions involves the native birds on Hawai'i. Berger (1981) reviews the role that such exotic diseases as bird pox virus and avian malaria played in the extinction of Hawaiian birds. In 1826, mosquitoes arrived in Hawai'i on a ship, and, although they may not have introduced avian malaria to Hawai'i, they—and, later, the hippobosid fly—have been blamed for the spread of the disease. Just how susceptible the native birds are is a matter for some debate. Warner (1968), on the basis of very small sample sizes, claimed that honeycreepers were very susceptible to avian malaria and bird pox. At present, the Hawaiian honeycreepers (Drepanididae) occur at elevations generally above about 600 m on the main islands and on several small islands in the remote northwest of the archipelago. Mosquitoes occur at low elevations in the wet parts on the main islands, and the range of overlap between mosquitoes and birds is small. Very little native vegetation remains below 600 m on the main Hawaiian islands (much of the lowland habitat is sugarcane, towns, and golf courses). The scarcity of native birds at lower elevations is mainly due to the lack of suitable habitat. In a considerably more detailed study, with large sample sizes, van Riper et al. (1986) found that, at the lower end of their altitudinal ranges, 2%–7% of four native Hawaiian birds were infected with malaria—and that, in one highly mobile species, 29% of the birds were infected.

If the current distribution of native honeycreepers is a reflection of the presence of introduced diseases, the birds must be susceptible to malaria; and just how susceptible they are is not clear. Some native birds, such as the 'amakihi on the island of O'ahu and several species in a small patch of native forest at 300 m on the island of Hawai'i (personal observations, 1988), are found at very low elevations. An unpublished report, by R. Nakamura, widely circulated among Hawaiian conservation biologists, found that substantial numbers of quite healthy native birds were infected with malaria and that an individual bird had lived for over five years with an infection. As van Riper et al. (1986) point out, we would expect the native species to be much more sensitive to malaria on first exposure to it. They found that the honeycreepers from remote islands in the Hawaiian chain were indeed highly susceptible to malaria and that these species probably have not been exposed to malaria. In sum, the role of malaria in determining the current range of Hawaiian birds is not clear, and the present may provide few clues to what happened historically.

Even less certain is the role of malaria in causing the widespread historical extinctions of Hawaiian birds. Many extinctions occurred before malaria became widespread on the islands. Some extinctions may have been caused by rats (Atkinson 1977); other causes will be discussed later in this chapter. Of the nine postmalaria extinctions, six were on Lana'i, Moloka'i, and O'ahu, (van Riper et al. 1986); these are relatively small islands where the native forest has been considerably reduced by human activity. Habitat destruction is a probable explanation for these six extinctions.

Introduced Animals as Prey Species.   Little is known about the effects that exotic prey species have on native communities, yet most introduced animals are potential prey for native predators. (Ebenhard (1988) gives some examples of the effects of exotic species as prey. In Ireland, several breeding pairs of the shorteared owls were found for the first time in 1977 and 1979 (Jones 1979). Since over 70% of the owls diet consisted of the introduced bank voles, establishment of the owls as a breeding species seems to be partially dependent on the successfully spreading bank voles.

The Effects of Introductions

The preceding survey shows that introduced species have many effects on communities, including sometimes devastating extinctions of native species and density changes substantial enough to be noticed by casual observers. The effects are clearly quite common—and often are important, for they are the basis of many anecdotal reports. That not all introduced species have devastating effects is equally clear. Determining what proportion of introductions have effects on density or what proportion cause extinctions is not easy. The literature must often fail to report effects, simply because no one was there to record them. When ecologists have examined the literature carefully, it has become clear that a large proportion of introductions have apparently caused effects—and that many have been so obvious that they were probably severe. It seems most reasonable to conclude that communities are sometimes resistant to introduced species but most often are not.

Introductions into Disturbed Habitats

Herbold and Moyle's (1986) second criticism of Simberloff's (1981) conclusions considers the habitats into which many introduced species are placed. They point out that many of the apparently benign introductions occurred in highly modified habitats from which native species had already been lost for other reasons. Using California fishes as examples, Herbold and Moyle found that 48 of the 137 species of freshwater fishes came from outside the state. Analysis of the habitat of these introduced species shows that 21 occur mainly in such highly modified habitats as reservoirs and farm ponds, that another 21 occur mainly in moderately modified habitats such as streams with altered temperature and flow regimes, and that 6 occur in near-pristine habitats; this latter group includes 4 species of trout and salmon that, when introduced into lakes and streams that previously had not had any fish species, altered invertebrate and amphibian populations. In all, twenty-four of forty-eight introductions have been documented as having a negative impact on native fauna. The effects of the remaining 22 species are not known, but only 2 species appear to be benign. These (probably) benign species occur in limited artificial habitats such as farm ponds and sewage-treatment plants.

THEORIES OF RESISTANCE: VACANT NICHES

Which introduced species should cause marked changes in abundance, and which communities are the most vulnerable to introduced species? Perhaps, in resistant communities, an invading species merely occupies a niche that was previously "vacant" and has little or no effect on most of the other species in the community. We can equate "resistance" to "vacant niche" and ask how we can identify vacant niches. Do vacant niches really exist? Historically, the concept of communities containing vacant niches has been used to justify numerous species introductions, many of which were later seen to have devastating effects (Herbold and Moyle 1986).

Herbold and Moyle (1986) considered the various definitions of "niche," noting that the definitions emphasize either an organism's individual characteristics or its relationships within a community. They note that much of the disagreement about niches (vacant or filled) reflects a failure to recognize that there are several definitions of niche. Different definitions support ideas that vacant niches are impossible, unknown, or arbitrarily defined.

If the niche is defined on the basis of an organism's individual characteristics, then a species brings its own niche with it into a community. Trivially, a persistent community has no vacant niches, while a community which does not persist *does* have vacant niches. If a niche is defined by relationships within a community, then an invading species cannot merely occupy a vacant niche, because the niche is inseparable from the community. For example, if niches are defined in terms of energy flows through the community, then an invading species sequesters a portion of energy that previously went to some other species. Neither of these definitions is very helpful in determining whether niches are vacant. In both definitions, vacant niches either do or do not exist—and which is the case depends on the definition and not on the ecological circumstances.

An alternative definition of niche involves arbitrary subdivisions grouping similar species; these are sometimes called "habitat" or "trophic niches." An example is provided by Lawton's study of the herbivorous insects on bracken fern (figure 9.5). Trophic niches may be either locally or globally vacant, and their precise definition is determined subjectively by the investigator There would seem to be no limit to the number of habitat niches, and how useful the concept is depends on the investigator's experience and judgment. A species introduced into a vacant habitat niche might share with native flora and fauna other resources, predators, or prey.

This is the only definition of niche that holds any promise for our being able to equate a vacant niche with resistant community. Even here, the notion of a vacant niche is almost certainly of no use in predicting resistance to the effects of an invading species. (In contrast, trophic niches are a useful concept in predicting whether a community might be invaded or not; see chapter 8.) Consider Lawton's discussions of the herbivores on bracken. His considerable experience has shown how these specialized herbivores partition the bracken according to where and how they feed (figure 9.5). Different continents not only have different species, but these species occupy different trophic niches. What would the effects of introducing a species into a vacant habitat niche be? While one might predict that such an introduction should be possible, what effects the introduction might have are not quite clear. Indeed, Lawton's attempts to introduce certain species into Britain aim to substantially reduce bracken abundance in potential pasture lands. If these introductions succeed, they will provide evidence of how vacant niches permit invasions. If they really succeed, the introductions will substantially reduce the plant; thus, we shall not be able to equate vacant niches with resistance.

In sum, the various ideas concerning vacant niches seem unlikely to help us predict the effects that introduced species have on community composition. If anything, the ideas seem to have been used to justify introductions by denying that they would have effects, when repeated experience has shown just how often these introductions did have effects.

THE FOOD-WEB THEORY OF RESISTANCE: ITS RELEVANCE TO SPECIES INTRODUCTIONS AND SECONDARY EXTINCTIONS

A Review of the Theory

The consequences of species introductions and extinctions depend critically on both which species are removed or added and the complexity of the food web (chapter 14). Removing a plant species from the base of a simple food chain destroys the entire community (figure 14.5a). The loss of one of several plant species utilized by a generalized herbivore in a more complex community (figure 14.5b) would cause few or no extinctions, because the herbivore is not so dependent on one species. These effects are obvious and well known. Less obvious are the effects of removing species from the tops of food chains. Removing a species that preys on a monophagous herbivore (figure 14.5c) probably reduces the density of the plant the herbivore eats; but the plant is likely to survive. Special conditions are required if a predator is to eliminate its sole prey before the predator itself becomes extinct. In the more complex community (figure 14.5d), the predator's absence may lead to the herbivore exterminating all but the one resistant plant species, which then regulates the herbivore's numbers.

Secondary extinctions are most likely after the loss of plants are from simple communities, because the remaining species in the food chains supported by these plants have no alternative food supplies. In complex communities, the remaining species have alternative food supplies. Yet, after the loss of top-predators, secondary extinctions are *more* likely in complex communities than in simple ones, because cascading extinctions can propagate more widely in complex communities.

These results have simple extensions to discussion involving competitors. Removing the predator of a species which competes with many other species may cause many changes. Removing a predator of a species having more specialized needs may have fewer effects. What is a generalized competitor? An obvious example is a species that competes for a resource that every species needs, such as space in communities where access to space guarantees access to food. For example, intertidal communities are well known as places where most of the species are limited by the space that gives them access to abundant food in plankton.

From these results involving species removals, we may be able to make predictions about the impacts of introduced species. A species which successfully invades a community may do so precisely because it lacks enemies in its new environment. The community may not persist (it may be invaded) and may lack resistance (the invader causes extinctions) for the same reason. Adding an herbivore to an island lacking in predators may be equivalent to removing the herbivore's predator in the herbivore's native community.

What kinds of introduced species are likely to cause the greatest changes? Highly polyphagous introduced species are likely to cause more damage than are monophagous species, because the previous results suggest that predator removals are more severe when their prey are polyphagous. A parallel argument suggests that polyphagous predators should cause more damage if they are introduced into areas containing neither competitors nor diseases.

Finally, the combination expected to produce the most profound changes will be the introduction of generalized herbivores, without their predators, into relatively simple communities. In simple communities and without their predators, these herbivores may attain high densities, their generalized feeding habits may exterminate many species of plants, and the removal of a few plant species may cause the collapse of entire food chains.

Testing the Theory. To test these food web–based theories will not be easy, because the food web's and the community's status, before as well as after an introduction or extinction, must be known. The best I can offer, first, is a reexamination of the literature on the effects of species introductions and, second, some interpretations of the data, in light of food-web theory. In part, this is a process of carefully selecting anecdotes that fit the theory. Such a selection may add a certain credibility to the theories but can hardly be said to test them.

Resistance to Plant Extinctions

The Loss of the American Chestnut Tree. The American chestnut was once an important component of the deciduous forests of eastern North America. It ranged from Georgia in the south and from Illinois in the west and was found as far north as Maine. In some forests it made up more than 40% of the overstory trees (Krebs 1978, 44). Yet, in the early part of this century, chestnuts were driven almost to extinction by a fungal disease, chestnut blight, first noticed in an area near New York City. The fungus was thought to have been introduced on nursery stock from Asia, where the fungus is endemic. It is found on other species of trees but appears to kill only chestnuts. These polyphagous habits are what one might expect from a species capable of devastating a species at the trophic level beneath it, but this is not the point I wish to make.

What interests me is the fate of the animal communities that depended on chestnuts. Many species might have been expected to use chestnuts, including one which enjoyed roasting them over open fires at Christmas.

Despite the importance of chestnuts, there appear to have been no extinctions of vertebrates, for there have been no extinctions of terrestrial vertebrates in eastern North American either during or since the chestnut's decline. Then what of the insects? Opler (1977) discusses the status of Lepidoptera known to have fed on chestnuts. Seven species of Lepidoptera (all of them possibly extinct) apparently fed exclusively on chestnuts. Forty-nine other species of Lepidoptera fed on chestnuts, but the tree constituted only part of their diets. Two of these forty-nine species fed primarily but not exclusively on chestnuts, and they still survive. Hence, only 12% of those Lepidoptera reported to have fed on chestnuts were so specialized that they may have become very rare if not extinct. Apparently, none of these seven species of Lepidoptera supported very specialized species of insect predators. The chestnut food web would seem to be like the one shown in figure 14.5c; most of the herbivores are polyphagous, and the food web is therefore fairly resistant to the removal of one of its plant species. That the loss of such an important species produced so few effects strikes me as interesting and unusual, whatever the explanation.

Trees in Hawai'i.　If there were no vertebrate extinctions after the loss of the chestnut, then the contrast with the Hawaiian situation is sharp. Hawai'i lost approximately half of its species of birds after the islands' discovery by Polynesians over a millenia ago. Approximately half the remaining bird species have been lost since Europeans colonized the islands in the early 1800s, and about half of that remnant are now endangered (Freed et al. 1988). There are many causes of these avian extinctions, but the patterns shown by the species which became extinct and by those which survived are fascinating and seem to indicate an unusual sensitivity to losses of plant species.

First, let me discuss the two families of Hawaiian nectarivores—the Meliphagidae and the endemic Drepanididae. Three of the five Meliphagid species are certainly extinct, and the remaining two species are probably extinct. With one exception, the declines or extinctions of these species occurred before the introduction of malaria (see above and Van Riper et al. 1986). Among the Drepanididae, two of the species which became extinct were nectarivorous. Two more nectarivorous species, the 'akoekoe and the 'i'iwi, have become extinct to some islands but not on others. Only two of these extinctions might be attributable to malaria and, if this attribution is correct, would be the *first* extinctions caused by malaria (van Riper et al. 1986). The birds' extinctions may have followed the destruction of a few important, nectar-producing plants such as various lobelliods and the native hibiscus (*Hibiscadelphus*), by introduced goats and pigs. If so, why these extinctions and not others? After all, other nectar-feeding species are the most common native birds in Hawai'i?

The extinct species were larger than the 'i'iwi or the 'akoekoe, which are the largest surviving nectar-feeding drepanidids. From the sparse field observations of these extinct species and their surviving relatives, it seems the large species were typically aggressive toward smaller species. Of the Meliphagids, at least two of the four species were observed to have been aggressive toward the 'i'iwi (Berger 1981). Pimm and Pimm (1982) found that the two largest surviving species are also behaviorally aggressive toward smaller species and are more specialized in terms of the nectar resources they exploit. These two species have been lost from some but not all of the islands, while the smallest species that exploit nectar are the most common native species on every island. We speculated that the large, aggressive species specialized on the richest sources of nectar, defending them against intruders. However, the large, aggressive species would probably be unable to profitably use or defend poorer-quality resources. When forests were reduced in extent and rich nectar sources were exterminated by introduced mammals, it was these larger, more specialized species that were lost. The smaller, less aggressive species were more generalized in their diets, and they survived.

Other Hawaiian birds also seem to be unusually specialized species. Another rare drepanidid is the 'akiapola'au, an insectivore. The 'akiapola'au may be endangered, because it is found mainly in areas where there are large koa trees. Not only is the koa forest restricted, but the trees are often felled as a source of wood for art objects intended for tourists. A granivore, the palila, is endangered because it too depends almost exclusively on the seeds of one tree, the mamane, that is sensitive to the effects of introduced goats and sheep (see above). The ranges of the frugivorous thrushes on the islands may be now restricted because the loss of a small number of fruiting species may prevent the thrushes from having a year-round food supply (van Riper and Scott 1979).

The unusual vulnerability of Hawaiian bird species seems to be a consequence of their dietary or habitat specializations. Forest communities in both eastern North America and Hawai'i have been greatly modified, yet it is Hawai'i in which the extinctions are so pronounced. When the number of species is the same, communities with specialized species are less complex than communities with species having more generalized diets—the food-web connectance is less. Islands probably have simple food webs, and their species appear to be less resistant to the loss of species at lower trophic levels than are species in the complex food webs of continental communities.

Resistance to Species Changes at Higher Trophic Levels

The food-web theory makes its more unusual predictions when it considers the way in which removals or additions of higher trophic levels affect species at lower trophic levels. Removing predators of generalized herbivores can effect great changes in plant communities, while removing predators of specialized herbivores should have relatively little effect. Similarly, adding generalized predators or herbivores can wreak havoc on their prey when such predators or herbivores are novelties in the community and when they lack competitors and predators. In contrast, adding specialized predators or herbivores should have relatively little effect other than reducing the densities of limited numbers of prey species. What is the evidence for these predictions?

Removing Top-Predators. Large predators have been lost from many communities. Their size and predatory habits make them rare and therefore vulnerable to extinction (chapter 7). To compound the problem, these species have been selectively removed by humans, either because they were thought to be dangerous, because they feed on livestock or other desirable herbivores, or simply for their skins. Consultation with my colleagues suggests that conventional wisdom views such losses as leading to an increased number of generalized herbivores, which in turn leads to substantial changes in plant communities. I have found this wisdom hard to document. Obvious sources, such as Jewell and Holt's (1981) book on the problems of abundant mammals, contain the same wisdom but few examples.

Even so, examples *do* exist, and perhaps the most entertaining is Alston Chase's pointed and sometimes idiosyncratic tale of the account of elk management in Yellowstone National Park (Wyoming) by the United States National Park Service (NPS) (Chase 1987). From 1886 to 1916, the park was guarded by the United States Cavalry. Although troopers were replaced by the NPS, they started a long tradition of "protection" that persists to this day—protection of tourists from game, protection of game from tourists and poachers, and protection of large herbivores from predators. Throughout the park's history, there have been continuous programs of predator control. The major result was the extinction, by 1930, of both cougars and wolves in the park. The result of the decline of these predators was predictable. Elk, the largest of the herbivores, multiplied rapidly, greatly degraded the range, and outcompeted deer, pronghorn antelope, and beaver. Even the grizzly bear suffered as the elk devoured tender vegetation and berries on which the bears also depended. The conclusion of such anecdotal studies is that it is the complexity of the food webs that make them vulnerable to the loss of top-species. As

Chase documents, the effects of changing the density of the elk propagated widely throughout the community.

Introductions of Generalized Mammals to Places where Enemies Are Absent. There is another way to test the prediction that introduced species with generalized diets will have severe impacts in places where they lack predators, competitors, and diseases. The test is to contrast the effect of introducing terrestrial mammals to oceanic islands with the effect of introducing terrestrial mammals to continents. As we saw earlier, Ebenhard has done exactly this.

Ebenhard's (1988) compilations show that a greater proportion of mammalian predators and herbivores caused obvious effects on oceanic islands than did so on continents. Ebenhard realized that it was important to separate oceanic islands from land-bridge islands, for land-bridge islands, by definition, were once part of a continent (chapter 7). Like continents, land-bridge islands will often have terrestrial mammals, their predators, and other enemies, all of which could not easily have dispersed to oceanic islands. The responses of land-bridge islands to introduced species should be more similar to the responses of continents than to those of oceanic islands. Indeed, the proportion of mammals having effects when they are introduced to land-bridge islands is comparable to the proportion of mammals producing such effects when they are introduced to continental communities. The addition of trophically generalized species causes the great changes in the community, again because the effects of the introduction propagate widely. However, oceanic islands may be vulnerable to introductions because their plants, for example, may be "ecologically naive"; they may never have encountered a large grazing mammal before. I shall presently return to the role played by ecological history in determining resistance.

The Introduction of Trophic Specialists versus Trophic Generalists. The most difficult theoretical prediction to test is that introductions of specialized species should have fewer effects than do the introductions of generalized species and similarly, that the removal of predators, competitors, and diseases from specialized species should also result in fewer effects. I know of no species-removal data that could be used to test the second half of this prediction, but there are data on species introductions. The following anecdotes are examples.

We have already encountered studies of chestnut blight in eastern North America and of the snake *Boiga* on Guam; in both studies the extent of the introduction's effect has been attributed to trophic generalization. The blight infects other species of trees, and the snake feeds on lizards; both were able to eliminate vulnerable species—chestnut trees and birds, respectively—without themselves becoming rare.

Ebenhard (1988) compared the effects of introductions of generalized mammalian herbivores with those of introductions of trophically more specialized mammalian herbivores. There were hundreds of the former type of introduction and only five of the latter, but these latter, trophically specialized species were not recorded as ever having effects. In contrast, Ebenhard found that omnivorous mammals had fewer effects on communities than did strictly predatory mammals.

Next, compare the effects of bird introductions with those of mammal introductions; birds rarely have produced documented effects, whereas mammals usually do. Birds are usually introduced for aesthetic reasons. Of the nine most commonly introduced bird species, only the pheasant is introduced as a game bird (often into farmland habitats), and the jungle fowl is often an escapee from chicken coops. The rest are human commensals—birds of our towns and parks. Birds have generally been introduced into habitats already highly modified and very simplified, for example, such as those introduced into lowland Hawai'i. There are few certain effects of these introductions. This result contrasts sharply with the obvious effects of introduced mammals. In short, birds are more commonly introduced into simpler communities than are mammals, and their effects are less dramatic. Herbold and Moyle make similar arguments about the effects of introduced fish in California—the fish appear to have fewer effects when introduced into highly modified and simplified habitats.

The problem with this argument is how to separate "modified" from "simplified." I would like to think that it is the communities' simplicity that makes introductions into human-modified habitats relatively benign. It is possible that the effects of introducing species into highly modified habitats may be reported less often, because we naturally notice the loss of some rare endemic species from a native forest more than we would notice the loss of some weed from only one of the communities in which it occurs almost worldwide.

Finally, we have the contrast between introductions of herbivorous mammals and introductions of insects introduced for biological control. Insect introductions have often been carefully screened to ensure that the species is trophically specialized, in order to avoid damage to nontarget organisms. There are several examples, such as *Cactoblastis* eating *Opuntia* cactus in Australia, in which these trophically specialized herbivores have greatly reduced their hosts and apparently left the rest of the community intact. Of course, these comparisons suffer from the obvious difficulty of comparing insects with mammals—the taxa differ in many respects apart from their degree of trophic specialization.

These anecdotes are at least consistent with the idea that adding trophically generalized predators or herbivores causes greater changes in a community than does adding trophically specialized species. I do not feel comfortable with evidence that relies on comparisons of birds and mammals, insects with mammals, and five trophically specialized mammals with several hundred trophically generalized mammals; neither does Ebenhard (personal communication). More detailed comparisons are needed.

THE ROLE OF HISTORY

Communities on oceanic islands are much less resistant to species introductions than are communities on continents. One explanation is that oceanic island floras are not resistant to introduced herbivores because the islands lack predators that would limit the herbivores. Adding herbivores to islands may be comparable to removing predators from herbivores in continental communities. Moreover, island communities are species poor and may therefore be more sensitive to the loss of plant species because the species dependent on the plants are trophically specialized. All these explanations for the lack of resistance are expressed in terms of species interactions. How else might the data be explained?

Unlike the continental community, an island flora may never have encountered large herbivores, even at low densities under tight predator control. An island community may lack resistance, either because it has never encountered a large grazing mammal before *or* because the large grazing mammal is freed from limits imposed by the predators which controlled its density on the continent. A continental community may lack resistance only for the second of these reasons. When both are subjected to the ravages of a grazing herbivore, how much less resistant will an island plant community be than a continental plant community? The island community experiences the *introduction* of the herbivore, while the continental community experiences the *increased abundance* of the herbivore once the latter's predators are removed. The difference, in change of species composition, between continental and island communities measures the effect of the latter's naivete. This is no more than a thought experiment; contrasting the effects of removing predators from a continent with the effects of introducing herbivores to an island would involve too many uncontrolled variables for the comparison to be meaningful.

These arguments raise the question of whether there is *any* way to evaluate the role of ecological history in shaping a community's vulnerability or resistance to species invasions. The most compelling study involves a comparison of continental floras.

The Resistance of Temperate Grasslands

Mack (1981, 1986) has studied continental grasslands worldwide, has contrasted their changes due to grazing by domestic animals, and has related these changes to the prior history of natural grazers. Grassland plant communities differ greatly in the extent of their resistance to plant invasions, even though the native communities share similar climates and life forms. Grasslands in Australia, South American, and western North America (the intermountain West and California's Central Valley) have suffered "massive changes in the species composition of once vast communities" (Mack 1986, 156). Not all temperate grasslands have proved to be so vulnerable. The grasslands in Europe, Asia, southern Africa, and central North America have also received many alien plants, but these communities have changed little, except for those areas brought under cultivation. Mack shows that the two characteristics that temperate grasslands have that make them vulnerable to plant invasions are (1) the absence of large, hoofed, congregating mammals in the last ten thousand years or so and (2) the dominance of caespitose grasses—those species growing in tussocks or bunches.

In each area of vulnerable grassland, tussock grasses persisted because large mammalian grazers were either rare or missing altogether. Australia had only marsupial grazers, the largest of which were the red and grey kangaroos, weighing less than 100 kg. The native South American fauna included placental mammals but lacked bovids and sheep. The recent distribution of large grazers into North America is more complex. There were enormous herds of bison on the Great Plains, but these large (approximately 500 kg) animals occurred only in small, isolated herds in the intermountain West.

Why are tussock-grass communities so vulnerable to invasion by alien plants? Tussock grasses are vulnerable to grazers which remove their erect flowering tillers, whereas rhizomatous or turf grasses develop a sprawling habit and have tillers that remain prostrate. Moreover, large grazers trample plants; the damage to areas of turf grasses can be repaired by vegetative propagation, an ability tussock grasses usually lack. As both adults and young plants, tussock grasses display little tolerance for grazing or trampling and are quickly lost when disturbance is increased. Mack argues that, once the native tussock grasses are removed by the large grazing mammals introduced by man, alien grasses quickly spread and come to predominate.

Tussock grasslands lacked large grazers, and the phenology of the tussock grasses may account for this scarcity. For example, in the intermountain West of the United States, these grasses do not grow in the early summer, when the lactating bison need maximum green forage. Tussock grasslands could not have supported the invasion into North America—or the evolution in Australia and South America—of large grazers. We have changed this vulnerability by artificially maintaining our domestic herbivores on the grasslands. In contrast to oceanic islands, where physical barriers prevent colonization by large herbivores, Mack's "vulnerable" grasslands lack resistance because of their ecological history. The combination of tussock grasses and the absence of large grazers meant that the grasslands were vulnerable to introduced grazers and alien grasses.

A COMMENT ABOUT HABITAT ISLANDS

One of the most obvious effects that humans have on natural ecosystems is the fragmentation of these ecosystems (Harris 1984) Much of the world is now a sea of highly modified habitats surrounding islands of natural habitat. What will be the fate of the animal and plant communities in these habitat islands? They are likely to be as vulnerable as oceanic islands, for similar reasons. As habitat fragments become isolated, they lose species, often from the highest trophic levels, because these species are the rarest and the slowest to recover from disasters (chapter 7). These losses will likely cause great changes in densities of species at lower trophic levels, for all the reasons I have been discussing. As the communities become highly simplified, they are likely to lose the species at lower trophic levels that are critical to the survival of species at higher trophic levels. This is only the start of the bad news.

Fragmented communities are surrounded by highly modified and often species-poor communities, which are readily invaded as a consequence (chapter 9). Fragmented communities will also progressively lose species; in turn, they too will become more easily invadable. Oceanic island communities may be vulnerable to introduced large herbivores and predators because they have never encountered these species before. Similarly, habitat fragments may be vulnerable to the invading species from surrounding, highly modified habitats because these species may also pose novel threats.

In short, species extinctions create more species extinctions, the potential for species invasions, and yet more species extinctions. There seem to be no theoretical arguments or empirical data to suggest that, as community composition collapses, further change becomes less likely. Indeed, just the opposite seems to be true; as community composition changes, further changes seem inevitable. I am reminded of Paul Ehrlich's "thirteenth-rivet" metaphor for community change (Ehrlich and Ehrlich 1981). A madman removes rivets from an airplane's wing while it is in flight, proclaiming that nothing has happened as a consequence. Indeed, nothing does happen until the thirteenth rivet is removed—and then the wing falls off.

SUMMARY

Species introductions and extinctions provide, on very large spatial scales, an opportunity to test theories of resistance. I argue that species introductions generally have effects on community composition; sometimes those effects are wholesale extinctions of native species.

The food-web theories developed in chapter 14 make predictions about the extinctions and introductions that should result in the greatest change in community composition. Complex communities should be most sensitive to the loss of species from the top of the food web, because secondary extinctions propagate more widely in complex than in simple communities. Simple communities should be more sensitive to the loss of plant species than complex communities, because, in simple communities, the consumers are dependent on only a few species and cannot survive their loss. The introduction of trophically generalized species should have profound effects on community composition, because such species can eliminate many of the species on which they feed. Specialized species should produce fewer changes.

The evidence for these theories is anecdotal, and I have consciously selected examples which fit the theories. I first contrasted the effects of species losses from low in the food chain. There have been few if any extinctions after the loss of the chestnut tree from North America, yet many bird extinctions in Hawai'i seem to have followed the loss of only a few keystone plant species. Thus the simpler, island communities seem less resistant to plant losses. The effects of removing species from higher trophic levels were examined by examining the effects of species which, when introduced to oceanic islands, were probably freed from their normal enemies. Mammal introductions to oceanic islands have had profound effects in a majority of cases; effects of mammal introductions to continents are less noticeable, probably because, on continents, enough enemies were present to control the numbers of the introductions. Birds are usually introduced into very highly modified habitats and are rarely reported as having effects, perhaps because these habitats are already so simplified.

16

Multispecies Models and
Their Limitations

The general thesis of this book is that ecologists have given relatively little attention to long-term, large-scale processes, and that, when they do, the importance of multispecies dynamics becomes increasingly important. Over the long term, population changes have to be understood in the context of the community structure in which each species is embedded, and community structure itself changes over time. Now that the material of the previous chap ters has been presented, it is time to look at this thesis critically. I shall do this in two chapters. In my final chapter, I shall deal with the empirical results. Here I shall deal with the more technical issues involving the ways in which I and like-minded colleagues have built multispecies models to deal with the various issues of stability. Critics are quick to point out the imperfections of these models; I have not missed them either. The purpose of this chapter is to look at the models, their form, their history, and their limitations and to suggest how the models may be improved.

There are both general and specific difficulties with these models. The three general criticisms, in increasing order of sophistication, are that (1) the models are wrong, (2) the models are worse than others, and (3) there has to be something better! I will deal with these criticisms first and then get on to more technical issues.

Are the Models Wrong?

I think it is pointless to try to justify models' equations biologically—their assumptions are almost bound to be wrong. That the assumptions are almost certainly wrong troubles many ecologists. Their concerns are misplaced. All models are simpler than nature, by their very definition; every model assumes that a wide array of natural effects can be ignored. The concern should not be whether the assumptions are wrong (they are!) but *whether it matters that they are wrong*.

We all use models that have indefensible assumptions. For example, how many times have you seen a linear regression fit through a set of points? Did

357

you really believe that the underlying relationship between those points was linear? Linearity is a convenient underlying model of something we often know is more complex; a straight line provides a simple and, one hopes, reasonable approximation to reality. Lotka-Volterra models are an equivalent in population dynamics. Of course, when we move away from the limits of the data on which we estimate the regression line, the assumption of linearity will become more tenuous; and some data are manifestly nonlinear anyway. The more important question is not whether we believe the assumptions, but whether we find them useful. The more we stretch the model, the more serious will be the failure to meet the assumptions.

How do we evaluate the failure of the assumptions? Endless discussions about the model's assumptions seem less persuasive to me than do tests against the data. After all, if the data look linear (and if nonlinear models do not improve the statistical fit), would you worry about my questioning whether the data are *really* linear? Similarly, the *utility* of dynamical models is established best by testing their predictions against the data. Notice I said "utility" and not "validity."

A slightly more refined version of "the models are wrong" criticism is that models—or, more exactly, the theoreticians who build them—have a more simple-minded view of nature than do those who study ecology in the field. The latter group of ecologists, goes the argument, appreciate nature's complexity because of their daily exposure to it, while their theoretical colleagues, sitting in their traditional armchairs, fail to see the subtleties. This hypothesis is testable; it is also known to be false. In an unusual study, Caswell (1988) surveyed a wide range of empirical and theoretical studies presented at a large annual ecology meeting. He counted how many potentially predictive factors were included in each study and found that theoreticians included more factors in their models than field ecologists did in their experiments. It was the theoreticians who had the more complex view of nature.

Are These Models Worse than Other Models?

Even those who accept that models are legitimate tools have difficulties with the kind of models used here. For one thing, they seem much less *real* than the models used by population ecologists. I have used Lotka-Volterra models as a starting point—I use them to explore some of the potential mechanisms that might determine how communities work. For other ecologists, the model of how a population's density changes is an end point, and they seek more mechanistic explanations for those changes. For example, in Lotka-Volterra models, encounters between predators and prey are modeled in the same way as the collisions of atoms in an "ideal" gas (that is, a gas for which physicists must make unrealistic assumptions; Royama 1971). This model form will seem hopelessly unrealistic to a behavioral ecologist. Predators seek out their prey in unique ways, solving along the way such complex optimizations as how to gain the most energy over some interval of time in a spatially patchy, temporally varying environment with different kinds of prey, different kinds of competitors, and variable threats from predators. There is something very compelling about watching an osprey crash into the water and struggle back into the air with its fish. Writing down that interaction as $a_{ij}X_jX_i$ seems a travesty!

Although the encounter between predator and prey may be the most compelling feature of predation, there are others. The physiologist may wonder how the osprey sees the fish, controls its dive, and solves those complex optimizations. The detailed behavioral model will seem mechanistic to the population ecologist but entirely phenomenological to the physiologist. What is mechanistic for one scientist is phenomenological for another scientist working at a different organizational level. In presenting seminars on the material in this book, I am often confronted by population ecologists who feel that community ecologists are not doing things the right way. The reasons, I am sure, have to do with this issue of what constitutes a mechanism.

The confrontation is not one-sided. Conceptually, the usual approaches to population biology have their difficulties as well. How can one expect to understand the behavior, genetics, or ecology of a single species in isolation—or, at best, in terms of interactions with one or two of the many species with which it interacts?

Perhaps no other topic dominates traditional population ecology more than does the discussion of whether density changes are density dependent. Should a population usually increase (density dependence) or not increase (density independence) when the population is subject to experimental or accidental decreases in density? Every ecologist has discussed this question; most ecologists have discussed it in print; and some influential ecologists have written, at book length, about it. The topic is compelling, important to both applied and theoretical ecology, and the mechanisms involved are *real*—we can go out and measure brood sizes to watch how quickly individuals are dying. And yet, as a community ecologist, I find the usual discussions of these issues incredibly circumscribed. As we saw in the chapters on resistance, whether the population will increase, stay the same, or even decrease *initially* is only a tiny part of the story. In the very long term, the density changes will depend on the arrangement and magnitude of interactions in the entire food web; and, in the middle term, whether the population increases or decreases will depend on how long we mean by "middle-term." The hypothetical population may indeed first increase in density, and we may study those increases in exquisite detail. In the longer term, the population may decrease and increase in an apparently random, yet actually just very complicated, fashion. Those "arrangements of interactions"—the various food-web patterns—may not seem as real as watching adults rear larger broods, and they may not be so immediately accessible to measurement, but their long-term importance is probably far greater.

Population biology "works"—that is, is useful—because it chooses scales of time and space where it is possible to excise a species from the matrix of species around it. Studying food webs and the dynamics of their constituent species is a different simplification, likely also to "work"—but at different scales.

Are There *no* Better Models?

Those for whom population models are reasonably mechanistic—but for whom community modes are too unrealistic—may wonder why community theorists do not build very complicated (but realistic!) models incorporating many parameters to describe the interactions between species. Should this not give us the best of both worlds—population realism plus the multispecies effects that I argue are necessary for long-term insight?

I built my first models this way, as a graduate student supported by the U.S. part of the International Biological Program (IBP) investigating desert and grassland biomes. I did not find insights in the richly parameterized IBP models, one of which had about six hundred parameters. Nothing I have done or seen since then has altered my perceptions, though insight is a personal matter and others may have different experiences. These models are often of such bewildering complexity that they are as hard to understand as nature itself. Large-scale simulation models are not discussed in this book because I did not see in them results that I considered relevant. (Important results may be there, hidden among the vast output these models produce). In fairness, however, the models most like these large-scale simulation models are those concerned with community assembly (Post and Pimm 1983; also see chapters 8–11 above). These simulations were so complex that it literally took years for us to understand what was going on, but it was these results that first made me think about the different meanings of ecological stability.

While insight is a very personal matter, prediction is more tangible. Complex, multiparameter models are not necessarily better and may be much *worse* at prediction than are simple models. Fisheries biologists deal with multispecies predictions over the long-term and on large spatial scales, and they find that models with a modest number of parameters are the most effective. The more parameters there are, each to be estimated and few known with any precision, the less likely the models will make accurate predictions (Ludwig 1989).

We are left with one obvious criticism: why not use better simple models? Almost certainly there must be better models "out there." Finding them is not trivial. The easy way would be to look at what others use and to follow them. It is not an accident that nearly all of the models in this book are of the Lotka-Volterra variety—they dominate the literature. Just because these models are the convention does not mean they are the best possible. I shall discuss presently a simple alternative to the Lotka-Volterra models, one that makes some surprisingly good predictions. In what follows, I shall provide details of the models used in this book and shall attempt to suggest what are their principal limitations.

RECIPES: GETTING COMMUNITY DYNAMICS FROM FOOD-WEB DIAGRAMS

Chapter 8 introduced the problem of how food-web structures modify persistence. This is one example of a broad class of problems encountered in all subsequent chapters—that is, the understanding of community dynamics in terms of food-web structure. All problems in this class have an obvious similarity: to investigate them theoretically requires moving from a food web to a model of interacting populations. We must graft a set of *dynamic* equations onto a *static* food-web framework.

I will now consider the details of this process of moving from food webs to community dynamics—and some of the many weaknesses of the recipes for doing this. Within ecology, this process has a history dating to at least May (1973) (see chapter 1). The history can be conveniently divided into two stages that reflect not just the increasing sophistication of the theories but a fundamental division in the kinds of questions asked. Taking a historical approach is conceptually sensible and shows why ecologists have taken each particular step. Each step has been determined by the previous one, rather than creating totally fresh procedures. This creates problems, though they are hardly unique to ecology.

The first stage in the recipes is to go from a food web to a set of equations that describe the species dynamics. The second stage is to alter these dynamics by altering species densities by adding or subtracting individuals or entire species. Most of the literature in the first stage I detail elsewhere (Pimm 1982), and I shall not repeat all the details here. Most of the models of the second stage have been developed since 1982, and they raise issues that have not been fully discussed elsewhere.

Stage 1: Getting a Working Model of Interacting Species

The first step is to create a set of equations that describe how the density of each species in a community changes over time. The Lotka-Volterra equations are one, but not the only, set of equations that can describe such interactions:

$$\frac{dX_i}{dt} = X_i(b_i + \Sigma a_{ij} X_j). \tag{16.1}$$

There will be one of these equations for every one of the n species in the community to be modeled. Each equation has both a term (in b_i) that describes how the density of each species grows in the absence of all others and as many as n terms that describe how each of the other species in the community modifies that growth. Some of these n terms may be zero when species do not directly affect each other, and one of these terms represents the effect that a species' own density has on its own population growth rate.

There are numerous extensions to these models, and generally they add one or more parameters to each interaction, a process which rapidly leads to incomprehensibly complicated models when more than a few species are involved. However, there has recently been a compelling attempt to replace these models with some simple alternatives.

I have already discussed the fact that it is the terms describing the interactions between predators and prey which seemed least convincing. In particular, notice that, per predator, the number of prey killed is proportional only to prey density, X_{prey}, and not to predator density, $X_{predator}$. Arditi and Ginzburg (1989) have discussed this assumption at length. They note that the rationale for excluding an effect of predator density was made explicit by Rosenzweig and MacArthur, who stated that "the greater the number of predators, the faster the [prey] density is reduced; still the *instantaneous* rate of change of the predator population depends on the *instantaneous* rate of kill, which depends on the *instantaneous* density of the prey" (1963, 315) [emphasis added]. This means that predator density has an effect on predator growth but that it is indirect yet mediated *instantaneously* by the prey density. What this is saying is that the behavior and population dynamics all occur on the same time scale. Arditi and Ginzburg argue that this will only be true in certain situations. Where it will not be true, we might expect the interaction terms, per predator, to depend on both prey density *and* predator density—or, perhaps, on their ratio, so that the terms would be proportional to $X_{prey}/X_{predator}$.

As an example, they asked whether we would really expect the number of rabbits consumed per coyote per unit time to remain the same if the number of coyotes were to be increased tenfold. Lotka-Volterra dynamics require us to answer yes; Arditi and Ginzburg's alternative formulation matches what I suspect is more likely—that is, that the number would go down because of the way the increased density of coyotes would affect each coyote's behavior. This effect would be on a time scale very much shorter than that needed to reveal the effect that the high numbers coyotes had on the density of rabbits and that the subsequent decline in rabbit density had on the coyotes.

Arditi and Ginzburg continue by suggesting a continuum: the per-predator effect may be proportional to $X_{prey}/(X_{predator})^m$, where m is the interference constant. If $m = 0$, we retrieve the Lotka-Volterra models, and, if $m > 0$, the predators behaviorally interfere with each other.

This is a very simple model; it seems plausible, makes some interesting predictions about communities, and brings into focus the difficulties of building models intended to make predictions over very different time scales. In a Lotka-Volterra model of a simple linear food chain, an increase in the availability of resources for the species at the base of the food chain causes an increase in the density of *only* the top-predator. From the results of chapter 14, we know that real systems do not behave this way. Increases in the density of species at one trophic level accompany increases in the density of species at other trophic levels. Moreover, the closer the trophic levels, the better the correlations in density. For example, increases in primary productivity often clearly accompany increases in plant density, yet there may be no discernible correlation with the densities of the top-predators. The ratio models of Arditi and Ginzburg make the correct predictions.

In fairness to Lotka-Volterra models, however, the consequences that increasing the density of all the species at one trophic level has on all the species in another trophic level is not nearly so clear (and not so clearly wrong) when we consider complicated food webs rather than simple chains. What constitutes a minimal model of how one trophic level affects another is an open question. What the Arditi and Ginzburg models *do* show, however, is that there are some simple alternatives to Lotka-Volterra models, alternatives that would permit exploration of multispecies effects. Moreover, these alternatives correctly make predictions about nature where at least the simple-minded Lotka-Volterra models fail. These alternative models are not considered in this book, for the simple reason that their potential significance for community ecology have only recently been recognized.

Choosing the Parameters. Given the choice of Lotka-Volterra models, we face the problem of choosing the parameters in those models—the b_i and the

a_{ij}. The method I have used has three steps that combine what we know about the parameters in general with our uncertainty about their exact values.

First, each food web contains a great deal of information about which species interact, for it indicates which species are prey to which predators. The food web also gives information about which species compete. Predators that share prey will compete—indirectly, of course—for the shared prey species. Exploitive competition is always an indirect interaction. (Species that share predators may also adversely affect each other; see chapter 8.) Most food webs contain only trophic information, although some contain information about both direct (or interference) competition and mutualistic interactions. (I still call these webs "food webs" [Pimm 1982]; not everybody agrees, but I prefer stretching the definition to inventing a piece of jargon such as "interaction webs.")

With this information, we can fill in the signs of the b_i and the a_{ij}. Species limited by resources outside the food web (such as either plants limited by nutrients or detritivores limited by detritus) will grow in the absence of other species: their b_i will be positive. Predators that depend entirely on species in the food web will die in the absence of other species: their b_i will be negative. Predators will have a negative effect on the growth rates of their prey, and prey will have a positive effect on the growth rates of their predators. Thus, if species i feeds on species j, then a_{ij} will be positive and a_{ji} will be negative. Directly mutualistic species will increase each other's growth rates, and directly competitive species will decrease each other's growth rates. Where species i does not directly affect species j, a_{ij} and a_{ji} will be zero.

The second step is to choose the relative and approximate magnitudes of the parameters. Because the variables, the X_i, are densities, nothing is lost if we choose one species' equilibrium density to be unity. I usually select the species at the base of the food web to have equilibrium densities on the order of unity. In doing so, I am saying that such species have a density of about one per unit area. After all, the choice of units for area (square centimetres, square kilometres, or perches) is an arbitrary one.

In the same way, how quickly a population density returns to equilibrium can also be set to unity, because the units of time are also arbitrary. In a simple, one-species model, such as the general form of the familiar "r and K" model,

$$\frac{dX_1}{dt} = X_1(b_1 - a_{11}X_1), \qquad (16.2)$$

there is no reason not to make the ratio b_1/a_{11} unity (which forces the equilibrium density to be one organism per unit area) and not to make b_1 unit (which forces the resilience, as defined in chapter 2, to also be unity).

Once some of the parameters have been established thus, the approximate magnitudes of the others are derived from what we know about (*a*) the relative densities of predators and prey and (*b*) the relative effects that predators have on prey and that prey have on predators. On the same per-area, per-unit-time basis, predators are generally rarer than their prey and typically are much rarer when they are very much larger than their prey. The choice of equilibrium densities for predators and prey should reflect this. The effect that a large predator has on the growth rate of its smaller prey will often be very large: predators kill many prey per unit time. The reverse effect will be much smaller, because the predator will need to consume many prey before it produces its own offspring. In contrast, the effect that a small insect parasitoid has on its host may be of approximately the same order of magnitude as the effect that the host has on it, for many parasitoids may emerge from each parasitized host. Moreover, host and parasitoid may have synchronized generation times. Thus, for the interaction between a host and a parasitoid, we would need parameter ranges different than those for the interaction between a large predator and a much smaller prey. These and similar considerations of natural history enable us to make sensible guesses about the approximate magnitudes of the parameters.

In the third and final step, we choose the actual parameters. We are very unlikely to know even one parameter, let alone all of them. My choice has been to select the parameters randomly over intervals that reflect the order-of-magnitude differences discussed in the previous paragraph. Thus, for a model of a predator and its prey, we could choose both that the prey have an equilibrium density, in the absence of its predator, of about unity and that the resilience be about unity. We could also choose that the predator population have an equilibrium density perhaps one to two orders of magnitude less than that of its prey and that the predator's (negative) effect on its prey be at least an order of magnitude greater than the prey's (positive) effect on the predator's growth rate.

Because the parameters are chosen randomly, comparisons of models should be made on a group of the parameters. Typically, this process of randomly selecting parameters over ecologically sensible intervals is repeated a hundred or more times, and comparisons are made between groups of models.

This process of parameter selection has some obvious limitations. Not only must the parameters be selected over a given interval, but we must select a statistical distribution for the parameters. The choice has always been a uniform distribution—that is, if we choose a parameter between 1 and 100, the number of parameters picked between 1 and 11 will be the same, on average, as that between 85 and 96 or any other interval of 10. One can raise two objections to this. First, I cannot think of any biological variables that are distributed this way; often, variables are approximately normally distributed—

values near the mean are found much more commonly than those farther away from it. Second, unlike the uniform distribution, in which, in my example, one may find values between 1 and 2 but never between 0 and 1, the normal distribution admits any value.

There are two reasons why the normal distribution is not used. First, it doubles the number of parameters to be specified; in addition to the mean values (which would be given by the orders-of-magnitude considerations discussed above), we would need to specify the variance of the parameters. I have no idea how to even guess the variances. Second, all values of parameters are possible with a normal distribution. Therefore, using the normal distribution would admit negative values for the effect that a prey has on its predator's growth rate—and this is hardly a realistic assumption. There are many other distributions, of course, but all seem equally arbitrary.

Selecting Models that Satisfy Ecologically Sensible Conditions. Once we have produced a collection of models, with the parameters selected randomly over sensible intervals, each such model must be scrutinized to see whether it satisfies ecologically sensible conditions. The first of these conditions is that the equilibrium densities of the species must be positive. This condition, called "feasibility" (Roberts 1974), is far from inevitable. Indeed, given a collection of models that have randomly selected parameters, we find that the chance of finding feasible models declines rapidly as the food web becomes more connected or has more species (Pimm 1982). For very large models—say, as many as twelve species—the computer programs spend most of their time trying to find the rare, feasible models. This is one reason why twelve-species models are considered "very large"!

The second condition is for the feasible equilibrium to be stable. Many of the feasible models have equilibrium densities that are unstable; that is, for even the smaller disturbance, densities will move away from equilibrium. Often the densities in such models move to another equilibrium that is stable but in which one or more species have become extinct in the process. For some models, species do not become extinct, however, and the densities continue to change in chaotic ways—such as the spiral chaos model or Gilpin (figure 6.2).

Deciding which models are unfeasible and which, although feasible, are unstable is an interesting problem in its own right, for communities cannot persist if some of their species densities are driven to extinction—and are unlikely to persist if some of their species have very large fluctuations in density (chapter 7). The models most likely to be able to describe persistent communities are those with stable equilibria. Using this constraint, one can predict a number of food-web patterns in the real world (Pimm 1982). Yet, some real populations do have dynamics more complex than a locally stable and fixed equilibrium (chapter 6).

Stage 2: Modifying the Models

The second stage of going from food webs to community dynamics is more recent in origin and involves taking food-web models and modifying them. These modifications may be either small changes in the density of a species or large changes caused by the removal or addition of species.

To enter this stage, the recipes reject both unfeasible and unstable models and examine the remainder. Rejecting unstable models is often justified, because simulations of these models will show that they lose species, although some of these unstable models show the kind of complex behaviors discussed in chapter 6. Unfortunately, there are no methods, other than simulation of the species' densities, to decide which unstable models yield extinction and which yield species persistence. For many models, simulation is such a computationally intensive process that one often has little choice but to select only feasible and stable models. In rejecting models in which species may persist (but not at a stable equilibrium point), I am failing to investigate a class of dynamical behaviors shown, in chapter 6, to be quite important.

Small Changes in Density.   In chapter 13, I presented a discussion of Yodzis' work on the effects that changing the densities of each species in a model food web had on every other species in the food web. He produced a set of model food webs by using the recipe I have discussed under the heading "stage one." Then, for each species in each food web, he determined the effects of small changes in its equilibrium density. This is conceptually simple, and the mathematics is straightforward though computationally tedious. From a knowledge of the parameters and the equilibrium densities they determine, the program calculates the *community matrix* that summarizes the effects that one species' *density* has on the *growth rate* of another species when the two species are at their equilibrium densities.

With some further computation, one obtains the *inverse* of this matrix. This inverse determines how small changes in the equilibrium *density* of one species will affect the equilibrium *density* of the other species in the food web. Because only very small changes in density are invoked, the models are not being stretched to implausible limits, and relatively few assumptions are likely to be violated. Of course, not only are all populations never at their equilibrium densities, but many populations fluctuate over many orders of magnitude during the time scales over which Yodzis considers them. We do not know what effects these fluctuations should have on Yodzis' results.

Large Changes in Density. I: Invasions.   In the chapters dealing with persistence, feasible and stable models were repeatedly tested to see whether a
species could successfully invade the model community. These tests are also
conceptually and mathematically simple—they involve deciding whether a
very rare putative invader can increase in density when all the other species in
the system are at their equilibrium densities. If the putative invader can increase, it is added to the community.

The assumptions needed to calculate invasiveness have some obvious difficulties; again, these difficulties are related to population fluctuations. While a
species may not be able to invade when all the species are at equilibrium, half
the species in the community are likely to be above their equilibrium density,
half are likely to be below it, and many species are likely to be far from equilibrium. All this may make it harder for a species to invade, but it is more
likely to mean that many species may have the chance to invade intermittently.
Furthermore, a community may be subject not to an isolated invasion attempt
but to a steady trickle of potential colonists. Such effects may lead to the models underestimating many species' chances for colonization. Just how population variability will effect community persistence through altered invasion rates
is a fascinating and clearly important question. Yet, again, the only way to
answer the question would be through massive simulations.

There are two ways to follow what happens when a species is added; one is
conceptually simple and computationally difficult, and the second is the reverse. The first and obvious way is to simulate the population densities of the
invading species and of the other species in the community. As the invader
increases in density, some of the former residents may increase in density,
others may decrease, and some may become extinct. The simulations stop
when a new equilibrium is reached, and the program then tries to add other
species to the community. Unfortunately, this process has proved to be too
time-consuming, and most modeling efforts have taken short cuts.

The computationally simple way of adding species to a community is to first
see whether the invading species and the resident species can all coexist at a
feasible, stable equilibrium. If so, the invading species is allowed to join the
community without further ado. This addition may not be justified, because
there is nothing to guarantee that the species can increase from being rare to
its new equilibrium density without this trajectory causing some species to
become extinct along the way.

Often, the equilibrium involving the invader and the residents is *not* feasible. Using the standard, simple numerical methods, we calculate that one or
more of the residents have negative equilibrium densities. The computer recipes generally throw out the species with the most negative equilibrium density,

calculate the new equilibrium and check its feasibility, remove the species with the most negative density, and so on, until a feasible equilibrium is reached. Although this may often give the equilibrium that would be reached by simulation, it is easy to come up with examples where it does not. Ted Case has drawn my attention to the obvious example of simple, two-species competition models, where the competitor with the more negative equilibrium density can be the species that wins, rather than loses, the competition.

Finally, there is a small class of models in which a rare species can invade where the equilibrium is feasible but not stable. The recipes generally reject such invaders. Again, in the absence of simulation, I know of no way to decide whether this is justified.

The process of successively adding species to a model food web is known to be flawed, but how important are these flaws? Post and I recognized the potential problems of our numerical recipes (Post and Pimm 1983). We felt that the problems were probably a minor source of error, because, by far, the most common reason why a species failed to join a model community was that it could not increase when rare. When a species did enter, by far the most common reason that it caused extinctions was that it fed on several species in the community and drove one or more of them to ever lower levels.

Large Changes in Density II: Species Removals. The final way in which food-web models can be altered is by removing species from them (chapters 12 and 13). Once stable, feasible models are produced, each species is removed in turn, and the resulting equilibrium is checked to see whether the adjusted model is both feasible and stable. Again there is the problem that, even if such an equilibrium is feasible and stable, the population densities may not be able to reach that equilibrium without incurring some species losses along the way. If the equilibrium is *not* stable, then the species might still persist—by showing complex dynamics—rather than go extinct as we would expect them to do.

In some simulations, Anthony King and I addressed these problems by simulating the densities of the species, from one equilibrium to the next, to track exactly which species become extinct along the way (King and Pimm 1983; also see chapter 14 above). Even in this study, the recipe has some limitations which are shared with all the other procedures involving large changes to the food web. The values of the parameters for the species not removed were the same in both the complete model and the model with the species deleted. This is probably the least realistic assumption in these analyses. For example, suppose predator species P feeds on prey species A and B. P may feed preferentially on A in the presence of B but, if A is removed, may have no choice but to switch its attentions to B and, in doing so, may increase its per-predator consumption rate. In short, the interaction coefficients between P and B may depend on whether A is present. With large perturbations, such as species deletions, the changes in the magnitude of the interaction coefficients could be large.

Equilibrium Models in a Variable World?

The obvious objection to all the models is that species densities vary greatly through time. In contrast, the recipes in stage two *all* start with species *exactly* at their equilibrium densities. Does this not completely invalidate the modeling process? As I have indicated, I think that examining such phenomena as the invasiveness of communities in terms of how variable are the densities of their constituent species should be a priority. I do think, however, that the existing models have something to tell us. The conclusions that I and others have drawn from these models are not about a *particular* model with *particular* parameters. Rather, the conclusions are about quite large numbers of models—typically in the hundreds—that share complexity, species richness, or some more specific feature of food-web structure.

For example, complex food webs are harder to invade than simple ones (chapter 10). Such a result comes from a comparison of a large number of simple models versus a large number of complex models, and, within each group, the parameters and equilibrium densities vary greatly. We would predict that, in the real world, a group of different complex communities would be harder to invade than a group of different simple communities. Yet, because the species densities in any one community are changing over time, a single community could be viewed, over the long term, as a collection of communities that have different parameters.

Many of the conclusions I have drawn in this book acknowledge the great diversity of parameter values. Indeed, the conclusions have been drawn independently of the parameter values and so may be relatively insensitive to the changing population densities. One may object that a collection of models with species exactly at their equilibrium densities is not the same as a collection of models with species densities frequently disturbed away from equilibrium, and I agree. There are, however, no shortcuts—the only way to calculate the similarities and differences is to use extremely time-consuming simulations. Since one must have a starting point, a sensible way to begin is with simple models and by remembering that no investigation is ever complete.

17

Conclusions and Caveats

In this final chapter, I shall review the principal results developed in chapters 2–15 and shall examine those results in two ways. After summarizing the results in each major section of the book, I shall consider both whether single-species or multispecies explanations are needed to interpret the phenomenon and how long it takes for various effects to become apparent. I shall end with some thoughts about why the results about multispecies, long-term processes often seem so contentious.

RESILIENCE

How fast a population recovers from a severe decline in density depends on the species' reproductive rate: the more young produced, the faster the population can recover its former level. In birds, those species that produced more eggs per year had densities that recovered more quickly from the effects of a severe winter. Across all animals, small-bodied species tend to have more young per year than do large-bodied species. Population densities of small-bodied species can be shown to increase more rapidly from year to year than do large-bodied species. Small-bodied species also decrease more from year to year than do large-bodied species. Although I have not directly related year-to-year changes in density to reproductive output, it seems likely that high reproductive rates will translate into high population resilience.

Resilience depends not just on the characteristics of individual species but also on the species' interactions with other species in the community. No species exists in isolation, so no species can recover until all the other species to which it is dynamically linked (directly *and* indirectly) have also recovered. I explored just one aspect of this—the length of food chains—in chapter 2. Theoretically, long food chains may be expected to reduce the resilience of their constituent species, and I presented evidence, from aphid populations in agricultural ecosystems, to support this prediction. Finally, theories also predict that resilience depends on how quickly the nutrients necessary for a species' growth become available. How quickly the nutrients become available will depend partially on food-web structure but also on abiotic processes.

Issues of Scale

The data that show that resilient populations are those with high reproductive rates are for relatively short time periods. The measures of resilience I used were either relative changes from one year to the next or measures of return time for bird populations, most of which were calculated to be five years or less. The aphid study demonstrated the effect of the presence or absence of predators in determining the rate of increase of the aphids and so provided an example of how community structure affects resilience. The choice of aphids was not accidental. These species have very short generation times, and so we could look at changes over many generations within one season, thereby studying an ecologically longer time span—albeit a shorter one in real time—than that considered in previous studies.

What these results show is that population effects on resilience can be detected in the short term and that multispecies effects can be detected in the long term. The results do not show that single-species effects predominate in the short term and that multispecies effects predominate in the long term, although they are consistent with this idea. If these short-term effects seem very much more *real* than the long-term effects, then surely it is because, over short scales, we can so easily see the changes. Two years after the very cold British winter of 1962–63, some bird species' densities were clearly back to what we remembered them as being in 1962; others were not. Before we can detect multispecies effects, we will probably need long-term data, and we will also need different kinds of data than those required to test single-species effects. Yet, just because the data may be harder to collect does not mean that the community effects are less important. In all likelihood, there is no single answer to what determines resilience. Rather, we need to recognize that which factors predominate will likely depend on the spatial and temporal scales we are considering.

VARIABILITY

Single-Species Effects

Like resilience, the temporal variability of population density depends on different factors that are likely to predominate at different spatial and temporal scales. Single-species effects on variability are complicated both empirically and theoretically. Across all animal species, densities of small-bodied species tend to be more variable than those of large-bodied species. Ecologists seem to have expected this result, but the available data have not always supported that expectation, and, in any case, its explanation is not obvious. Moreover, what the data appear to show is that, while the result is particularly true for species completing a generation within a year or less, for long-lived species, variability actually *increases* with increasing body size. Body size may affect variability directly, but large body size also tends to be associated with individuals living longer and reproducing more slowly than do species with small body sizes. Longevity and resilience also affect variability.

During inclement conditions, the densities of small-bodied vertebrates are likely to decline more than the densities of large-bodied species. At least in mammals, the differential declines may occur because, as body size increases, so do the species' available fat stores relative to their energetic needs. Data on bird species' densities after cold winters show that it is indeed the small-bodied species that decrease the most.

Large-bodied species live longer, so we may be counting the same individuals from year to year. Thus, long life may act to reduce population variability, but this effect cannot be that important, because, for the longest-lived species (those weighing more than 100 grams), large-bodied animals vary more than small-bodied animals.

Resilience is greater in small-bodied species, but the effects that resilience has on variability are complicated. Highly resilient species may have high variability, because highly resilient species can track environmental fluctuations more closely than can less resilient species. In contrast, highly resilient species can recover more quickly from brief but severe declines in density, and this will make their densities *less* variable. Detailed examinations of the relationship between resilience and variability provide empirical evidence for both these diametrically opposite effects.

Multispecies Effects

In discussing multispecies effects on population variability, I introduced MacArthur's now classic argument that food-web structure should modify temporal variability. Species that feed on few species should be more likely to become very rare after the loss of a single prey species than should species that are more polyphagous. In contrast, Watt argued that, after the loss of the predators that controlled prey densities, polyphagous species would be able both to increase faster and to become more abundant than monophagous species. For the few data that seem adequate to test these ideas, polyphagous aphids are less variable than species that take fewer host plants; but, in some moths, just the opposite relationship is obtained.

The diversity of predators is also likely to affect the variability of their prey; many predator species may provide more reliable control than do few predator species. For voles in the far north of Europe, where there is only one important vole predator, vole densities cycle and, as a consequence, are relatively more variable. Farther south, there is a diversity of relatively more polyphagous predators, and the vole densities do not cycle and are less variable. Similarly, in two insect groups, the more parasitoid species that each species suffers, the less that species varies from year to year. Moreover, there is the suggestion that, the more parasitoids, the less likely the species is to reach unusually high densities.

Finally, some environments are more variable than others, and, to a limited extent, these ecosystem differences are reflected in differences in population variability. There is need for caution, however. Ecologists have rather freely hypothesized about how ecosystems differ in environmental variability, and these differences are not always substantiated by careful analyses of the data.

Single-Species versus Multispecies Explanations of Variability. The data show quite clear differences in variability, which are due to differences in the life-history characteristics of the species. Detecting multispecies effects ought to be very difficult, for we need not only data on population changes over long periods but also such community statistics as the number of species of prey the species exploits and the number of species of predator the species suffers. Yet, despite these difficulties and despite the shortcomings of the data, we can demonstrate the effects that community structure has on variability. Perhaps we have just been lucky in the systems chosen for analysis, but I think it is much more likely that the effects are important and ubiquitous, so that we shall find them almost wherever we look for them.

Perhaps no other result demonstrates the importance of temporal scale more than does the empirical result that variability increases with the length of time over which it is calculated. The longer we measure a population, the more we uncover long-term, large-amplitude changes in abundance. Over the short term, a population-level explanation for this result comes from the consequences of a population's dependence on the densities of previous years. Recovery from unusually low densities, for example, will take time, and so the population in subsequent years will still be dependent on that initial low density. During this recovery period, the population will show an upward trend in densities and a consequent increase in variability. There are also community-level explanations for increasing variability. Each species' dynamics depend on many other species in the community. So the effect of waiting for all the other species to recover their former densities (after some perturbation) is to create a yet longer trend—and a similarly long time period over which variabilities will increase. Over the shortest times we detect how long individuals live and how quickly they reproduce, and over progressively longer times we detect the effects of other species progressively more indirectly related to the species we are measuring. Over yet longer periods, long-term changes in such physical variables as temperature and rainfall will drive similarly long-term changes in population densities. Very long time series of these physical variables are known to have reddened spectra and so will impose an increasing variance on population densities.

The role of multispecies interactions also appears in the considerations of complex dynamics. Even the most simple models of population changes show a variety of behaviors that include cycles, cycles on cycles, strange attractors, and chaos. The difficulties of detecting complex dynamics in the field data are not trivial, and ecologists are only just starting to employ existing—or to develop new—techniques for handling the very limited data available to them. In lynx, use of these new techniques suggests the existence of something more complex than a simple limit cycle, and Schaffer and Kot (1986a) have shown the existence of an attractor in chicken pox data that closely resembles the attractor obtained by periodically forcing a simple model of disease transmission. Analyses of these and other long-term population data sets suggest that complex dynamics are a more common feature in natural populations than has previously been expected.

Quite what causes these complicated dynamics in nature is not clear. One-species models incorporating one or more time delays can generate them—but what causes those time delays if not interactions with other species? It is quite likely that complicated dynamics are a simple consequence of interspecific interactions, interactions that are inevitable because each species interacts with others in its community. If complex dynamics are a consequence of interspecific interactions, why do such examples as limit cycles appear to be most common in the species-poor arctic? The answer may be that, in communities with more species, the dynamics could become so complicated that, with the limited data available, we are unable to distinguish the patterns from random changes.

EXTINCTION

The process of extinction is central to both population and community ecology. It is the ultimate population catastrophe, and, when species are lost, community composition changes. It is not surprising that it is small populations that are most prone to extinction, and there is abundant evidence that small populations persist for shorter periods of time than do large populations. The length of time a population lasts before going extinct depends on many more factors than just its average size, however.

When demographic accidents are the only cause of extinctions, the time a population lasts should increase very rapidly as population size increases. When environmental disturbances are the principal cause of extinction, the time to extinction will increase only slowly with increasing population size. Indeed, for the most extreme environmental catastrophes, large population size may offer no protection against extinction. Slowly growing populations should be more likely to become extinct than more resilient populations, because they recover less quickly from environmental disturbances. Species in which the individuals live a long time should be less likely to become extinct in any given interval, because demographic accidents are less likely to exterminate a population of long-lived individuals. Across all animal species, small-bodied species tend to have fast-growing populations and short-lived individuals, and these features affect the chance of extinction in opposite ways. At relatively high densities, the cause of extinction will be environmental disasters, and so small-bodied species with their fast-growing populations will persist longer than large-bodied species. At very low densities, demographic accidents are the more likely cause of extinction, and large-bodied species with their long-lived individuals will most likely persist longer. For birds on small islands off the coast of Britain, these theoretical predictions are supported. Below the crossover density of about seven pairs, large-bodied species persisted longer than small-bodied species with the same average density. At this crossover population size, small-bodied species and large-bodied species survive equal periods.

Other things being equal, species whose population densities fluctuate greatly should be more likely to become extinct than species which vary less. Recall, however, from chapters 2–5, that small-bodied species have more variable densities than do large-bodied species (making the formers' extinction *more* likely) but that small-bodied species also have more resilient populations (making their extinction *less* likely). For birds, however, more resilient populations are less variable. So, for birds, more variable populations should always be more likely to become extinct. Analyses of the birds on British islands support this prediction. The "other things being equal" clause means that variability is independent of population abundance and that populations do not differ greatly in their skewness. If rare populations are much less variable than are abundant populations, then they may become extinct more often. There is also the possibility that extremely variable populations may be so variable because of their rare excursions to very high densities and consequently be less prone to extinction than are less variable populations which more regularly encounter small, yet fatal, population crashes.

A population's spatial and demographic structure can influence its chance of extinction, in either direction. Spatially partially isolated populations may be more likely to become extinct through demographic accidents: a population of forty is in little danger from demographic accidents, but eight isolated populations of five individuals will likely persist for only a short period of time. In contrast, a spatially partially isolated population will be better protected against environmental disasters, which typically may be geographically restricted or may affect only some age classes of the population. Not all environmental disasters are geographically restricted, of course. If the populations are very isolated from one another, then a local extinction may not be quickly followed by a recolonization.

Small populations become inbred; inbreeding tends to reduce growth rate; and a reduced growth rate in turn may be expected to increase the chance of extinction. Moreover, species that have lost genetic variability may be less able to exploit new ecological opportunities. Inbred species may be particularly vulnerable to disasters such as epidemics, since different genotypes often differ in their susceptibility to diseases. The inability of inbred species to exploit new opportunities, as well as their increased susceptibility to disease, may be very important in the long term, but they are difficult features to demonstrate in field populations. If inbreeding is important in the wild, then species must loose genetic variability even at quite large population sizes.

Extinction and Time Scales

The discussion of what effects extinction clearly demonstrates how the relative importance of different factors change over progressively longer time scales. In the very short term, it is demographic accidents that doom populations. What makes species vulnerable to this kind of extinction is their own life-history characteristics—for instance, their vulnerability to inbreeding or the small body size that goes hand in hand with short generation time. Species that live longer as individuals persist longer as populations. In contrast, when we see extinctions at longer time scales, it is environmental perturbations that are likely to be their cause, for more persistent populations typically tend to be larger—and therefore less susceptible to demographic accidents—than are very small populations. What determines these longer-term extinctions is how variable the populations are from year to year, for this assays how vulnerable are populations to their varying environments. Variability depends on a species' life-history characteristics, of course, but also on how those interact with the environment and on how the species interacts with other species in the community. In sum, what determines whether and how fast extinctions occur during a five-year period does not necessarily give us the answer to these questions when they are considered a fifty-year period.

PERSISTENCE

How long community composition lasts depends, in part, on both how fast the community is loosing species and how fast it is gaining species. Why do communities gain species at different rates? Physical features of the environment—such as how isolated the habitat is—play a part in determining the rate of invasion. There are also many population-level explanations for why a community is not being invaded. Species differ greatly in their individual abilities to reach communities, and many species introductions may fail because they land in places where the environment is unsuitable for them. Even when a few colonists find a potentially suitable environment, this introduction may fail for one of the many reasons that very small populations become extinct. Analyses of game-bird data have shown that most introductions failed even for species that eventually were successfully introduced.

Persistence depends on features of the community as well. Models of how diverse communities assemble from a small number of species suggest that food-web structure plays an important role in determining persistence. Some aspects of this idea are quite familiar: the more species in a community, the more likely the invading species will encounter a competitor that prevents it from being successful. Other aspects are much less obvious. Finally, these models show that the assembly process itself plays a role. For, even when the model communities are not increasing their numbers of species, each species replacement makes a subsequent replacement less likely. Separating the various explanations of why communities persist is difficult and comes from several different sources, which I now shall consider.

Evidence from Community Composition

Elton (1946) argued that communities are "structured" in that there are relatively few representatives of each genus in any community. He suggested that communities exclude many species through ecological interactions. Elton's ideas about the role of limited taxonomic similarity were extended to limited morphological similarity by Lack (1947). In contrast, Williams (1964) argued that for a given number of species in a community, there are actually *more* species per genus than one would expect. If communities persist, Williams argued, it is because there is a dearth of species able to reach the community— or to survive when they get there. The debate about these hypotheses continues to this day. For example, Grant (1986) observed that morphologically similar species of Darwin's finches rarely occur on the same islands in the Galapagos archipelago. Simberloff (1984) argued that this is because two of those species are rare and, by this fact alone, cannot be found on many islands. Grant replies that such arguments are self-defeating, for the two rare species are probably rare because competition from morphologically similar species restricts their distribution.

Whether the patterns of community composition provide evidence for persistence, they certainly generate interesting hypotheses about it. The patterns suggest that community composition may be limited taxonomically, morphologically, geographically, and temporally. Thus, communities may show geographical patterns, such as "checkerboard" distributions. Some islands in an archipelago may have one species while other islands may have another species. The two species may only very rarely occur on the same island because they cannot coexist. Alternatively, communities may show temporal patterning, with species' activities staggered to reduce overlap.

The most compelling evidence for interspecific competition restricting the species which can invade comes from the known failed breeding attempts of the species. It is this actual demonstration of failure that is the most direct evidence. An obvious way to study persistence is to contrast the failed invasion attempts with those that were successful and to look at how community structure alters the proportion of successes. Studies of deliberately introduced species provide these contrasts. Studies of plants, insects, and birds show that the *proportions* of introduced species are greatest in communities with fewest species, a result consistent with the hypothesis that increased species richness makes invasion more difficult. Such studies, however, typically find that the *numbers* of introduced species are relatively constant across different communities. So there might be a fixed number of species available to invade, and species-poor communities would have a higher proportion of introduced species just because they have fewer native species.

The most convincing evidence for invasions being more difficult in species-rich communities comes from studies of birds introduced on islands. The proportion of failed introductions is greatest on islands that have more species and is lower on islands than on adjacent mainlands (which also have more species).

Moreover, at least for the introduced Hawaiian birds, species fail more often when the island avifauna already contains morphologically similar species.

Predation has a complex role in persistence. While a predator may make it impossible for a species to invade, it may also facilitate invasion, by reducing the densities of the native species, thereby making those species less effective competitors. Thus, communities rich in predators could be easier *or* more difficult to invade than are species-poor communities. In one study, an introduced species of fish was more common in lakes with unusually large numbers of predators. In another study, the presence of rats on islands appeared to have little effect on the number of species of introduced lizards, even though the rats probably had a major effect on the population densities of the lizards.

Evidence from Food-Web Patterns

In nature, food webs show a number of patterns that are statistically unusual when compared with model food webs. All these patterns may be related to persistence, for each pattern implies that the ways in which a species may enter a community are limited. For instance, the limitation on food-chain lengths may suggest that species cannot simply join a community by feeding on all the species below them in the food chain and so avoiding predation and competition. Alternatively, the limitation suggests that species cannot easily insert between trophic levels, because such species would suffer simultaneous competition and predation. If these suggestions were not true, food chains would be unlimited. Similarly, except where prevented by habitat boundaries, a species should feed across rather than within food chains, so that, within a habitat, compartmented food webs should be more easily invaded than food webs that are reticulate. Furthermore, because predatory species can more readily invade communities that have more prey species, and because, under some circumstances, prey species more readily invade communities that have more predatory species, we observe, across different communities, a close correlation between the numbers of species of prey and the number of species of predator.

Although these food-web patterns may indicate how difficult it is for a community to be invaded, what is to stop a potential invader from violating the food-web patterns and, in doing so, making the extinction of other species in the community more likely? The existence of food-web patterns may indicate persistence, but we do not know for certain. Testing these ideas by experimental introductions of species to communities has yet to be carried out.

The Role of Community Assembly

Model communities become progressively harder to invade as assembly proceeds, even when they are not changing either their average numbers of species or their connectance. Assembly itself increases community persistence—therefore, "old" communities are harder to invade than "new" ones. Testing to see whether old communities are easier to invade than new ones is clearly going to be very difficult. Ecologists readily accept that disturbed communities are more readily invaded than undisturbed ones. While "disturbed" means several things, one meaning uses the term to describe communities that are recent assemblages of introduced species. In Hawai'i, relatively species-poor upland, native (hence old) bird communities appear harder to invade than the relatively species-rich lowland communities composed largely of introduced species (hence new communities).

A ubiquitous feature of community-assembly models is the existence of alternative persistent states—different combinations of species, each of which is persistent in the presence of the other combinations. In the laboratory, these alternative communities seem very easy to obtain, and, in the field, there is abundant, albeit circumstantial, evidence of their existence. Another theoretical possibility—that changes in composition constantly cycle appears to be rare in models. Cycles occur in nature, and the patterns of competition between species growing on coral reefs shows how complex they can be. The conditions needed to generate them are special ones, however, and cycles probably are rare in nature.

A final feature of the models is that it may not always be possible to create persistent communities from just their constituent species: other species, which become extinct in the process, may be needed. There is a humpty-dumpty effect: even with all the pieces, we cannot put the community back together again.

Single-Species and Multispecies Explanations for Persistence

There is no point in reiterating the debate about whether the failure of a species to invade a community is determined principally by the species itself or by the community it is trying to invade. This debate is as old as modern ecology. Rather, I would like to speculate on the future of this debate, for the topic of persistence is a crucial one in applied ecology. I think the most compelling studies of persistence have been those that examine deliberately introduced species, for they are the most directly experimental studies available. They provide a unique opportunity to test the community-based hypotheses, and, while patterns in more or less pristine communities are important sources of hypotheses about persistence, the results of analyzing patterns are often ambiguous. Species are being moved around the world ever more quickly, genetically new organisms are being "engineered," and many ecologists are involved in attempts to restore highly modified communities to something like their original states. If we are to prevent accidental introductions, we need to predict which species are likely to be the successful ones. If we are to restore communities, we need to evaluate the importance of alternative persistent states that may prevent us from obtaining the desired persistent state. If there are humpty-dumpty effects, then, when we destroy communities, we may not be able to reassemble them, even if we have kept all the species alive in botanical gardens and zoos.

The rather more academic efforts of detecting patterns in communities are likely to be replaced by these persistence experiments, which will increase in number in the future, both despite our efforts to prevent accidental introductions and because of our restoration efforts. Not only will this increased use provide tests for the theories, but the theories may provide crucial clues in the interpretation of the empirical studies.

RESISTANCE

In discussing resistance, I have focused on the extent to which species densities are changed when the densities of other species in the community are altered either by additions or removals of individuals. Even this restricted topic is complex, for additions or removals may be at very different spatial scales, and their consequences may be examined at different levels of biological organization. I have examined two topics: (1) the effects that changing the density of one species has on the density of another target species ("one-on-one" resistance) and (2) the effects that changing the density of one species has on both the composition and total density of a group of species ("one-on-many" resistance).

One on One

Are There Effects?   It is not obvious that removing one species, A, *should* have effects on the density of another, B, even when A is B's predator or putative competitor. Yet another species, C, may be responsible for determining the density of B. If only some predator removals cause an increase in the density of their prey, or if only some competitor removals produce an increase in the density of competing species, can we predict which species will increase and by how much? Several surveys have reviewed studies on predator and competitor removals. The great majority of these removals produce the expected increases in the density of some species, suggesting that species are generally not resistant to the removals of other species. Yet, studies are biased in many ways, for those that find statistically nonsignificant results may be less likely to be published than those are that find significant results. Furthermore, studies that *do* find significant results were probably motivated by prior simple natural history observations that indicated the likely effects to be found by the experiments.

Single-Species Explanations of Resistance.   What, then, should determine the degree of resistance? Certainly the degree of change in one species may depend solely on its interaction with the one species that is removed. Insect hosts and their parasitoids may be special cases in which we can predict the changes in prey (host) density that are caused by a predator (parasitoid). The introductions of highly specific parasitoids to control often accidentally introduced hosts may occur in relative dynamical isolation from species in the rest of the community. Comparisons of different models, as well as comparisons of host densities in the laboratory versus the field suggest that both the dispersal behavior of the host and the searching behavior of the parasitoid are important in determining the host population's resistance to the parasitoid.

Multispecies Explanations of Resistance 1: Trophic-Level Effects.   There are a number of theories of resistance involving multispecies effects. Perhaps the most famous hypothesis that tries to predict the species that are most resistant to removals is that of Hairston et al. (1960). They imagined a three-trophic-level world in which predators should limit the density of their herbivorous prey and in which, consequently, these herbivores should not be able to limit the density of the plants on which they feed—that is, the world should be "green." The empirical evidence from competitor-and-predator removals does not support this hypothesis. The data also reject an alternative argument—that is, that the world is green because plants are so well protected by their physical and chemical defenses that herbivores can have little impact on them. There is an intriguing extension to "the world is green" hypothesis proposed by Fretwell and Oksanen (Fretwell 1977; Oksanen et al. 1981), who imagined one-, two-, or three-trophic-level ecosystems. They argue that, at the lowest levels of plant productivity, there will not be enough food for effective herbivores. At higher levels of plant productivity, effective herbivory will be possible and will greatly reduce the density of the plants. At yet higher levels, predators will begin to control the herbivores. Most predator-and-competitor removal studies have not been examined in the light of this hypothesis, but some limited data on plant competition support it.

Unfortunately, the data available in the surveys of species removals may not be appropriate for testing these hypotheses. The hypotheses make their most distinctive predictions about where removals should *not* have effects. Not only are ecologists unlikely to report negative studies, but they may be reluctant to even undertake studies that are unlikely to find negative results. Of more concern is that all the hypotheses predict changes of *all* the species at one trophic level when *all* the species at another trophic level are removed. In contrast, the data accumulated so far address the effects that removing one species has on the densities of (at most) a small number of other species. This is a startling mismatch of scale; any one of the theories could be correct, yet studies of small numbers of species might fail to verify it. In short, the surveys may deal with one-on-one questions of resistance while the theories deal with many-on-many effects.

Only in aquatic systems have ecologists manipulated entire trophic levels. There is abundant evidence that top-predators affect predators and that predators affect herbivores. The effects that herbivores have on phytoplankton are mixed; some studies find such effects, while others do not. There is some suggestion that the difference is due to the level of plant productivity, which would tend to support the hypothesis of Oksanen et al., and there is some evidence that, at high levels of productivity, inedible plants predominate.

Multispecies Explanations of Resistance II: Food-Web Effects. Food web–based ideas suggest that it may be extremely difficult to make any statements about resistance. Even if species A should increase if its predator B is removed, this result is likely to be complicated when there are indirect effects—for example, B may also feed on species C, which competes with species A. Perhaps, with a knowledge of the food web, we can predict what the effects of removing any given species will be. As it happens, food-web theory predicts that, even when the food web is known, slight changes in the parameters that describe interactions between species will be amplified by complex pathways that describe all the indirect effects between any pair of species. The result is a high degree of uncertainty: we cannot even be certain that removing a predator will generally cause an increase in the density of one of its prey species!

This theory may seem unlikely in view of the surveys of field experiments, yet I think it is not the theory but rather the field experiments that need reinterpretation. There is ample evidence of indirect effects in field studies, which also show a considerable degree of uncertainty in their outcomes. The same removal repeated several times often does not yield the same result. Thus, while it may be possible to say that a predator reduces the densities of *some* of its prey species, it is generally not possible to say *which* species. The prey species that increase may differ from experiment to experiment.

Finally, the directional uncertainty inherent in complex food webs only applies to the final change in a species' density—that is, the change that appears over a very long time. When we examine the transient changes—for example, whether a prey species will increase in density when its predator is removed—there should be no uncertainty. The prey species in the example should increase. At intermediate time scales, we should see effects rippling through the community, as the change in one species affects more and more species and as these indirect effects begin to alter the density of the species we are measuring with progressively increasing complexity.

It follows that, when we *do* see expected changes in density, it may only be because of the short duration of the field study. The importance both of indirect effects and of how long it takes for the effects of a species removal to be manifested are best illustrated by studies of the removals of rodents from desert communities. For example, seed-eating ants and rodents may be indirect mutualists rather than competitors, because they take seeds from different species of plants that compete with one another. The pathway of effects from rodent to plant, from plant to competing plant, and so to the ants is a long one; it is not surprising that, when rodents were removed, decreases in ant numbers took some time to appear. The initial change in ant numbers, however, was an increase, as they began to take seeds once exploited by the rodents.

One on Many

Predicting the change in one species' density when another is removed may be straightforward in the short term, but, in the long term, the complexity of natural food webs will make predictions impossible, except, perhaps, in some special circumstances. What if we ask a different question: How much does species composition change when we add or subtract species? If we are unable to predict the individual species that will change, can we predict how many species will become extinct? In addition, can we predict how much the total density of a group of prey species will be changed when we either remove their common predator or add a species of predator?

Species Composition. The food-web theories predict that complex communities should be most sensitive to the loss of species from the top of the food web, because secondary extinctions propagate more widely in complex than in simple communities. In contrast, simple communities should be more sensitive to the loss of plant species than are complex communities, because, in simple communities, the consumers are dependent on only a few species and cannot survive their loss. Similarly, the introduction of trophically generalized species should have profound effects on community composition, because such species can eliminate many of the species on which they feed, while introductions of specialized species should effect fewer changes.

These predictions are not easy to test, for we must not only obtain data on more species than are required for the tests of the one-on-one theories of resistance, but we must ensure that we have looked at the systems long enough for transient changes to have disappeared. Compilations of data on the effects of species introductions and extinctions provide an opportunity to test these theories of resistance on very large spatial scales and over very long time periods. While the evidence is often only anecdotal, and while I have consciously selected examples to support those theories, the contrasts provided by the examples are a challenge to explain otherwise.

For example, few if any extinctions occurred after the loss of the chestnut tree from North America, yet many bird extinctions in Hawai'i seemed to follow the loss of a few "keystone" plant species. Presumably, the island communities had simpler food chains with more trophically specialized species and so were more vulnerable. Introductions of trophically generalized mammals to oceanic islands where the communities are relatively species poor have had profound effects in a majority of cases; effects of introductions to continents are less noticeable, probably because there the introduced mammals have been controlled by native enemies. In contrast to mammals, bird introductions are rarely reported as having effects. Birds are generally introduced for the purpose of brightening our lives, rather than as a source of meat. Consequently, birds are introduced into close proximity to us and thus are placed into highly modified, species-poor habitats where we should not expect them to have noticeable effects.

Total Species Abundances. Food-web theories predict that, when grazers are removed from a community, increased interspecific competition between the plants will lead to large changes in species composition but to relatively small changes in total density. Adding grazers to communities where some plant species are much more abundant than others can lead to larger changes in total density than when the plant species are more even in their densities. The reason for this is that the greatest changes in total density can only come when grazers increase evenness, by feeding heavily on the abundant plant species. If the grazers feed heavily on the relatively rare plant species, they cannot decrease their density by much before they eliminate them. So this result—of more plant communities with more even densities being more resistant—depends on the coexistence of the grazers and the plants. If the plants do not coexist when the grazers are present, then grazers may indeed have a substantial impact on the total density *and* on species composition of the plant communities.

Complex though these changes may be, the predictions are supported by field studies. In eastern Africa, when large mammalian herbivores were added, grassland plant communities with more even densities changed less in total density than did less even communities. Similar results were observed with a wide range of grazers in aquatic communities involving both plant-animal and animal-animal interactions. In these communities, grazers and their prey species have coexisted on a regional scale over geological time scales.

Issues of Scale

The species-removal experiment may be the most common kind of experiment conducted by ecologists. My entirely subjective impressions of removal experiments are these: "unexpected" results—that is, those that come from indirect effects—though quite common (see chapter 13), still strike the experimenters as noteworthy. (Why else would they be called unexpected?) Results that change direction in different replicates of the experiment are definitely considered special. Recognition that results will change over time is rare, and the appreciation that the changing results are the manifestations of progressively more indirect effects is very rare. Indeed, explicit recognition that time may be a critical variable is often lacking: many removal experiments do not discuss in detail whether enough time has elapsed for effects, expected or otherwise, to appear.

I hope that my impressions are wrong. For what most studies show is both the ubiquity of indirect effects and how critical the time period of the experiments is in setting a limit to how many such effects will appear. To interpret species-removal experiments in terms of only the interaction between the species removed and its immediate prey or competitor is to impose a very short time period on that interpretation. To understand fully the consequences of species-removal experiments, full consideration of spatial, temporal, and organizational scales is essential.

THE DIFFICULTIES: INEVITABLE APPROACHES TO LONG-TERM, LARGE-SCALE, AND HENCE MULTISPECIES STUDIES

The summaries provided so far in this chapter provide many examples of how the various meanings of "stability" can be related to different aspects of community structure. The results can be criticized in a number of ways, and I shall finish this chapter by trying to address some of these difficulties.

Some of the examples I have used are detailed studies of either a single system or a small group of systems. (One such example is the discussion of the role of predator diversity in the variability of vole densities; chapter 4.) For such examples, the details of the studies are often appealing; but the examples are carefully selected, so it is not surprising that they meet the predictions of the theory. (I have not deliberately excluded counterexamples, however; but inconclusive, statistically nonsignificant results are easy to dismiss as being the consequence of poor data and are easy to overlook, as they generally take up less journal space than do significant results.) Surely, goes the criticism, there is some group of organisms in which we will find the result predicted by the theories, if only because there are so many different groups of organisms that the assumptions of the theories must surely match at least one of them! Only when we find predicted relationships in most of the groups for which we have reasonable data will we conclude that the theories are compelling. So, selected examples may support the theories, but they certainly do not provide adequate tests of them. Only many new studies or compilations of existing studies will do that.

Wherever possible, I have tried to use large-scale compilations of studies to test theories; but large-scale compilations have their critics too. There are concerns over both the data and the methods used to analyze them.

The data needed to test ideas in community ecology almost always involve a large number of variables. The obvious danger in analyzing such data without having any prior ideas of what one should find is the "night-sky effect": if you look at random data long enough, all kinds of patterns can be imagined. Modern computers and their statistical programs can find relationships in data sets very quickly indeed. At the 5% significance level, by an amazing coincidence, one in twenty results will be considered significant, if each analysis is considered in isolation. Community ecology is so conceptually complex that we rarely have any difficulty in explaining any result we encounter, have very little difficulty believing the explanations, and may feel that we even *expected* the result. Consequently, relationships in the data gain considerable credibility when they match theoretical expectations—and especially when they do so consistently. Theories tell us where to look, and, when we readily find what we are looking for, we gain confidence in both the theories, for which we must make unreasonable assumptions (see previous chapter) and in the data, which we know are less than ideal. For this reason, I have quite deliberately presented data and theories together in this book.

By their very nature, broad surveys require assembling data from the literature, rather than conducting field studies. Not only is searching archives nowhere near as much fun as field work, but it means both using data for purposes to which they were never originally intended and omitting all the intricacies of the original work. Worse, it means using data of less than sterling quality when one wants to test theories on large temporal and spatial scales. To test theories of persistence and resistance, I have used assembled anecdotes about whether

introduced species "succeeded" of "failed" or did or did not have "effects,"—
and have hoped that the hundreds of naturalists who used those terms meant
more or less the same things by them.

Certainly, poor data may lead to false results—for example, in analyzing
food-web patterns, we find that, the poorer the quality of the data, the more
prominent some features become (Pimm 1982). One must be ever vigilant for
such spurious results. When the data contain the phenomena we expect, however, it is likely that the phenomena are so pervasive that they appear despite
the fact that the data are poor. It is not that the data are particularly good but
that they are good enough.

I have a feeling of déjà vu about these arguments concerning limited and
poor-quality data. When ecologists started to compile studies of food webs,
only about thirty were available for study (Cohen 1978; Pimm 1982). There
was no doubt that the data were poor, yet they readily demonstrated a dozen or
more features that differed from what we might expect from random food
webs. The number of food webs available for study is now almost an order of
magnitude larger. Most of those food-web features are still intact. More important, the claims made on the basis of those original thirty webs were the
reason why (a) we now have so many more data available to us and (b) many
studies were conducted to directly test the existence of those features. It would
be surprising if there were not many studies whose compilation and comparisons would both support and sometimes reject many of the theories and empirical relationships claimed here. How many other studies are there that show
the relationship between polyphagy or predator diversity and population variability or between the species richness of a community and the ease with which
it can be invaded by a deliberately introduced species? If this book forces new
analyses of these studies or more recognition for the ones which have already
been published, it will be a success.

Literature Cited

Addicott, J. F. 1974. Predation and prey community structure: an experimental study of the effect of mosquito larvae on the protozoan communities of pitcher plants. *Ecology* 55:475–92.

Alford, R. A., and H. M. Wilbur. 1985. Priority effects in experimental pond communities: competition between *Bufo* and *Rana*. *Ecology* 66:1097–1105.

Andrewartha, H. G., and L. C. Birch. 1954. The distribution and abundance of animals. University of Chicago Press, Chicago.

Anonymous. 1944. British Ecological Society symposium on the ecology of closely allied species. *Journal of Animal Ecology* 13:176–78.

Anonymous. 1974. Proceedings of the First International Congress of Ecology. Centre for Agricultural Publishing and Documentation, Wageningen, The Netherlands.

Arditi, R., and L. R. Ginzburg. 1989. Coupling in predator-prey dynamics: ratio-dependence. *Journal of Theoretical Biology* 139:311–26.

Ashwood, T. L., and J. L. Gittleman. 1991. Behavioral and ecological factors influencing critical density in cheetahs *Acinonyx jubatus*. In H. Genoways, ed. Current Mammalogy 3. Plenum Press, New York (in press).

Atkinson, I. A. E. 1977. A reassessment of factors, particularly *Rattus rattus* L., that influenced the decline of endemic forest birds in the Hawaiian Islands. *Pacific Science* 31:109–33.

Ballou, J., and K. Ralls. 1982. Inbreeding and juvenile mortality in small populations of ungulates: a detailed analysis. *Biological Conservation* 24:239–72.

Banks, R. C. 1981. Summary of foreign game bird liberations 1969 to 1978. Special Scientific Report, Wildlife no. 239. Bureau of Sport Fisheries and Wildlife, Washington, D.C.

Barkai, A., and C. McQuaid, 1988. Predator-prey reversal in a marine benthic ecosystem. *Science* 242:62–64.

Beaver, R. A. 1985. Geographical variation in food web structure in *Nepenthes* pitcher plants. *Ecological Entomology* 10:241–48.

Beddington, J. R., C. A. Free, and J. H. Lawton. 1978. Modelling biological control: on the characteristics of successful natural enemies. *Nature* 273:513–19.

Beirne, B. P. 1975. Biological control attempts by introductions against pest insects in the field in Canada. *Canadian Entomology* 107:225–36.

Bell, B. D. 1977. The Big South Cape Islands rat irruption. Pages 33–45 in P. R. Dingwall, I. A. E. Atkinson, and C. Hay, eds. The ecology and control of rodents in New Zealand nature reserves. Information Ser. 4. Department of Lands and Survey, Wellington.

Bellows, T. S. 1981. The descriptive properties of some models of density dependence. *Journal of Animal Ecology* 50:139–56.

Belovsky, G. E. 1987. Extinction models and mammalian persistence. Pages 35–58 in M. E. Soulé, ed. Viable populations for conservation. Cambridge University Press, Cambridge.

Bender, E. A., T. J. Case, and M. E. Gilpin. 1984. Perturbation experiments in community ecology: theory and practice. *Ecology* 65:1–13.

Bentham, G. 1924. Handbook of British flora: a description of the flowering plants and ferns indigenous to, or naturalized in the British Isles. 7th ed., revised by J. D. Hooker. L. Reeve and Co., Ashford, Kent.

Berger, A. J. 1981. Hawaiian birdlife. 2d ed. University of Hawaii Press, Honolulu.

Bird, R. P. 1930. Biotic communities of the aspen parkland of central Canada. *Ecology* 11:356–442.

Blackman, R. L., and V. F. Eastop. 1984. Aphids on the world's crops: an identification and information guide. John Wiley and Sons, Chichester, England.

Blueweiss, L., H. Fox, V. Kudzma, D. Nakashima, R. Peters, and S. Sams. 1978. Relationship between body size and some life history parameters. *Oecologia* 37:257–72.

Bohl, W. H., and G. Bump. 1970. Summary of foreign game bird liberations 1960 to 1968. Special Scientific Report, Wildlife no. 130. Bureau of Sport Fisheries and Wildlife, Washington, D.C.

Bonnell, M. L., and R. K. Selander. 1974. Elephant seals: genetic variation and near extinction. *Science* 184:908–9.

Bonner, J. T. 1965. Size and cycle: an essay on the structure of biology. Princeton University Press, Princeton, N.J.

Bradley, R. A. 1983. Complex food webs and manipulative experiments in ecology. *Oikos* 41:150–52.

Brown, J. H. 1971. Mammals on mountaintops: non-equilibrium insular biogeography. *American Naturalist* 105:467–78.

Brown, J. H., D. W. Davidson, J. C. Munger, and R. S. Inouye. 1986. Experimental community ecology: the desert granivore system. Pages 41–62 in J. Diamond and T. J. Case, eds. Community ecology. Harper and Row, New York.

Brown, J. H., and A. Kodric-Brown. 1977. Turnover rates in insular biogeography: effect of immigration on extinction. *Ecology* 58:445–49.

Brown, C. J., I. A. W. MacDonald, and S. E. Brown. 1985. Invasive alien organisms in South West Africa/Namibia. South African National Scientific Programmes Report 119.

Brown, V. K. 1985. Insect herbivores and plant succession. *Oikos* 44:17–22.

Buss, L. W. 1986. Competition and community organization on hard surfaces in the sea. Pages 517–36 in J. Diamond and T. J. Case, eds. Community ecology. Harper and Row, New York.

Buss, L. W., and J. B. C. Jackson. 1979. Competitive networks: nontransitive competitive relationships in cryptic coral reef environments. *American Naturalist* 113:223–34.

Calder, W. A. 1984. Size, function, and life history. Harvard University Press, Cambridge, Mass.

Carpenter, S. R., J. K. Kitchell, and J. R. Hodgson. 1985. Cascading trophic interactions and lake productivity. *Bioscience* 35:634–39.

Carpenter, S. R., J. F. Kitchell, J. R. Hodgson, P. A. Cochran, J. J. Elser, M. M. Elser, D. M. Lodge, D. Kretchmer, X. He, and C. N. von Ende. 1987. Regulation of lake primary productivity by food web structure. *Ecology* 68:1863–76.

Case, T. J., and D. T. Bolger. 1991. The role of introduced species in shaping the distribution and abundance of island reptiles. *Evolutionary Ecology* 5:272–290.

Caswell, H. 1988. Theory and models in ecology: a different perspective. *Bulletin of the Ecological Society of America* 69:102–8.

Cawthorne, R. A., and J. H. Marchant. 1980. The effects of the 1978/79 winter on British bird populations. *Bird Study* 27:163–72.

Chase, A. 1987. Playing God in Yellowstone. Harcourt Brace Jovanovich, New York.

Chatfield, C. 1984. The analysis of time series: an introduction. 3d ed. Chapman and Hall, London.

Clark, J. S. 1988. Effect of climate change on fire regimes in northwestern Minnesota. *Nature* 334:233–35.

Coblentz, B. E. 1978. The effect of feral goal (*Capra hircus*) on island ecosystems. *Biological Conservation* 13:279–86.

Cohen, J. E. 1977. Ratio of prey to predators in community webs. *Nature* 270:165–67.

————. 1978. Food webs and niche space. Princeton University Press, Princeton, N.J.

Cohen, J. E., F. Briand, and C. M. Newman. 1990. Community food webs: data and theory. Springer-Verlag, New York.

Cole, B. J. 1981. Overlap, regularity and flowering phenologies. *American Naturalist* 117:993–97.

————. 1983. Assembly of mangrove ant communities: patterns of geographic communities. *Journal of Animal Ecology* 52:349–55.

Colwell, R. K., and D. W. Winkler. 1984. A null model for null models in biogeography. Pages 344–59 in D. R. Strong, D. Simberloff, L. G. Abele, and A. B. Thistle, eds. Ecological communities: conceptual issues and the evidence. Princeton University Press, Princeton, N.J.

Connell, J. H. 1978. Diversity in tropical rain forests and coral reefs. *Science* 199:1302–10.

————. 1983. On the prevalence and relative importance of interspecific competition: evidence from field experiments. *American Naturalist* 122:661–96.

Connell, J. H., and W. P. Sousa. 1983. On the evidence needed to judge ecological stability or persistence. *American Naturalist* 121:789–824.

Coulson, J. C., G. R. Potts, I. R. Deans, and S. M. Fraser. 1968. Exceptional mortality of shags and other seabirds caused by paralytic shellfish poison. *British Birds* 61:381–404.

Crawley, M. J. 1987. What makes a community invasible? Pages 429–54 in M. J. Crawley, P. J. Edwards, and A. J. Gray, eds. Colonization, succession and stability. Blackwell Scientific Publications, Oxford.

Crowell, K. L. 1973. Experimental zoogeography: introductions of mice to small islands. *American Naturalist* 107:535–58.

Davidson, D. W., D. A. Samson, and R. S. Inouye. 1985. Granivory in the Chihuahuan Desert: interactions within and between trophic levels. *Ecology* 66:486–502.

Davidson, J., and H. G. Andrewartha. 1948a. Annual trends in a natural population of *Thrips imaginis* (Thysanoptera). *Journal of Animal Ecology* 17:193–99.

Davidson, J., and H. G. Andrewartha. 1948b. The influence of rainfall, evaporation and atmospheric temperature on fluctuations in the size of a natural population of *Thrips imaginis* (Thysanoptera). *Journal of Animal Ecology* 17:200–22.

Day, D. 1981. The doomsday book of animals. Ebury Press, London.

Dayan, T., D. Simberloff, E. Tchernov, and Y. Yom-Tov. 1989. Inter- and intra-specific character of displacement in mustelids. *Ecology* 70:1526–39.

————. 1990. Feline canines: community-wide character displacement in the small cats of Israel. *American Naturalist* 136:39–60.

Dayton, P. K. 1971. Competition, disturbance and community organization: the provision and subsequent utilization of space in a rocky intertidal environment. *Ecological Monographs* 41:351–89.

DeAngelis, D. L. 1991. Dynamics of nutrient cycling and food webs. Chapman and Hall, London.

Delcourt, P. A., and H. R. Delcourt. 1987. Ecological studies. Vol. 53: Long-term forest dynamics in the temperate zone. Springer-Verlag, New York.

den Boer, P. J. 1977. Dispersal power and survival: carabids in a cultivated countryside. Miscellaneous papers 14. Landbouwhogeschool, Wageningen, The Netherlands.

Dennis, B. 1989. Allee effects: population growth, critical density and the chance of extinction. *Natural Resource Modelling* 4:481–538.

Diamond, J. M. 1975. Assembly of species communities. Pages 342–44 in M. L. Cody and J. M. Diamond, eds. Ecology and evolution of communities. Harvard University Press, Cambridge, Mass. © 1975 by the President and Fellows of Harvard College.

———. 1984a. Historic extinctions: a rosetta stone for understanding prehistoric extinctions. Pages 824–62 in P. S. Martin and R. G. Klein, eds. Quaternary extinctions: a prehistoric revolution. University of Arizona Press, Tucson.

———. 1984b. "Normal" extinctions of isolated populations. Pages 191–246 in M. H. Nitecki, ed. Extinctions. University of Chicago Press, Chicago.

Diamond, J. M., and R. M. May. 1977. Island biogeography and the design of nature reserves. Pages 228–52 in R. M. May, ed. Theoretical ecology. Sinauer Associates, Sunderland, Mass.

Dinerstein, E., and G. F. McCracken. 1990. Endangered greater one-horned rhinoceros carry high levels of genetic variation. *Conservation Biology* 4:417–23.

Dobson, A. P., and R. M. May. 1986. Patterns of invasions by pathogens and parasites. Pages 58–76 in H. A. Mooney and J. A. Drake, eds. Ecology of biological invasions of North America and Hawaii. Springer-Verlag, New York.

Drake, J. A. 1985. Some theoretical and experimental explorations of structure in food webs. Ph.D. dissertation, Purdue University.

———. 1988. Models of community assembly and the structure of ecological landscapes. Pages 584–604 in T. Hallam, L. Gross, and S. Levin, eds. Mathematical ecology. World Press, Singapore.

———. 1989. Communities as assembled structures: do rules govern pattern? *Trends in Ecology and Evolution* 5:159–63.

————. 1990. The mechanics of community assembly and succession. *Journal of Theoretical Biology* 147:213–34.

————. 1991. Community assembly mechanics and the structure of an experimental species ensemble. *American Naturalist* 137: 1–26.

Drake, J. A., H. A. Mooney, F. di Castri, F. Kruger, R. Groves, M. Rejmanek, and M. Williamson, eds. 1989. Biological invasions: a global perspective. John Wiley and Sons, Chichester, England.

Ebenhard, T. 1987. An experimental test of the island colonization survival model: bank vole (*Clethrionomys glareolus*) populations with different demographic parameters. *Journal of Biogeography* 14:213–23.

————. 1988. Introduced birds and mammals and their ecological effects. *Swedish Wildlife Research* 13:1–107.

Ehler, L. E., and R. W. Hall. 1982. Environmental entomology (forum): evidence for competitive exclusion of introduced natural enemies in biological control. *Entomological Society of America* 11:1–4.

Ehrlich, P. R. 1965. The population biology of the butterfly, *Euphydras editha*. II. The structure of the Jasper Ridge colony. *Evolution* 19:327–36.

Ehrlich, P. R., and A. Ehrlich. 1981. Extinction: the causes and consequences of the disappearance of species. Random House, New York.

Ehrlich, P. R., D. D. Murphy, M. C. Singer, C. B. Sherwood, R. R. White, and I. L. Brown. 1980. Extinction, reduction, stability and increase: the responses of checkerspot butterfly (*Euphydras*) populations to the California drought. *Oecologia* 46:101–5.

Ehrlich, P. R., R. White, C. Singer, W. McKechinie, and E. Gilbert. 1975. Checkerspot butterflies: a historical perspective. *Science* 188:221–28.

Elton, C. S. 1946. Competition and the structure of ecological communities. *Journal of Animal Ecology* 15:54–68.

————. 1958. The ecology of invasions by animals and plants. Chapman and Hall, London.

Erlinge, S., G. Göransson, L. Hansson, G. Hogstedt, O. Liberg, I. N. Nilsson, T. Nilsson, T. von Schantz, and M. Sylvén. 1983. Predation as a regulating factor on small rodents populations in southern Sweden. *Oikos* 40:36–52.

Evans, D. O., and D. H. Loftus. 1987. Colonization of inland lakes in the Great Lakes region by rainbow smelt, *Osmerus mordax:* their freshwater niche and effects on indigenous fishes. *Canadian Journal of Fisheries and Aquatic Sciences* 44:249–66.

Ewens, W. J., P. J. Brockwell, J. M. Gani, and S. I. Resnick. 1987. Minimum viable population size in the presence of catastrophes. Pages 59–68 in M.

Soulé, ed. Viable populations for conservation. Cambridge University Press, Cambridge.

Fenchel, T. 1974. Intrinsic rate of natural increase: the relationship with body size. *Oecologia* 14:317–26.

Feeny, P. 1976. Plant apparency and chemical defence. *Recent Advances in Phytochemistry* 10:1–40.

Finerty, J. P. 1980. The population ecology of cycles of small mammals. Yale University Press, New Haven, Conn.

Fox, L. R., and P. A. Morrow. 1981. Specialization: species property or local phenomenon? *Science* 211:887–93.

Fox, M. D., and B. J. Fox. 1986. The susceptibility of natural communities to invasion. Pages 57–66 in R. H. Groves and J. J. Burdon, eds. Ecology of biological invasions: an Australian perspective. Australian Academy of Sciences, Canberra.

Freed, L. A., S. Conant, and R. C. Fleischer. 1988. Evolutionary ecology and radiation of Hawaiian passerine birds *Trends in Ecology and Evolution* 2:196–202.

Fretwell, S. D. 1977. The regulation of plant communities by food chains exploiting them. *Perspectives in Biology and Medicine* 20:169–85.

Gardner, M. R., and W. R. Ashby. 1970. Connectance of large, dynamical (cybernetic) systems: critical values for stability. *Nature* 228:784.

Gaston, K. J. 1988. Patterns in the local and regional dynamics of moth populations. *Oikos* 53:49–59.

Gaston, K. J., and J. H. Lawton. 1988a. Patterns in body size, population dynamics and regional distribution of bracken herbivores. *American Naturalist* 132:662–80.

———. 1988b. Patterns in the distribution and abundance of insect populations. *Nature* 331:709–12.

Gause, G. F. 1934. The struggle for existence. Williams and Wilkins, Baltimore.

Gibb, J. A., and J. E. C. Flux. 1973. Mammals. Pages 334–71 in G. R. Williams, ed. The natural history of New Zealand. Reed Press, Wellington.

Gilpin, M. E. 1979. Spiral chaos in a predator-prey model. *American Naturalist* 113:306–8.

———. 1987. Spatial structure and population vulnerability. Pages 125–40 in M. Soulé, ed. Viable populations for conservation. Cambridge University Press, Cambridge.

Gilpin, M. E., M. P. Carpenter, and M. J. Pomerantz. 1986. The assembly of a laboratory community: multi-species competition in *Drosophila*. Pages

23–40 in J. M. Diamond and T. J. Case, eds. Community ecology. Harper and Row, New York.

Gilpin, M. E., and M. Soulé. 1986. Minimum viable populations: processes of species extinctions. Pages 19–34 in M. Soulé ed., Conservation biology, Sinauer Associates, Sunderland, Mass.

Gleick, J. 1988. Chaos: making a new science. Heinemann Press, London.

Goodman, D. 1987. The demography of chance extinction. Pages 11–34 in M. Soulé, ed. Viable populations for conservation. Cambridge University Press, Cambridge.

Grant, B. R., and P. R. Grant. 1989. Evolutionary dynamics of a natural population. The University of Chicago Press, Chicago.

Grant, P. R. 1975. The classical case of character displacement. Pages 237–70 in T. Dobzhansky, M. K. Hecht, and W. C. Steele, eds. Evolutionary ecology. Vol. 8. Plenum Press, New York.

———. 1986. Darwin's finches. Princeton University Press, Princeton, N.J.

Grant, P. R., and D. Schluter. 1984. Interspecific competition inferred from patterns of guild structure. Pages 201–233 in D. R. Strong, D. Simberloff, L. G. Abele, and A. B. Thistle, eds. Ecological communities: conceptual issues and the evidence. Princeton University Press, Princeton, N.J.

Greenway, J. C., Jr. 1967. Extinct and vanishing birds of the world. Dover Books, New York.

Groves, R. H., and J. J. Burdon. 1986. Ecology of biological invasions. Cambridge University Press, Cambridge.

Hairston, N. G., F. E. Smith, and L. B. Slobodkin. 1960. Community structure, population control and competition. *American Naturalist* 94:421–25.

Hale, R. L., and H. H. Shorey. 1971. Effect of foliar sprays on the green peach aphid on peppers on southern California. *Journal of Economic Entomology* 64:547–49.

Hall, D. J., W. E. Cooper, and E. E. Werner. 1970. An experimental approach to the production dynamics and structure of freshwater animal communities. *Limnology and Oceanography* 15:839–928.

Hall, R. W., and L. E. Ehler. 1979. Rate of establishment of natural enemies in classical biological control. *Entomological Society of American Bulletin* 25:280–82.

Hallett, J. G. 1982. Habitat selection and the community matrix of a desert small-mammal fauna. *Ecology* 63:1400–10.

Hambright, K. D., R. J. Trebatoski, R. Drenner, and D. Kettle. 1986. Experimental study of the impacts of bluegill (*Lepomis macrochirus*) and largemouth bass (*Micropterus salmoides*) on pond community structure. *Canadian Journal of Fisheries and Aquatic Sciences* 43:1171–76.

Hanski, I. 1987. Populations of small mammals cycle—unless they don't. *Trends in Ecology and Evolution* 2:55–56.

Hansson, L. 1984. Predation as the factor causing extended low densities in microtine cycles. *Oikos* 43:255–56.

———. 1987. An interpretation of rodent dynamics as due to trophic interactions. *Oikos* 50:308–18.

Hansson, L., and H. Henttonen. 1985. Gradients in density variations of small rodents: the importance of latitude and snow cover. *Oecologia* 67:394–402.

Hansson, L., and T. Larsson. 1980. Small rodent damage in Swedish forestry during 1971–79. *Rapport Institutionen for Viltekologie* 1.

Hansson, L., and J. Zejda. 1977. Plant damage by bank voles (*Clethrionomys glareolus* Schreber) and related species in Europe. *EPPO Bulletin* 7:223–42.

Harris, L. D. 1984. The fragmented forest: island biogeography theory and the preservation of biotic diversity. University of Chicago Press, Chicago.

Harrison, D. L. 1968. The mammals of Arabia. Vol. 2. Ernest Benn, London.

Hassell, M. P., J. Lato, and R. M. May. 1989. Seeing the wood for the trees: detecting density dependence from existing life-table studies. *Journal of Animal Ecology* 58:883–92.

Hassell, M. P., J. H. Lawton, and R. M. May. 1976. Patterns of dynamical behaviour in single-species populations. *Journal of Animal Ecology* 45:471–86.

Hawkins, B. A., and J. H. Lawton. 1987. Species richness for parasitoids of British phytophagous insects. *Nature* 326:788–90.

Heads, P., and J. H. Lawton. 1983. Studies on the natural enemy complex of the holly leaf-miner: the effects of scale on the detection of aggregative responses and the implications for biological control. *Oikos* 40:267–76.

Hedrick, P. W. 1987. Genetic bottlenecks. *Science* 237:963.

Henttonen, H. 1985. Predation causing extended low densities in microtine cycles: further evidence from shrew dynamics. *Oikos* 45:156–57.

———. 1986. Causes and geographic patterns of microtine cycles. Ph.D. dissertation. Helsinki, Finland.

Henttonen, H., A. D. Macquire, and L. Hansson. 1985. Comparisons of amplitudes and frequencies (spectral analyses) of density variations in long-term data sets of *Clethrionomys* species. *Annales Zoologici Fennici* 22:221–27.

Henttonen, H., T. Oksanen, A. Jortikka, and V. Haukisalmi. 1987. How much do weasels shape microtine cycles in northern Fennoscandian taiga? *Oikos* 50:353–65.

Herbold, B., and P. B. Moyle. 1986. Introduced species and vacant niches. *American Naturalist* 128:751–60.

Holdren, C. E., and P. R. Ehrlich. 1981. Long-term dispersal in checkerspot butterflies: transplant experiments with *Euphydras gilletti*. *Oecologia* 50: 125–29.

Holdgate, M. W. 1960. The fauna of the mid-Atlantic islands. *Proceedings of the Royal Society [B]* 152:550–67.

Holling, C. S. 1959. Some characteristics of simple types of predation and parasitism. *Canadian Entomologist* 91:385–98.

———. 1973. Resilience and stability of ecological systems. *Annual Reviews of Ecology and Systematics* 4:1–23.

Holt, R. D. 1977. Predation, apparent competition, and the structure of prey communities. *Theoretical Population Biology* 12:197–229.

———. Prey communities in patchy environments. *Oikos* 50:276–90.

Holt, R. D., and J. Pickering. 1985. Infectious disease and species coexistence: a model of Lotka-Volterra form. *American Naturalist* 126:196–211.

Hope, J. H. 1973. Mammals of the Bass Strait islands. *Proceedings of the Royal Society of Victoria* 85:163–96.

Hurd, L. E., M. V. Mellinger, L. L. Wolf, and S. J. McNaughton. 1971. Stability and diversity at three trophic levels in terrestrial successional ecosystems. *Science* 173:1134–36.

Hutchinson, G. E. 1959. Homage to Santa Rosalia or why are there so many kinds of animals? *American Naturalist* 93:145–59.

Jackson, J. B. C. 1977a. Competition on marine, hard substrata: the adaptive significance of solitary and colonial strategies. *American Naturalist* 111:743–67.

———. 1977b. Habitat area, colonization, and development of epibenthic community structure. Pages 349–58 in B. F. Keegan, P. O'Ceidigh, and P. J. S. Boaden, eds. Biology of benthic organisms. Pergamon Press, Elmsford, N.Y.

Janzen, D. H. 1970. Herbivores and the number of trees in tropical forests. *American Naturalist* 104:501–28.

Järvinen, O. 1985. Predation causing extended low densities in microtine cycles: implications from predation on hole-nesting passerines. *Oikos* 45: 157–58.

Jeffries, J. J., and J. H. Lawton. 1985. Predator-prey ratios in communities of freshwater invertebrates: the role of enemy free space. *Freshwater Biology* 15:105–12.

Jewell, P. A., and S. Holt. 1981. Management of locally abundant wild mammals. Academic Press, New York.

Joemje, W., K. Bakker, and L. Vlijm. 1987. The ecology of biological invasions. *Proceedings of the Koninklijke Nederlandse Akademie van Wetenschappen,* Series C 90:1–80.

Jones, E. 1979. Breeding of the short-eared owl in south west Ireland. *Irish Birds* 1:77–80.

Jones, H. L., and J. M. Diamond. 1976. Short-time-base studies of turnover in breeding birds of the California channel islands. *Condor* 76:526–49.

Jordan, C. F., J. R. Kline, and D. S. Sasscer. 1972. Relative stability of mineral cycles in forest ecosystems. *American Naturalist* 106:237–53.

Karr, J. R. 1982a. Avian extinction on Barro Colorado Island Panama: a reassessment. *American Naturalist* 119:220–39.

———. 1982b. Population variability and extinction in the avifauna of a tropical land-bridge island. *Ecology* 63:1975–78.

Kauffman, S. 1992. The origins of order: self-organization and selection in evolution. Oxford University Press, Oxford.

Keller, M. A. 1984. Reassessing evidence for competitive exclusion of introduced natural enemies. *Environmental Entomology* 13:192–95.

King, A. W. and S. L. Pimm. 1983. Complexity and stability: a reconciliation of theoretical and experimental results. *American Naturalist* 122:229–39.

Kitching, R. L. 1986. The ecology of exotic animals and plants. John Wiley and Sons, Brisbane.

Kitching, R. L., and S. L. Pimm. 1986. Food chain lengths: phytotelmata in Australia and elsewhere. *Proceedings of the Ecological Society of Australia* 14:123–39.

Kneib, R. T. 1988. Testing for indirect effects of predation in an intertidal soft-bottom community. *Ecology* 69:1795–805.

Kneidel, K. A. 1983. Fugitive species and priority during colonization in carrion-breeding Diptera communities. *Ecological Entomology* 8:163–69.

Kohn, A. 1959. The ecology of *Conus* in Hawaii. *Ecological Monographs* 29:47–90.

Kornberg, H., and M. Williamson, eds. 1986. Quantitative aspects of the ecology of biological invasions. Cambridge University Press, Cambridge.

Krebs, C. J. 1978. Ecology: the experimental analysis of distribution and abundance. Harper and Row, New York.

Kremmer, J. N., and S. W. Nixon. 1978. A coastal marine ecosystem. Springer-Verlag, Berlin.

Lack, D. L. 1947. Darwin's finches. Cambridge University Press, Cambridge.

———. 1966. Population studies of birds. Oxford University Press, Oxford.

———. 1969. The number of bird species on islands. *Bird Study* 16:193–209.

————. 1976. Island biology illustrated by the land birds of Jamaica. Blackwell Scientific Publications, Oxford.

Lande, R. 1988. Genetics and demography in biological conservation. *Science* 241:1455–60.

Langeland, A., J. I. Koksvik, Y. Olsen, and H. Reinertsen. 1987. Limnocorral experiments in a eutrophic lake—effects of fish on the planktonic and chemical conditions. *Polskie Archiwum Hydrobiologii* 34:51–65.

Lawton, J. H. 1976. The structure of the arthropod community on bracken (*Pteridium aquilinum* (L.) Kuhn). *Botanical Journal of the Linnean Society* 73:187–216.

————. 1984. Non-competitive populations, non-convergent communities, and vacant niches: the herbivores of bracken. Pages 67–100 in D. R. Strong, D. Simberloff, L. G. Abele, and A. B. Thistle, eds. Ecological communities: conceptual issues and the evidence. Princeton University Press, Princeton, N.J.

————. 1988. More time means more variation. *Nature* 334:563.

————. 1989. Food webs. Pages 43–78 in J. M. Cherrett, ed. Ecological concepts. Blackwell Scientific Publications, Oxford.

Lawton, J. H., and K. C. Brown. 1986. The population and community ecology of invading insects. *Philosophical Transactions of the Royal Society of London* 314:607–17.

Lawton, J. H., and D. R. Strong. 1981. Community patterns and competition in folivorous insects. *American Naturalist* 188:317–38.

Leigh, E. G. 1975. Population fluctuations, community stability and environmental variability. Pages 51–73 in M. L. Cody and J. M. Diamond, eds. Ecology and evolution of communities. Harvard University Press, Cambridge, Mass.

Leigh, E. G., Jr. 1981. The average lifetime of a population in a varying environment. *Journal of Theoretical Biology* 90:213–39.

Levitan, C., W. C. Kerfoot, and W. R. DeMott. 1985. Ability of *Daphnia* to buffer trout lakes against periodic nutrient inputs. *Internationale Vereinigung für Theoretische und Angewandte Limnologie, Verhandlungen* 22:3076–82.

Lewin, R. 1983a. Santa Rosalia was a goat. *Science* 221:213–39.

————. 1983b. Predators and hurricanes change ecology. *Science* 221:737–40.

Lewontin, R. C. 1969. The meaning of stability. Pages 13–24 in G. M. Woodwell and H. H. Smith, eds. Diversity and stability in ecological systems. Brookhaven Symposium in Biology 22. Brookhaven National Laboratory, Upton, N.Y.

Linden, H. 1988. Latitudinal gradients in predator and prey interactions, cyclicity and synchronism in voles and small game populations in Finland. *Oikos* 52:341–49.

Lindstedt, S. L., and M. S. Boyce. 1985. Seasonality, fasting endurance, and body size in mammals. *American Naturalist* 125:873–78.

Long, J. 1981. Introduced birds of the world. David and Charles, London.

Lotka, A. J. 1925. Elements of physical biology. Williams and Wilkins, Baltimore.

Lovelock, J. 1982. Gaia: a new look at life on Earth. Oxford University Press, Oxford.

Lovejoy, T. E., J. M. Rankin, R. O. Bierregaard, K. S. Brown, L. H. Emmons, and M. E. Van der Voort. 1984. Ecosystem decay of Amazon forest remnants. Pages 295–72 in M. H. Nitecki, ed. Extinctions. University of Chicago Press, Chicago.

Luckinbill, L. S., and M. Fenton. 1978. Regulation and environmental variability in experimental populations of Protozoa. *Ecology* 59:1271–76.

Ludwig, D. 1989. Small models are beautiful; efficient estimators are even more beautiful. Pages 274–84 in C. Castillo-Chavez, S. A. Levin, and C. A. Schoemaker, eds. Mathematical approaches to problems in resource management and epidemiology. Lecture notes in biomathematics 81. Springer-Verlag, New York.

Lynch, M. 1979. Predation, competition, and zooplankton community structure: an experimental study. *Limnology and Oceanography* 24:253–72.

Lynch, M., and J. Shapiro. 1981. Predation, enrichment and phytoplankton community structure. *Limnology and Oceanography* 26:86–102.

McArdle, B. H. 1989. Bird population densities. *Nature* 338:628.

McArdle, B. H., K. J. Gaston, and J. H. Lawton. 1990. Variation in the size of animal populations: patterns, problems, and artefacts. *Journal of Animal Ecology* 59:439–54.

MacArthur, R. H. 1955. Fluctuations of animal populations and a measure of community stability. *Ecology* 36:533–36.

———. 1972. Geographical ecology. Harper and Row, New York.

MacArthur, R. H., and E. O. Wilson. 1967. The theory of island biogeography. Princeton University Press, Princeton, N.J.

McBrien, H., R. Haraman, and A. Crowder. 1983. A case of insect grazing affecting plant succession. *Ecology* 64:1035–39.

McCune, B., and T. F. H. Allen. 1985. Will similar forests develop on similar sites? *Canadian Journal of Botany* 63:367–76.

MacDonald, I. A. W., and M. L. Jarman. 1985. Invasive alien plants in the terrestrial ecosystems of Natal, South Africa. South African National Scientific Programmes Report 118. South African National Scientific Programmes, Pretoria.

MacDonald, I. A. W., F. J. Kruge, and A. A. Ferrar. 1985. The ecology and management of biological invasions in southern Africa. Oxford University Press, Capetown.

McGowan, J. A., and P. W. Walker. 1985. Dominance and diversity maintenance in an oceanic ecosystem. *Ecological Monographs* 55:103–18.

McGugan, B. M. 1958. Forest lepidoptera of Canada. Vol. 1. Publication 1034. Canada Department of Agriculture, Ottawa.

Mack, R. N. 1981. Invasion of *Bromus tectorum* L. into western North America: an ecological chronicle. *Agro-Ecosystems* 7:145–65.

———. 1986. Temperate grasslands vulnerable to plant invasions: characteristics and consequences. Pages 155–79 in H. Mooney and J. A. Drake, eds. Ecology of biological invasions of North America and Hawai'i. Springer-Verlag, Berlin.

McNaughton, S. J. 1968. Structure and function in California grasslands. *Ecology* 49:962–72.

———. 1977. Diversity and stability of ecological communities: a comment on the role of empiricism in ecology. *American Naturalist* 111:515–25.

———. 1985. Ecology of a grazing ecosystem: the Serengeti. *Ecological Monographs* 55:259–94.

McNaughton, S. J., M. Oesterheld, D. A. Frank, and K. J. Williams. 1989. Ecosystem productivity and herbivory in terrestrial habitats. *Nature* 341:142–44.

McQueen, D. J., D. O. Evans, and J. A. Downing. 1991. Productivity mediation of predation cascades in freshwater pelagic communities (submitted).

McQueen, D. J., M. R. S. Johannes, N. R. LaFontaine, A. S. Young, E. Longbotham, and D. R. S. Lean. 1990. Effects of planktivore abundance on chlorophyll a and Secchi depth. *Hydrobiologia* 200/201:337–43.

McQueen, D. J., M. R. S. Johannes, J. R. Post, T. J. Stewart, and D. R. S. Lean. 1989. Bottom-up and top-down impacts on freshwater pelagic community structure. *Ecological Monographs* 59:289–309.

McQueen, D. J., J. R. Post, and E. L. Mills. 1986. Trophic relationships in freshwater pelagic ecosystems. *Canadian Journal of Fisheries and Aquatic Sciences* 43:1571–81.

Magurran, A. E. 1988. Ecological diversity and its measurement. Princeton University Press, Princeton, N.J.

Malicky, H. 1976. Trichopteran-Emergenz in zwei Lunzer Bachen 1972–1974. *Archiv für Hydrobiologie* 77:51–65.

Marchant, J. J., R. Hudson, S. P. Carpenter, and P. Whittington. 1990. Population trends in British breeding birds. British Trust for Ornithology, Tring, Hertfordshire.

Margalef, R. 1965. Ecological correlations and the relationship between primary productivity and community structure. Pages 355–64 in C. R. Goldman, ed. Primary productivity in aquatic environments. Memoires Istituto Italiano di Idrobiologia. University of California Press, Berkeley.

———. 1969. Diversity and stability: a practical proposal and a model of interdependence. Pages 25–37 in G. M. Woodwell and H. H. Smith, eds. Diversity and stability in ecological systems. Brookhaven Symposium in Biology 22. Brookhaven National Laboratory, Upton, N.Y.

May, R. M. 1971. Stability in multi-species community models. *Mathematical Biosciences* 12:59–79.

———. 1972. Will a large complex system be stable? *Nature* 238:413–14.

———. 1973. Stability and complexity in model ecosystems. Princeton University Press, Princeton, N.J.

———. 1974. Stability and complexity in model ecosystems. 2d ed. Princeton University Press, Princeton, N.J.

———. 1984. Prehistory of Amazonian Indians. *Nature* 312:19–20.

———. 1987. Chaos and the dynamics of biological populations. *Proceedings of the Royal Society [A]* 413:27–44.

May, R. M., J. R. Beddington, J. W. Horwood, and J. G. Shepherd. 1978. Exploiting natural populations in an uncertain world. *Mathematical Biosciences* 42:219–52.

May, R. M., and M. P. Hassell. 1989. Parasitoid theory: against Manichaeism. *Trends in Ecology and Evolution* 4:20–21.

Mellinger, M. V., and S. J. McNaughton. 1975. Structure and function of successional vascular plant communities in central New York. *Ecological Monographs* 45:161–82.

Menge, B. A., and J. P. Sutherland. 1976. Species diversity gradients: synthesis of the roles of predation, competition and temporal heterogeneity. *American Naturalist* 110:351–69.

Miller, D. J. 1989. Introductions and extinction of fish in the African great lakes. *Trends in Ecology and Evolution* 4:56–59.

Mithen, S. J., and J. H. Lawton. 1986. Food-web models that generate constant predator-prey ratios. *Oecologia* 69:542–50.

Mooney, H., and J. A. Drake, eds. 1986. Ecology of biological invasions of North America and Hawaii. Springer-Verlag, Berlin.

Moreau, R. E. 1948. Ecological isolation in a rich tropical avifauna. *Journal of Animal Ecology* 17:113–26.

———. 1952. Africa since the Mesozoic: with particular reference to certain biological problems. *Proceedings of the Zoological Society of London* 121:869–913.

Morin, P. J. 1986. Interactions between intraspecific competition and predation in an amphibian predator-prey system. *Ecology* 67:713–20.

Morin, P. J., H. M. Wilbur, and R. N. Harris. 1983. Salamander predation and the structure of experimental communities: responses of *Notophthalmus* and microcrustacea. *Ecology* 64:1430–36.

Moulton, M. P. 1985. Morphological similarity and the coexistence of congeners: an experimental test with introduced Hawaiian birds. *Oikos* 44:301–5.

Moulton, M. P., and S. L. Pimm. 1983. The introduced Hawaiian avifauna: biogeographical evidence for competition. *American Naturalist* 121:669–90.

———. 1985. The extent of competition in shaping an experimental avifauna. Pages 80–97 in J. Diamond and T. Case, eds. Community ecology. Harper and Row, New York.

———. 1986. Species introductions to Hawaii. Pages 231–49 in H. Mooney and J. A. Drake, eds. Ecology of biological invasions of North America and Hawaii. Springer-Verlag, Berlin.

———. 1987. Morphological assortment in introduced Hawaiian passerines. *Evolutionary Ecology* 1:113–24.

Moyle, P. B., H. W. Li, and B. W. Barton. 1986. The Frankenstein effect: impact of introduced fishes on native fishes of North America. Pages 415–25 in R. Stroud, ed. The role of fish culture in fisheries management. American Fisheries Society, Bethesda, Md.

Muellar, L. D., and F. J. Ayala. 1981. Dynamics of single-species population growth: stability or chaos? *Ecology* 62:1148–54.

Murdoch, W. W. 1966. Population stability and life history phenomena. *American Naturalist* 100:5–11.

Murdoch, W. W., J. Chesson, and P. L. Chesson. 1985. Biological control in theory and practice. *American Naturalist* 125:344–66.

Neill, W. E. 1988. Complex interactions in oligotrophic lake food webs: responses to nutrient enrichment. Pages 31–43 in S. R. Carpenter, ed. Complex interactions in lake communities. Springer-Verlag, New York.

Newsome, A. E., and I. R. Noble. 1986. Ecological and physiological characters of invading species. Pages 1–20 in R. H. Groves and J. J. Burdon, eds. Ecology of biological invasions. Cambridge University Press, Cambridge.

Nicotri, M. E. 1977. Grazing effects of four marine intertidal herbivores on the microflora. *Ecology* 58:1020–32.

Norse, E. A., K. L. Rosenbaum, D. S. Wilcove, W. H. Romme, D. W. Johnston, and M. L. Stout. 1986. Conserving biological diversity in our national forests. The Wilderness Society, Washington, D.C.

O'Brien, S. J., and J. F. Evermann. 1988. Interactive influence of infectious disease and genetic diversity in natural populations. *Trends in Ecology and Evolution* 3:254–59.

O'Brien, S. J., D. E. Wildt, and M. Bush. 1986. The cheetah in genetic peril. *Scientific American* 254:84–92.

O'Connor, R. J. 1981. comparisons between migrant and non-migrant birds in Britain. Pages 167–95 in D. J. Aidley, ed. Animal migration. Cambridge University Press, Cambridge.

Odum, E. P. 1963. Ecology. Holt, Rinehart, and Winston, New York.

Odum, H. T., and R. C. Pinkerton. 1955. Time's speed regulator: the optimum efficiency for maximum power output in physical and biological systems. *American Scientist* 43:331–43.

Oksanen, L. 1983. Trophic exploitation and arctic phytomass patterns. *American Naturalist* 122:45–52.

———. 1988. Ecosystem organization: mutualism and cybernetics or plain Darwinian struggle for existence? *American Naturalist* 131:424–44.

Oksanen, L., S. D. Fretwell, J. Arruda, and P. Niemelä. 1981. Exploitation ecosystems in gradients of primary productivity. *American Naturalist* 118:240–62.

O'Neill, R. V. 1976. Ecosystem persistence and heterotrophic regulation. *Ecology* 57:1244–53.

O'Neill, R. V., D. L. DeAngelis, J. B. Waide, and T. F. H. Allen. 1986. A hierarchical concept of ecosystem. Princeton University Press, Princeton, N.J.

Opler, P. A. 1977. Insects of American chestnut: possible importance and conservation concern. The American Chestnut Symposium. West Virginia University Press, Morgantown.

Owen, J., and F. S. Gilbert. 1989. On the abundance of hoverflies (Syrphidae). *Oikos* 55:183–93.

Paine, R. T. 1980. Food webs: linkage, interaction strength and community infrastructure. *Journal of Animal Ecology* 49:667–85.

Patterson, B. D. 1984. Mammalian extinction and biogeography in the southern Rocky Mountains. Pages 247–93 in M. H. Nitecki, ed. Extinctions. The University of Chicago Press, Chicago.

Peitgen, H.-O., D. Saupe, and M. F. Barnsky. 1988. The beauty of fractal images. Springer-Verlag, New York.

Peters, R. H. 1983. The ecological implications of body size. Cambridge Studies in Ecology, Cambridge University Press, Cambridge.

Petersen, R. O., R. E. Page, and K. M. Dodge. 1984. Wolves, moose, and the allometry of population cycles. *Science* 224:1350–52.

Phillipi, T. E., M. P. Carpenter, T. J. Case, and M. E. Gilpin. 1987. *Drosophila* population dynamics: chaos and extinction. *Ecology* 68:154–59.

Pianka, E. R. 1966. Latitudinal gradients in species diversity: a review of concepts. *American Naturalist* 100:33–46.

Pimm, S. L. 1979a. Complexity and stability: another look at MacArthur's original hypothesis. *Oikos* 33:351–57.

———. 1979b. The structure of food webs. *Theoretical Population Biology* 16:144–58.

———. 1980a. The properties of food webs. *Ecology* 61:219–25.

———. 1980b. Species deletion and the design of food webs. *Oikos* 35:139–49.

———. 1982. Food webs. Chapman and Hall, London.

———. 1984a. The complexity and stability of ecosystems. *Nature* 307:321–26. © 1984 Macmillan Magazines Ltd.

———. 1984b. Food chains and return times. Pages 397–412 in D. R. Strong, D. Simberloff, L. F. Abele, and A. B. Thistle, eds. Community ecology: conceptual issues and the evidence. Princeton University Press, Princeton, N.J.

———. 1986a. Community structure and stability. Pages 309–29 in M. Soulé, ed. Conservation biology: the science of scarcity and diversity. Sinauer Associates, Sunderland, Mass.

———. 1986b. Putting the species back into community ecology. *Trends in Ecology and Evolution* 1:51–52.

———. 1987a. Determining the effects of introduced species. *Trends in Ecology and Evolution* 2:106–7.

———. 1987b. The snake that ate Guam. *Trends in Ecology and Evolution* 2:293–95.

———. 1988a. Energy flow and trophic structure. Pages 263–78 in J. Alberts and L. Pomeroy, eds. Concepts in ecosystem ecology. Springer-Verlag, Berlin.

———. 1988b. The geometry of niches. Pages 92–111 in A. Hastings, ed. Community ecology: lecture notes in biomathematics. Springer-Verlag, Berlin.

———. 1989. Theories of predicting success and impact of introduced species. Pages 351-368 in J. A. Drake, H. A. Mooney, F. di Castri, F. Kruger,

R. Groves, M. Rejmanek, and M. Williamson, eds. SCOPE 37, Biological Invasions: a global perspective. John Wiley and Sons, Chichester, England.

Pimm, S. L., and J. Gittleman. 1990. Carnivores and ecologists on the roads near Damascus. *Trends in Ecology and Evolution* 5:70–72.

Pimm, S. L., J. Gittleman, G. F. McCracken, and M. Gilpin. 1989. Genetic bottlenecks: alternative explanations for low genetic variability. *Trends in Ecology and Evolution* 4:176–78.

Pimm, S. L., H. L. Jones, and J. Diamond. 1988. On the risk of extinction. *American Naturalist* 132:757–85.

Pimm, S. L., and R. L. Kitching. 1987. The determinants of food chain lengths. *Oikos* 50:302–7.

Pimm, S. L., and J. H. Lawton. 1977. On the number of trophic levels. *Nature* 268:329–31. © 1977 Macmillan Magazines Ltd.

———. 1978. On feeding on more than one trophic level. *Nature* 275:542–44.

———. 1980. Are food webs compartmented? *Journal of Animal Ecology* 49:879–98.

Pimm, S. L., J. H. Lawton, and J. E. Cohen. 1991. Food web patterns and their consequences. *Nature* 350:669–674.

Pimm, S. L., and J. W. Pimm. 1982. Resource use, competition and resource availability in Hawaiian honeycreepers. *Ecology* 63:1468–80.

Pimm, S. L., and A. Redfearn. 1988. The variability of animal populations. *Nature* 334:613–14. © 1988 Macmillan Magazines Ltd.

———. 1989. Bird population densities. *Nature* 338:628.

Pimm, S. L., and J. A. Rice. 1987. The dynamics of multispecies, multi-life-stage models of aquatic food webs. *Theoretical Population Biology* 32:303–25.

Pitts, G. C., and T. R. Bullard. 1968. Some interspecific aspects of body composition in mammals. Pages 45–70 in National Academy of Sciences, ed. Body composition in animals and man. Publication 1598. National Academy of Sciences, Washington, D.C.

Pomeroy, L. R. 1970. The strategy of mineral cycling. *Annual Reviews of Ecology and Systematics* 1:171–90.

Post, J. R., and D. J. McQueen. 1987. The impact of planktivorous fish on the structure of a plankton community. *Freshwater Biology* 17:79–89.

Post, W. M., and S. L. Pimm. 1983. Community assembly and food web stability. *Mathematical Biosciences* 64:169–92.

Potts, G. R., J. C. Coulson, and I. R. Deans. 1980. Population dynamics and breeding success of the shag, *Phalacrocorax aristotelis,* on the Farne Islands, Northumberland. *Journal of Animal Ecology* 49:465–84.

Power, M. E., W. J. Matthews, and A. J. Stewart. 1985. Grazing minnows, piscivorous bass, and stream algae: dynamics of a strong interaction. *Ecology* 66:1448–56.

Prentice, R. M. 1962. Forest lepidoptera of Canada. Vol. 2. Bull. 128. Canada Department of Forestry, Ottawa.

———. 1963. Forest lepidoptera of Canada. Vol. 3. Bull. 1013. Canada Department of Forestry, Ottawa.

Quinn, J. F. 1982. Competitive hierarchies in marine benthic communities. *Oecologia* 54:129–35.

Quinn, J. R., and A. Hastings. 1987. Extinction in subdivided habitats. *Conservation Biology* 1:198–209.

Raitt, R. J., and S. L. Pimm. 1976. Dynamics of bird communities in the Chihuahuan Desert, New Mexico. *Condor* 78:427–42.

Ralls, K., J. Ballou, and R. L. Brownell, Jr. 1983. Genetic diversity in California sea otters: theoretical considerations and management implications. *Biological Conservation* 25:209–32.

Ralls, K., J. D. Ballou, and A. R. Templeton. 1988. Estimates of lethal equivalents and the cost of inbreeding in mammals. *Conservation Biology* 2:185–93.

Ralls, K., and P. H. Harvey. 1985. Geographic variation in size and sexual dimorphism of North American weasels. *Biological Journal of the Linnean Society* 25:119–67.

Ralls, K., P. H. Harvey, and A. M. Lyles. 1986. Inbreeding in natural populations of birds and mammals. Pages 35–56 in M. Soulé, ed. Conservation biology: the science of scarcity and diversity. Sinauer Associates, Sunderland, Mass.

Redfearn, A., and S. L. Pimm. 1987. Insect pest outbreaks and community structure. Pages 99–133 in P. Barbosa and J. C. Schultz, eds. Insect pests. Academic Press, Orlando, Fla.

———. 1988. Population variability and polyphagy in herbivorous insect communities. *Ecological Monographs* 58:39–55.

———. 1991. Predation and population variability. In M. Crawley, ed. Predation. Blackwell Scientific Publications, Oxford.

Rejmánek, M., and K. Spitzer. 1982. Bionomic strategies and long-term fluctuations in abundance of Noctuidae (Lepidoptera). *Acta Entomologica Bohemoslovaca* 79:81–96.

Richter-Dyn, N., and N. S. Goel. 1972. On the extinction of a colonizing species. *Theoretical Population Biology* 3:406–33.

Risch, S. J., D. Andow, and M. A. Altieri. 1983. Agroecosystem diversity and pest control: data, tentative conclusions and new research directions. *Environmental Entomology* 12:625–29.

Roberts, A. 1974. The stability of a feasible random ecosystem. *Nature* 251:607–8.

Roberts, A., and K. Tregonning. 1981. The robustness of natural systems. *Nature* 288:265–66.

Robinson, J. V., and J. E. Dickerson, Jr. 1987. Does invasion sequence affect community structure? *Ecology* 68:587–95.

Robinson, J. V., and M. A. Edgemon. 1988. An experimental evaluation of the effect of invasion history on community structure. *Ecology* 69:1410–17.

Robinson, J. V., and W. D. Valentine. 1979. The concepts of elasticity, invulnerability, and invadability. *Journal of Theoretical Biology* 81:91–104.

Root, R. B. 1973. Organization of a plant-arthropod association in simple and diverse habitats: the fauna of collards, *Brassica oleracea*. *Ecological Monographs* 43:95–124.

Root, T. 1988. Energy constraints on avian distributions and abundances. *Ecology* 69:330–39.

Roots, C. 1976. Animal invaders. Universe Books, New York.

Rosenzweig, M. L., and Z. Abramsky. 1986. Centrifugal community organization. *Oikos* 46:339–48.

Rosenzweig, M. L., and R. H. MacArthur. 1963. Graphical representation and stability conditions of predator-prey interactions. *American Naturalist* 97:209–23.

Rothrock, D. A., and A. S. Thorndike. 1980. Geometrical properties of the underside of sea ice. *Journal of Geophysical Research* 85:3955–63.

Roughgarden, J. 1989. The structure and assembly of communities. Pages 203–26 in J. Roughgarden, R. M. May, and S. A. Levin, eds. Perspectives in ecological theory. Princeton University Press, Princeton, N.J.

Royama, T. 1971. A comparative study of models for predation and parasitism. *Researches on Population Ecology,* suppl. 1:1–91.

Rummel, J. D., and J. Roughgarden. 1983. Some differences between invasion-structured and co-evolution-structured competitive communities: a preliminary theoretical analysis. *Oikos* 41:477–86.

Ryther, J. H. 1969. Photosynthesis and fish production in the sea. *Science* 166:72–76.

Sadovskaja, E. B., T. P. Povališina, M. A. Štern, J. A. Mjasnikov, and T. V. Panina. 1971. Mnogoletnie zoologičeskie nabljundenija v odnom iz očagov gemorragičeskoj lihoradki s počečnym sindromom v Tul'skoj Oblasti (Long-term zoological observations in a source of hemorraghic fever with renal syndrome in Tula Province). *Trudy Instituta Poliomielita i Virusnyh Encefalitov AMN SSSR* 19:314–21.

Sailer, R. I. 1978. Our immigrant insect fauna. *Entomological Society of America Bulletin* 24:3–11.

Sale, P. F. 1978. Coexistence of coral reef fishes: a lottery for living space. *Environmental Biology of Fishes* 3:85–102.

Savidge, J. A. 1987. Extinction of an island avifauna by an introduced snake. *Ecology* 68:660–68.

Schaffer, W. M. 1984. Stretching and folding in lynx fur returns: evidence for a strange attractor in nature? *American Naturalist* 124:798–820.

Schaffer, W. M., and M. Kot. 1985. Nearly one-dimensional dynamics in a simple epidemic. *Journal of Theoretical Biology* 112:403–27.

———. 1986a. Chaos in ecological systems: the coals that Newcastle forgot. *Trends in Ecology and Evolution* 1:58–63.

———. 1986b. Differential systems in ecology and epidemiology. Pages 158–78 in A. V. Holden, ed. Chaos. Princeton University Press, Princeton, N.J.

Schoener, T. W. 1974. Resource partitioning in ecological communities. *Science* 185:27–39.

———. 1983a. Field experiments on interspecific competition. *American Naturalist* 122:240–85.

———. 1983b. Rate of species turnover declines from lower to higher organisms: a review of the data. *Oikos* 41:372–77.

———. 1985. Patterns in terrestrial vertebrate versus arthropod communities: do systematic differences in regularity exist? Pages 556–86 in J. Diamond and T. J. Case, eds. Community ecology. Harper and Row, New York.

———. 1989. Food webs from the small to the large. *Ecology* 70:1559–89.

Schoener, T. W., and D. A. Spiller. 1992. Is extinction rate related to temporal variability in population size? an empirical answer for orb spiders. *American Naturalist* 139:1176–1206

Schoenly, K., R. A. Beaver, and T. A. Heumier. 1991. On the trophic relations of insects: a food web approach. *American Naturalist* 137:597–638.

Scott, J. M., S. Mountainspring, F. L. Ramsey, and C. B. Kepler. 1986. Forest bird communities on the Hawaiian Islands: their dynamics, ecology and conservation. Cooper Ornithological Society, Berkeley, Calif.

Shaffer, M. L. 1981. Minimum population sizes for species conservation. *Bioscience* 31:131–34.

Shapiro, J., and D. I. Wright. 1984. Lake restoration by biomanipulation. Round Lake, Minnesota—the first two years. *Freshwater Biology* 14:371–83.

Sih, A., P. Crowley, M. McPeek, J. Petranka, and K. Strohmeier. 1985. Predation, competition, and prey communities: a review of field experiments. *Annual Review of Ecology and Systematics* 16:269–311. © 1985 by Annual Reviews, Inc.

Simberloff, D. 1981. Community effects of introduced species. Pages 53–81 in H. Nitecki, ed. Biotic crises in ecological and evolutionary time. Academic Press, New York.

———. 1984. Properties of coexisting bird species in two archipelagos. Pages 234–53 in D. R. Strong, D. Simberloff, L. G. Abele, and A. B. Thistle, eds. Ecological communities: conceptual issues and the evidence. Princeton University Press, Princeton, N.J.

———. 1986. Introduced species: a biogeographic and systematic perspective. Pages 3–26 in H. Mooney and J. A. Drake, eds. Ecology of biological invasions of North America and Hawaii. Springer-Verlag, Berlin.

Simberloff, D., and W. Boecklen. 1981. Santa Rosalia reconsidered: size ratios and competition. *Evolution* 35:1206–28.

Simola, H. 1984. Population dynamics of plankton diatoms in a 69 year sequence of annually laminated sediment. *Oikos* 43:30–40.

Skinner, B. 1984. Colour identification guide to the moths of the British Isles. Viking, Harmondsworth, England.

Slobodkin, L. B., F. E. Smith, and N. G. Hairston. 1967. Regulation in terrestrial ecosystems, and the implied balance of nature. *American Naturalist* 101:109–24.

Smithers, R. H. N. 1971. The mammals of Botswana. Memoire 4. National Museum of Rhodesia, Salisbury, Rhodesia.

Soulé, M. 1987. Viable populations for conservation. Cambridge University Press, Cambridge.

Southwood, T. R. E. 1977. Habitat, the templet for ecological strategies. *Journal of Animal Ecology* 46:337–65.

———. 1978. Ecological methods. Chapman and Hall, London.

Spitzer, K., M. Rejmánek, and T. Soldan. 1984. The fecundity and longer-term variability in abundance of noctuid moths (Lepidoptera, Noctuidae). *Oecologia* 62:91–93.

Sprules, W. G., and J. E. Bowerman. 1988. Omnivory and food chain length in zooplankton food webs. *Ecology* 69:418–26.

Steele, J. H. 1986. A comparison of terrestrial and marine ecological systems. *Nature* 313:355–58.

Stiles, G. F. 1977. Coadapted competitors: the flowering seasons of hummingbird-pollinated plants in a tropical forest. *Science* 198:1177–78. © 1977 by the AAAS.

Stiling, P. 1987. The frequency of density dependence in insect host-parasitoid systems. *Ecology* 68:844–56.

Strong, D. R. 1986. Density vagueness: adding the variance in the demography of real populations. Pages 257–68 in J. Diamond and T. J. Case, eds. Community ecology. Harper and Row, New York.

——. 1988. Parasitoid theory: from aggregation to dispersal. *Trends in Ecology and Evolution* 3:277–80.

——. 1989. Reply. *Trends in Ecology and Evolution* 4:21–22.

Strong, D. R., Jr., J. H. Lawton, and T. R. E. Southwood. 1984. Insects on plants: community patterns and mechanisms. Blackwell Scientific Publications, Oxford.

Sugihara, G. 1984. Graph theory, homology and food webs. *Proceedings of Symposia in Applied Mathematics* 30:83–101.

Sugihara, G., and R. M. May. 1990. Nonlinear forecasting as a way of distinguishing chaos for measurement error in time series. *Nature* 344:734–41.

Sutherland, J. P. 1974. Multiple stable points in natural communities. *American Naturalist* 108:859–73.

Takens, F. 1981. Detecting strange attractors in turbulence. Pages 366–81 in D. A. Rand and L. S. Young, eds. Dynamical systems and turbulence. Springer-Verlag, New York.

Tanner, J. T. 1966. Effects of population density on the growth rates of animal populations. *Ecology* 47:733–45.

Tatchell, G. M., S. J. Parker, and I. P. Woiwod. 1983. Synoptic monitoring of migrant insect pests in Great Britain and Western Europe. IV. Host plants and their distribution for pest aphids in Great Britain. *1982 Rothamsted Experimental Station Report,* Part 2:45–159. Rothamsted Experimental Station, England.

Taylor, L. R., I. P. Woiwod, and J. N. Perry. 1978. The density dependence of spatial behaviour and the rarity of randomness. *Journal of Animal Ecology* 47:383–406.

——. 1979. The negative binomial as dynamic ecological model for aggregation, and the density dependence of k. *Journal of Animal Ecology* 48:289–304.

——. 1980. Variance and the large scale spatial stability of aphids, moths, and birds. *Journal of Animal Ecology* 49:831–54.

Terborgh, J., and B. Winter. 1980. Some causes of extinction. Pages 119–34 in M. E. Soulé and B. A. Wilcox, eds. Conservation biology. Sinauer Associates, Sunderland, Mass.

Testa, J. W. 1986. Electromorph variation in Weddell seals (*Leptonychotes weddelli*). *Journal of Mammalogy* 67:606–10.

Tietz, H. M. 1972. An index to the described life histories, early stages, and hosts of the Macrolepidoptera of the continental United States and Canada. Allyn Press, Sarasota, Fla.

Toft, C. A., and T. W. Schoener. 1983. Abundance and diversity of orb spiders on 106 Bahamian islands: biogeography at an intermediate trophic level. *Oikos* 41:411–26.

Tregonning, K., and A. Roberts. 1978. Ecosystem-like behaviour of a random interaction model. Part I. *Bulletin of Mathematical Biology* 40:513–24.

———. 1979. Complex systems which evolve towards homeostasis. *Nature* 281:563–64.

Turcek, F. J. 1951. Effect of introduction on two game populations in Czechoslovakia. *Journal of Wildlife Management* 15:113–14.

Turchin, P. 1990. Rarity of density dependence or population regulation with lags? *Nature* 344:660–63.

Turelli, M. 1978. A reexamination of stability in randomly versus deterministic environments with comments on the stochastic theory of limiting similarity. *Theoretical Population Biology* 13:244–67.

Vanni, M. J. 1987a. Effects of food availability and fish predation on a zooplankton community. *Ecological Monographs* 57:61–88.

———. 1987b. Effects of nutrients and zooplankton size on the structure of phytoplankton community. *Ecology* 68:624–35.

van Riper, C., III, and J. M. Scott. 1979. Observations on distribution, diet, and breeding of the Hawaiian thrush. *Condor* 81:65–71.

van Riper, C., III, S. G. Van Riper, M. L. Goff, and M. Laird. 1986. The epizootiology and ecological significance of malaria on the birds of Hawaii. *Ecological Monographs* 56:327–44.

Virnstein, R. W. 1977. The importance of predation by crabs and fishes on benthic infauna in Chesapeake Bay. *Ecology* 58:1199–1218.

Vitousek, P. 1986. Biological invasions and ecosystem properties: can species make a difference? Pages 163–79 in H. A. Mooney and J. Drake, eds. Biological invasions of North America and Hawaii. Springer-Verlag, New York.

Volterra, V. 1926. Fluctuations in the abundance of species, considered mathematically. *Nature* 118:558–60.

Wallace, A. R. 1876. The geographical distribution of animals. Macmillan, London.

Walters, S. M., A. Brady, C. D. Brickell, J. Cullen, P. S. Green, J. Lewis, V. A. Matthews, D. A. Webb, P. F. Yeo, and J. C. M. Alexander. 1984. The European garden flora: a manual for the identification of plants cultivated in Europe, both out-of-doors and under glass. Vol. 2. Monocotyledons (part 2). Cambridge University Press, Cambridge.

Ware, F. J., and R. D. Gassaway. 1976. Effects of grass carp on native fish populations in two Florida lakes. *Proceedings of the Annual Conference of Southeastern Game and Fish Commissioners* 30:324–25.

Warner, R. E. 1968. The role of introduced diseases in the extinction of the endemic Hawaiian avifauna. *Condor* 70:101–20.

Wasserman, S. S., and C. Mitter. 1978. The relationship of body size to breadth of diet in some Lepidoptera. *Ecological Entomology* 3:155–60.

Watt, K. E. F. 1964. Comments on long-term fluctuations of animal populations and measures of community stability. *Canadian Entomologist* 96:1434–42.

———. 1965. Community stability and the strategy of biological control. *Canadian Entomologist* 97:887–95.

———. 1969. A comparative study of the meaning of stability in five biological systems: insect and furbearer populations, influenza, Thai hemorrhagic fever, and plague. Pages 142–50 in G. M. Woodwell and H. H. Smith, eds. Diversity and stability in ecological systems. Brookhaven Symposium in. Biology 22. Brookhaven National Laboratory, Upton, N.Y.

———. 1973. Principles of environmental science. McGraw Hill, New York.

Watt, K. E. F., and P. P. Craig. 1986. System stability principles. *Systems Research* 3:191–201.

Whittaker, R. H. 1975. Communities and ecosystems, 2d ed. Macmillan, New York.

Whittaker, R. H., and D. Goodman. 1978. Classifying species according to their demographic strategy. I. Population fluctuations and environmental heterogeneity. *American Naturalist* 113:185–200.

Wilcove, D. S. 1987. Public lands management and the fate of the spotted owl. *International Council for Bird Preservation* 41:361–67.

Wildt, D. E., M. Bush, K. L. Goodrowe, C. Packer, A. E. Pusey, J. L. Brown, P. Joslin, and S. J. O'Brien. 1987. Reproductive and genetic consequences of founding isolated lion populations. *Nature* 329:328–30.

Williams, C. B. 1958. Insect migration. Collins Press, London.

———. 1964. Patterns in the balance of nature. Academic Press, New York.

Williamson, M. H. 1957. An elementary theory of interspecific competition. *Nature* 180:422–25.

———. 1972. The analysis of biological populations. Edward Arnold, London.

———. 1981. Island populations. Oxford University Press, Oxford.

———. 1983. The land-bird community of Skokholm: ordination and turnover. *Oikos* 41:378–84.

———. 1984. The measurement of population variability. *Ecological Entomology* 9:239–41.

———. 1987. Are communities ever stable? Pages 352–71 in A. J. Gray, M. J. Crawley, and P. J. Edwards, eds. Colonizations, succession, and stability. Blackwell Scientific Publications, Oxford.

Willis, E. O. 1974. The composition of avian communities in remanescent woodlots in southern Brazil. *Papeis Avulsos Museo Paulisto* 33:1–25.

———. 1980. Species reduction in remanescent woodlots in southern Brazil. *Proceedings of the 19th International Ornithological Congress,* 783–86.

Witteman, G. J., A. Redfearn, and S. L. Pimm. 1990. The extent of complex population changes in nature. *Evolutionary Ecology* 4:173–83.

Wolda, H. 1978. Fluctuations in abundance of tropical insects. *American Naturalist* 112:1017–45.

———. 1983. "Long-term" stability of tropical insect populations. *Researches in Population Ecology* 3:112–26.

Wolda, H., and P. Gulundo. 1981. Population fluctuations of mosquitos in the nonseasonal tropics. *Ecological Entomology* 6:99–106.

Woodwell, G. M., and H. H. Smith. 1969. Diversity and stability in ecological systems. Brookhaven Symposium in Biology 22. Brookhaven National Laboratory, Upton, N.Y.

Yodzis, P. 1988. The indeterminacy of ecological interactions as perceived through perturbation experiments. *Ecology* 69:508–15.

Zaret, T. M. 1980. Predation and freshwater communities. Yale University Press, New Haven, Conn.

Zaret, T. M., and R. T. Paine. 1973. Species introduction in a tropical lake. *Science* 182:449–545.

Species Index

Subject Index

434